Special cases. If the random variables are independent or uncorrelated ($\rho = 0$), then

$$\sigma_{X+Y}^2 = \sigma_X^2 + \sigma_Y^2 = \sigma_{X-Y}^2; \qquad \sigma_{aX+bY}^2 = a^2\sigma_X^2 + b^2\sigma_Y^2;$$

$$\sigma_{\Sigma X_i}^2 = \sum \sigma_i^2; \qquad \sigma_{\Sigma w_i X_i}^2 = \sum w_i^2 \sigma_i^2;$$

$$E(XY) = E(X) \cdot E(Y) = \mu_X \mu_Y.$$

If all correlations $\rho_{ij} = \rho$ and all variances $\sigma_i^2 = \sigma^2$, then

$$\sigma_{\Sigma X_i}^2 = n\sigma^2[1 + (n-1)\rho].$$

B. SAMPLING THEORY

Notation: $X_1, X_2, \ldots, X_n$ are uncorrelated random variables whose average is $\bar{X}$. X_i has mean μ_i and variance σ_i^2, $i = 1, 2, \ldots, n$.

1. Mean and Variance of Sample Average

$$\mu_{\bar{X}} = \frac{1}{n} \sum \mu_i; \qquad \sigma_{\bar{X}}^2 = \frac{1}{n^2} \sum \sigma_i^2.$$

Special case: identically distributed random variables. If all means $\mu_i = \mu$, and all variances $\sigma_i^2 = \sigma^2$, then $\mu_{\bar{X}} = \mu$,

$$\sigma_{\bar{X}}^2 = \sigma^2/n$$

for sampling with replacement, and

$$\sigma_{\bar{X}}^2 = \frac{\sigma^2(N-n)}{n(N-1)}$$

for sampling without replacement, population size N.

2. Binomial Distribution

If X is the number of successes and $\bar{p}$ is the proportion of successes ($\bar{p} = X/n$), then

$$E(X) = np, \qquad \sigma_X^2 = np(1-p);$$

$$E(\bar{p}) = p, \qquad \sigma_{\bar{p}}^2 = p(1-p)/n.$$

3. Central Limit Theorem
(for large samples, drawn with replacement)

If Z is the standard normal random variable, then

D1106139

$$P\left(\frac{\bar{X} - \mu}{\sigma/\sqrt{n}} \geq z\right) \approx P(Z \geq z).$$

4. Central Limit Theorem for Binomial Distributions

If Z is the standard normal random variable and np is far from 0 and n, then if X is the number of successes

$$P(X \geq x) \approx P(Z \geq z), \qquad \text{where } z = (x - \tfrac{1}{2} - np)/\sqrt{np(1-p)}.$$

Continued inside back cover

STURDY STATISTICS
Nonparametrics and Order Statistics

STURDY STATISTICS
Nonparametrics and Order Statistics

FREDERICK MOSTELLER
Harvard University

ROBERT E. K. ROURKE
*Formerly Kent School, Connecticut
and St. Stephen's School, Rome*

ADDISON-WESLEY PUBLISHING COMPANY
Reading, Massachusetts · Menlo Park, California
London · Amsterdam · Don Mills, Ontario · Sydney

ISBN 0-201-04868-X
BCDEFGHIJK-HA-798765

Preface

Why "sturdy"? Our choice of "sturdy" to describe the statistical methods in this book reflects the use of the technical terms "robust" and "resistant". Statisticians attach these labels to methods that retain their useful properties in the face of changing distributions and violations of standard assumptions. However, if we used a title like "robust statistics", many knowledgeable readers would naturally expect the book to focus mainly on the study of the comparative robustness of various statistics. Instead, we emphasize the statistics themselves and their applications, as is appropriate in an early course. The everyday adjective "sturdy" avoids the implied scope of the more technical terms.

What this book deals with. The methods of sturdy statistics, based on signs, counts, ranks, and order statistics, are easy to understand and simple to execute. But simplicity in itself is an empty goal. Life's problems are not always simple; sometimes complexity is the rule, not the exception. Yet our elementary statistical devices often help people solve complex problems. Indeed, the concepts underlying these techniques involve a mode of reasoning that extends far beyond the specific methods we discuss. What we are dealing with is a general pattern of devices for statistical inference.

What this book assumes. The authors assume that the reader has had two years of high school algebra, along with some exposure to statistics. To help recall, or reinforce, the statistical ideas needed in the exposition, the book includes seven appendixes, and we refer to these appendixes at appropriate points. "Appendix IV–2" refers to the second section in the fourth appendix at the back of the book. The appendixes allow the reader to study this book after a short exposure to probability or statistics, typically a quarter or a semester.

When the reader meets an idea with an appendix reference attached, he should:

1) ignore the reference, if the idea is clear, or
2) read the given appendix, if the idea needs freshening, or
3) consult both the appendix and the supplementary references at the beginning of the appendix, if the idea needs fuller explanation.

[Note: A reference to *PWSA* refers to *Probability with Statistical Applications*, 2nd edition, by F. Mosteller, R. E. K. Rourke, and G. B. Thomas, Jr., Addison-Wesley, 1970.]

Style. The text is written to be read. Most of the sections and chapters are short. The exposition extracts general ideas and methods from specific examples: the general flows from the particular and then returns to enrich it. Since this procedure slows the pace, a reader who wishes to go faster may focus his attention on the definitions and theorems. But a careful study of the illustrative examples usually pays off in deeper theoretical insights.

Attempting, in an introductory course, to get interest out of rigor is sometimes like trying to wring charity out of piety. For a book of this kind, the authors' attitude to proof parallels that so admirably set forth in an article by R. P. Boas, Jr.* Professor Boas wrote, in part:

> In mathematics . . . we believe things because *somebody* proved them; we have not necessarily studied the proof ourselves . . . only just so much time is available. In order to make the best use of it, . . . give proofs when they are easy and justify unexpected things; . . . omit tedious or difficult proofs, especially those of plausible things. . . . give easy proofs under simplified assumptions rather than complicated proofs under general hypotheses.

To this we would add: include proofs, even tedious or difficult ones, if their study will equip the student with tools of permanent value.

Payoff. Sturdy statistics offer the reader several worthwhile payoffs. Not only is he exposed to nonparametric and other methods in common use, but he is also offered a set of tools whose uses are much broader in scope than the problems that produced them. While the reader can gain a sure grasp of general methods by studying the specific examples of the text, the authors' basic objective is always to illuminate the big ideas behind the techniques.

Examples and exercises. The repeated use of some examples and exercises is intentional. This practice underlines the fact that often the data can and should be analyzed in various ways in order to give an informative report.

---

* R. P. Boas, Jr., "Calculus as an Experimental Science", *American Mathematical Monthly* (Vol. 78, No. 6, June-July, 1971), pp. 664–667.

Frequently the authors have been able to give both examples and exercises the tang of reality. To give the reader some sense of complicated real-life problems, the book occasionally offers large examples involving more extensive data than is usually found in such a text. A star, placed beside an exercise or other material, indicates the presence of a special challenge.

At some points, the reader is asked to put forth extra effort, to go beyond the text and supply some ideas of his own. The instructor should encourage these efforts, but he should recognize that they take time. The resulting opportunities for discussion are to be welcomed, for students have important contributions to make. Statistics is more than pure mathematics; it makes hard contact with the real world.

Tables and answers. The tables of the text are numbered in two ways. Table 2–1 is the first table in Chapter 2, and Table A–6 is the sixth table at the back of the book. The assortment of tables at the back of the book is rather generous. In most cases, examples at the beginning of a table illustrate its use.

A section at the back of the book gives answers for the odd-numbered exercises.

Some possible sequences. The book, designed for a one-quarter or one-semester course, lends itself to a number of possible sequences. A good basic sequence consists of Chapters 1 through 11. This material may be supplemented, according to the taste of the instructor and the time available, by adding one or more of the following self-contained units.

Chapter 12
The Kruskal-Wallis Statistic

Chapter 13
The Problem of m Rankings: The Friedman Index

Chapters 14 and 15
Order Statistics

Chapter 16
Designing Your Own Statistical Test or Measure

If necessary, the basic sequence may be shortened by deleting Chapters 10 and 11, the last two chapters on the chi-square distribution.

For well-prepared students, the instructor might skip, or pass quickly over, Sections 1–3, 1–4, 2–1, 2–2, and 2–3. Indeed, for students familiar with the normal, binomial, and multinomial distributions and the t-test, the instructor might deal with the permutation test (Sections 1–5, 1–6, and 1–7), and then proceed at once to Chapter 3.

Diagram 1 gives a schematic representation of the connections between chapters and reveals many other possible sequences (see page viii).

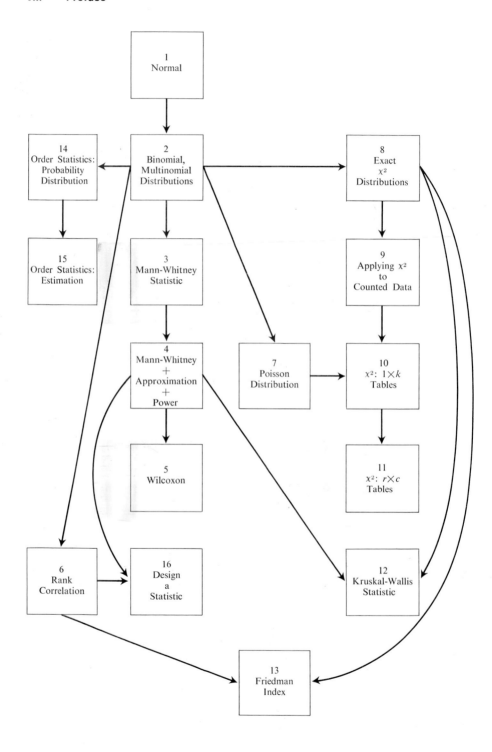

Acknowledgements. The authors are deeply grateful to many people for help and encouragement in the writing of this book.

Robert Berk, Bert Green, and Gus W. Haggstrom provided critiques and suggestions that improved the organization and exposition.

Joseph Gastwirth and Arthur P. Dempster discussed parts of the manuscript with us.

Michael Brown, Michael Sutherland, and Sanford Weisberg made clarifying suggestions.

Nathan Keyfitz helped us get the data required for women over 80 in Fig. 1–3.

In addition, we owe much to Holly Grano and Gale Robin Mosteller for typing and corrections, and to Cleo Youtz for computations, tables and graphs, and efficient organization of materials.

October 1972   F.M.
                                                                                      R.E.K.R.

Diagram 1. Connections between chapters. The arrows indicate possible sequences. Each chapter rests on those chapters whose arrows indicate an unbroken flow from Chapter 1 to the given chapter. Thus Chapter 12 rests on Chapters 1, 2, 3, 4, and 8.

Contents

Normal and Non-normal Data

1–1 WHAT ARE STURDY STATISTICS?

The best-known methods of statistical inference assume that we know the form of the probability distribution that the measurements come from. For example, we often assume that the measurements are drawn from a normal distribution and then proceed to test hypotheses about its parameters, such as the mean μ or the variance σ^2. Frequently, however, we cannot comfortably make an assumption about the form of the distribution under study. The evidence may not support the chosen form or, still worse, may sharply deny it.

In such circumstances, we prefer to use methods whose strengths do not depend much on the precise shape of the distribution. We may want to compare distributions when we know little about their forms. So we turn to methods based on *signs* of differences, *ranks* of measurements, and *counts* of objects or events falling into categories. Such methods may not rest heavily on the explicit parameters of the distribution, and for this reason they are called *nonparametric*.

The adjective "nonparametric" is somewhat misleading, because nonparametric statistics do in fact deal with parameters such as the median of a distribution or the probability of success p in a binomial. Indeed, the word "nonparametric" as commonly used does not lend itself to a precise definition. Its meaning will emerge from a study of the examples and discussions in the text.

The main point is that many of the methods we describe defend themselves against wild observations and stand up well against various shapes of distributions and failures of assumptions. Statisticians use such words as "robust" and "resistant" for methods that have these properties. Since these words have grown to have specific technical meanings, we use the everyday word "sturdy" to describe most of the statistics we present in this book.

1-2 WHAT DOES THIS BOOK DO?

For a student of applied statistics with a modest mathematical background, this book offers the following.

1. An opportunity to acquire a considerable kit of tools whose use can go well beyond the techniques discussed here. In particular, these tools can be used to develop new methods tailored to the students' own problems. Thus these sturdy statistics offer a vehicle for introducing the student to the more general techniques of intermediate statistics.

2. An introduction to nonparametric methods that are in common use, together with some theoretical background for the methods and some illustrations of their applications in important practical problems.

3. An opportunity to grasp the general pattern of devices for statistical inference by studying a modest number of them. For instance, we shall see that although most methods require a special table of exact probabilities for *small sample sizes*, a formula plus a standard table may be adequate *for large sample sizes* (indeed, often the table is not needed).

4. A few examples of problems that are slightly larger than the small ones that naturally predominate in the sets of exercises. These larger practical problems are intended to give the student a feeling for the way these sturdy statistics are used in real life, so that when he faces practical problems he will not be lost in a forest of uncertainties. Real-life problems have a way of being larger, more complicated, and less organized than the smaller textbook examples. Some exposure to such problems under textbook conditions may help the student get started more easily on his own statistical problems.

We meet two kinds of distributions—*discrete* distributions and *continuous* distributions. For those wishing to review the basic ideas about distributions, we provide, at the back of the book, Appendix I (discrete distributions) and Appendix II (continuous distributions). In the next two sections, we shall review the normal distributions, an important family of continuous distributions.

1-3 THE SHAPES OF NORMAL DISTRIBUTIONS

Because our work makes extensive use of the theory of the normal distribution, we need to review some of its properties. Many random variables found in practical work have frequency distributions whose shapes are approximately symmetrical, have a rounded top like an old mountain, and tail off to zero on each side of the center. Such frequency distributions are often called bell-shaped, and the graph of a normal distribution (Fig. 1–1b) does look a bit like a cross section of a bell. We shall give a concrete example of an approximately normal distribution using the heights of men.

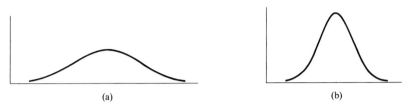

Fig. 1–1 Shapes of two normal probability density functions.

Example 1. *Distribution of heights of men* (R. W. Newman and R. M. White, 1951). Table 1–1 gives the frequency distribution of the heights in inches of 24,404 U.S. Army males at the time of leaving the service. It shows the percentage of individuals falling into various intervals of height, each interval being 0.8 inches wide. Figure 1–2 shows the histogram corresponding to this table. Note that the midpoints of the bar tops of the histogram appear to lie on a curve similar to the

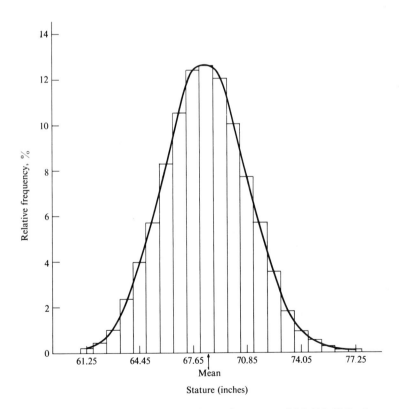

Fig. 1–2 Histogram of frequency distribution of stature of 24,404 U.S. Army males. Adapted from data of Newman and White.

4   Normal and non-normal data

normal curves in Fig. 1–1. Thus, if we approximate the upper boundary of the histogram by drawing a smooth curve through the midpoints of its bar tops, then the shape of the histogram resembles that of a normal distribution.

Table 1–1
Stature of 24,404 U.S. Army males

Midpoint of class interval (inches)	Relative frequency %	Midpoint of class interval (inches)	Relative frequency %
61.25	0.2	70.05	10.0
62.05	0.5	70.85	7.7
62.85	1.0	71.65	5.6
63.65	2.4	72.45	3.5
64.45	4.0	73.25	1.7
65.25	5.6	74.05	0.9
66.05	8.3	74.85	0.6
66.85	10.5	75.65	0.2
67.65	12.4	76.45	0.1
68.45	12.6	77.25	0.1
69.25	12.0		
		Total	99.9

Mean = 68.4 inches
Standard deviation = 2.5 inches

Adapted from data of R. W. Newman and R. M. White given in A. Damon, H. W. Stoudt, and R. A. McFarland, *The Human Body in Equipment Design*. Cambridge, Mass.: Harvard University Press, 1966, p. 24.

From the grouped data of Table 1–1, we find that the mean height is 68.4 inches and the standard deviation is 2.5 inches. (See Appendix I for discussion of mean and standard deviation.) These two numbers are sufficient to enable us to "fit" a normal distribution to these data, because a normal distribution is completely determined by its mean and standard deviation, as we shall see in Section 1–4.

We know from tables of the normal distribution (see Table A–1 at the back of the book) that approximately 68% of a normal distribution is within one standard deviation of the mean and approximately 95% is within two standard deviations. For Example 1, we find 73.5% within one standard deviation of the mean and 96.3% within two. The agreement between the theoretical and the observed curves is close in these comparisons.

Positive variables. As we shall soon see from its formula, a theoretical normal distribution spreads its probability over the whole real axis from minus to plus infinity. Yet the height of an individual must be positive. How, then, can the normal distribution be used as a model to approximate the distribution of heights? The answer is that the mean of the heights is many standard deviations away from zero, indeed about 27, and so very little probability lies to the left of zero. In Example 1, the smallest observed height was 3.6 standard deviations below the mean. Since physical quantities are ordinarily positive, and since the main body of such measurements is usually many standard deviations to the right of zero, a normal distribution fitted to the data would have a negligible part of its probability to the left of zero. This agrees well with the histogram of Example 1, which has no heights to the left of zero.

Non-normality. Naturally, many distributions of measurements are not well approximated by the normal, as the following example shows.

Example 2. *Distribution of age at death for U.S. females* (adapted from N. Keyfitz and W. Flieger). Figure 1–3 shows the histogram for the distribution of age at death for U.S. females based on experience up to 1965. The histogram for men has the same general shape, but more men than women die at the earlier ages.

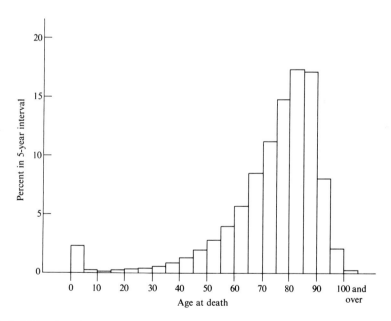

Fig. 1–3 U.S., female, 1965: percent dying in each 5-year age interval (the 100–105 interval includes all deaths after 100 rather than only those occurring in the interval). Data from N. Keyfitz and W. Flieger, *World Population: An Analysis of Vital Data.* Chicago: University of Chicago Press, 1968, p. 45.

Note that the distribution is not very much like a normal. It has two peaks (technically, *modes*) instead of one, it is not approximately symmetric, and it is not bell-shaped. The spike at the left is caused by infant mortality.

EXERCISES FOR SECTION 1–3

1. Use interpolation in Table 1–1 to estimate the proportion of men having heights more than one standard deviation above the mean height.

2. (Continuation) From Table A–1 at the back of the book, find the probability that a measurement drawn from a normal distribution exceeds one standard deviation above the mean. Compare the result with the proportion obtained in Exercise 1.

3. The text says that Example 1 has an observation as far out as 3.6 standard deviations, but Table 1–1 does not appear to extend nearly that far. Explain how that could happen.

1–4   THE FAMILY OF NORMAL DISTRIBUTIONS

In the histogram of the statures of Army men (Fig. 1–2), the midpoints of the bar tops fall on a curve whose general shape is that of a normal curve. We now review more formally the distributions associated with normal curves.

The normal probability distribution is the most important continuous distribution in statistics. The family of these distributions can be obtained from the distribution of the *standard* normal random variable Z, whose probability density function is defined by

$$f(z) = \frac{1}{\sqrt{2\pi}} e^{-z^2/2}, \qquad -\infty < z < +\infty. \tag{1}$$

It is shown in advanced calculus that the area between the graph of the density curve and the z axis is 1. Areas represent probabilities, and the 1 represents the total probability, as it must for f to be a density function. (See Appendix II, Sections 1 and 2.) As usual, $e = 2.71828 \ldots$ and $\pi = 3.14159 \ldots$.

The graphs of the density function f and the cumulative function F are shown in Figs. 1–4 and 1–5.

Let us review some properties of the standard normal distribution.

1. The mean of the distribution is 0.

2. The graph of $f(z)$ is symmetrical about the $f(z)$-axis, implying Property 1.

3. The variance of the distribution is 1.

4. The distribution extends without limit to both left and right, and $f(z) \to 0$ as $|z| \to \infty$.

Notation. In this book, the symbol "$P(\ )$" means "the probability that". Thus "$P(a < Z < b)$" means "the probability that Z lies between a and b".

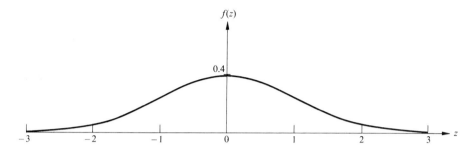

Fig. 1-4 Standard normal density function.

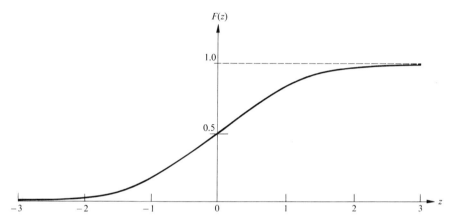

Fig. 1-5 The standard normal cumulative distribution function (cdf): $\mu = 0$, $\sigma = 1$.

5. At the origin, the ordinate is $1/\sqrt{2\pi} \approx 0.3989 \approx 0.4$. Hence, if we want to find $P(0 < Z < 0.1)$, we can approximate this probability by using $f(0) \times z \approx (0.4)(0.1) = 0.04$. A more precise value is 0.03983. Our approximation is close because the normal curve is nearly flat between 0 and 0.1.

Definition. *The family of normal distributions.* If the random variable Y is given by

$$Y = aZ + b, \qquad a \neq 0, \tag{2}$$

where Z is a *standard* normal random variable, then Y is said to be normally distributed.

If we denote the mean of Y by μ_Y and the standard deviation of Y by σ_Y, then this definition leads at once to the following theorem.

Theorem. *Mean and standard deviation.* The normal random variable $Y = aZ + b$, $a \neq 0$, has mean and standard deviation

$$\mu_Y = b \quad \text{and} \quad \sigma_Y = |a| .$$

Proof. We have, using the expected value notation, E (see Appendix I–4),

$$\mu_Y = E(aZ + b) = E(aZ) + E(b) = aE(Z) + b$$
$$= a(0) + b = b,$$

and

$$\sigma_Y^2 = \text{Var}(aZ + b) = \text{Var}(aZ) = a^2 \text{Var}(Z) = a^2(1) = a^2.$$

Hence

$$\sigma_Y = |a|.$$

For simplicity in the rest of this discussion, we shall drop the subscript Y from μ and σ.

It can be proved that the density function of Y is defined by

$$g(y) = \frac{1}{\sqrt{2\pi}\sigma} e^{-(y-\mu)^2/2\sigma^2}, \qquad -\infty < y < +\infty. \tag{3}$$

Formula (3) shows that the density function is symmetrical about the line $y = \mu$; the proof is that $g(\mu + k) = g(\mu - k)$ for any real k. It can also be proved that the area between the graph of function g and the y-axis is unity.

For arbitrary values of μ and σ, the basic parameters of the normal distribution, formula (3) defines the *two-parameter* family of normal distributions. *When specific values are assigned to μ and σ, g is completely determined.* Figure 1–6 shows three members of the normal family. The subscripts correspond to the variables. For example, Y_1 has mean $\mu_1 = -2$ and standard deviation $\sigma_1 = 1$; and Y_2 has mean $\mu_2 = 0$ and standard deviation $\sigma_2 = 0.5$.

Why is the family of normal distributions important? Here are five reasons.

1. Many populations of measurements are distributed normally or approximately normally.

2. Some non-normal random variables, such as the binomial, can sometimes be transformed into random variables that are approximately normally distributed.

3. Under very general conditions, *sums* of random variables, normal or non-normal, are approximately normally distributed.

4. The normal probability distribution is *tabulated*. Hence, when we know μ and σ, we can readily get approximate probabilities for a wide range of distributions.

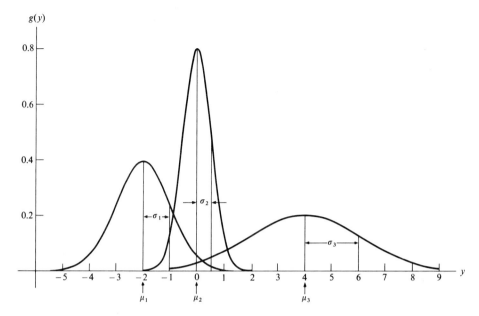

Fig. 1–6 A collection of three normal density functions with differing means and differing standard deviations: $\mu_1 = -2$, $\sigma_1 = 1$; $\mu_2 = 0$; $\sigma_2 = 0.5$; $\mu_3 = 4$, $\sigma_3 = 2$.

5. The normal is mathematically tractable, and so in a new problem we can often find out what would happen if the distribution were normal and thus get an idea of what happens for other distributions.

The following examples illustrate the use of the standard normal probability table.

Note. For a continuous random variable X, the probability that X takes any specific value x is 0. Thus $P(X = x) = 0$. Therefore, $P(X \leq x) = P(X < x)$ for a continuous random variable.

[*Proof:* Since $X < x$ and $X = x$ are mutually exclusive events, $P(X \leq x) = P(X < x \text{ or } X = x) = P(X < x) + P(X = x) = P(X < x)$.]

Example 1. For the standard normal random variable Z, find

   a) $P(-1 < Z < 1)$,   b) $P(-2.4 \leq Z \leq -0.5)$,   c) $P(Z > 1.2)$.

Solutions. We use Table A–1 at the back of the book.

a) Since the normal curve is symmetric about $z = 0$,

$$P(-1 < Z < 1) = \text{area above the interval from } z = -1 \text{ to } z = 1$$
$$= 2 \text{ (area from } z = 0 \text{ to } z = 1)$$
$$= 2(0.3413) = 0.6826.$$

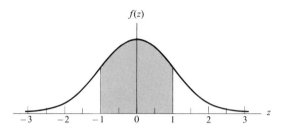

b) From symmetry,

$$P(-2.4 \le Z \le -0.5) = P(0.5 \le Z \le 2.4)$$
$$= \text{(area from } z = 0 \text{ to } z = 2.4\text{)}$$
$$- \text{(area from } z = 0 \text{ to } z = 0.5\text{)}$$
$$= 0.4918 - 0.1915 = 0.3003.$$

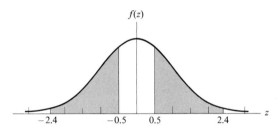

c) $$P(Z > 1.2) = \text{(area to the right of } z = 0\text{)}$$
$$- \text{(area from } z = 0 \text{ to } z = 1.2\text{)}$$
$$= 0.5000 - 0.3849 = 0.1151.$$

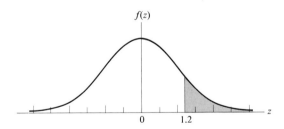

Nonstandard normal random variables. The normal distributions make up a two-parameter family, as we have noted. Fortunately, the table for the *standard* normal distribution enables us to find the probabilities for *any* normal random variable, X. The method is simple: by means of a transformation, we *standardize* X. Thus, if X is a normal random variable, then

$$Z = \frac{X - \mu_X}{\sigma_X} \tag{4}$$

is the standard normal random variable, which has $\mu = 0$ and $\sigma = 1$. (See Appendix III–2.)

Note that the transformation (4) in effect expresses each value of X in terms of its distance from its own mean, measured in standard deviations.

Because of (4), we have, for example,

$$P(X \le x) = P\left(Z \le \frac{x - \mu_X}{\sigma_X}\right),$$

where Z is the standard normal random variable.

Example 2. The results of a test are normally distributed with mean 500 and standard deviation 100. What is the probability that a score drawn at random is less than or equal to 740?

Solution. Let the normal random variable X denote a random score. We want $P(X \le 740)$. Since $\mu_X = 500$ and $\sigma_X = 100$, our standardized X is

$$Z = \frac{X - 500}{100},$$

where Z is the *standard* normal random variable. Therefore,

$$P(X \le 740) = P\left(Z \le \frac{740 - 500}{100}\right) = P(Z \le 2.4).$$

Table A–1 gives

$$P(Z \le 2.4) = 0.5000 + 0.4918 = 0.9918.$$

EXERCISES FOR SECTION 1–4

If Z is a standard normal random variable, use Table A–1 at the back of the book to evaluate the following.

1. $P(0 < Z < 1)$
2. $P(-2 < Z < 2)$
3. $P(0.3 < Z < 3.2)$
4. $P(-0.3 < Z < 3.2)$
5. $P(-3.2 < Z < -0.3)$
6. $P(|Z| < 0.4)$
7. $P(|Z| > 0.4)$
8. $P(Z < -0.3)$
9. If X is a normal random variable with $\mu_X = 100$ and $\sigma_X = 10$, evaluate:
   a) $P(X > 112)$,   b) $P(X < 85)$,
   c) $P(80 < X < 120)$.

10. For the numbers given in Example 2, find (a) $P(X > 640)$, (b) $P(X < 380)$.

11. If Z is the standard normal random variable and $P(|Z| < k) = 0.4$, find k. If $P(0 < Z < k) = 0.433$, find k.

12. From the normal table, find the area above the interval from 0 to h for $h = 0.01$, 0.1, 0.2, 0.3, 0.4, 0.5. Compare these values with $0.4h$. How big must h be before the difference is 10% of the correct value?

1–5 PERMUTATION TESTS FOR DIFFERENCES IN DISTRIBUTIONS

Suppose that we have samples from two distributions and that we wish to test whether the distributions are essentially identical, as in (a) below, or whether they have different means, as in (b). If, further, we are unwilling to make strong assumptions about the shapes or standard deviations of the underlying distributions, then what sort of test could we make?

(a)   (b)

No doubt, since the sample means of the two samples offer good estimates of the population means, we could continue to use the difference in sample means, $\overline{Y} - \overline{X}$, to estimate the difference in population means, $\mu_Y - \mu_X$. But how would we be able to assess the credibility of the difference? A small example will illustrate an approach.

Example 1. *Difference of means.* Without making assumptions about the shapes of the distributions of measurements, find how unlikely we are to get, from identical populations by chance alone, a difference in the means as large as that of the following two samples:

$$X\text{-sample: } 108, 96, 120 \qquad Y\text{-sample: } 150, 126$$

Solution. Taking the averages for each sample we get

$$\overline{Y} - \overline{X} = 138 - 108 = 30,$$

which appears to be a substantial difference. But we need some measure of the chance of getting such an extreme result when the original distributions are identical.

If the two sets of measurements are independently drawn from the same population, then the usual way to view the sampling is this:

Draw a first sample of 3 for the X-sample, and then draw a sample of 2 for the Y-sample.

For our purpose, a different but more useful viewpoint is this:

Draw a pooled sample of 5. Then from this sample draw at random 2 measurements for the Y-sample, and leave the remaining 3 measurements for the X-sample.

The two ways of getting the X- and Y-samples give identical long-run results, because the order of the drawings is independent of the size of the measurements and because the X-sample and the Y-sample are all independently drawn from the same distribution. Proving the equivalence formally requires a bit of notation that we shall not tool up for, since we merely want to describe the method for this problem.

If the numbers are all independently drawn from the same distribution, then the second way of looking at the sampling process implies that, once the original 5 numbers are drawn to make up the two samples, it is just a matter of chance which pair of numbers turns up in the Y-sample. There are 10 possible outcomes, and they are equally likely. We can therefore compute the difference of the means for each of the possible outcomes and see how extreme the observed difference is, given these 5 measurements and given that the populations are identical.

Table 1–2

The 10 possible samples and differences of means

Outcome	96	108	120	126	150	$\bar{Y} - \bar{X}$
1	X	X	X	Y	Y	$138 - 108 = 30$
2	X	X	Y	X	Y	$135 - 110 = 25$
3	X	Y	X	X	Y	$129 - 114 = 15$
4	Y	X	X	X	Y	$123 - 118 = 5$
5	X	X	Y	Y	X	$123 - 118 = 5$
6	X	Y	X	Y	X	$117 - 122 = -5$
7	Y	X	X	Y	X	$111 - 126 = -15$
8	X	Y	Y	X	X	$114 - 124 = -10$
9	Y	X	Y	X	X	$108 - 128 = -20$
10	Y	Y	X	X	X	$102 - 132 = -30$

If the distributions for X and Y are not identical to begin with, and if Y, say, is generally to the right, as in (b) above, then the Y-sample has a high probability of having the larger mean.

We list in Table 1–2 the 10 possible samples and the differences in sample means that occur with these 5 measurements.

Among all the differences that can occur, the one that we observed, 30, is the most extreme in the positive direction. We can say there is 1 chance in 10 of getting a result this extreme in the positive direction, if these 5 measurements were drawn and if the original populations were identical. In technical language, we say that "the descriptive level of significance is 1/10".

> **Definition.** The *descriptive level of significance* is the probability of getting a result as extreme as, or more extreme than, the observed result, under the given hypothesis.

For example, if we had observed a difference of 25 (outcome 2, Table 1–2), then we would report a descriptive level of significance of 2/10, because the chances are 2 in 10 of getting a result at least as extreme as 25.

   To this point, we have considered extremeness only in the positive direction. If we wish to consider results far from zero in either the positive or the negative direction as extreme, then the descriptive level of significance is the probability of getting results with absolute values at least as extreme as the *absolute value* of the observed result. Thus, if we consider extreme results in either the positive or the negative direction, and if the observed difference is 25, then we report that the descriptive level of significance is 3/10, because Table 1–2 shows that

$$P(|\text{observed difference}| \geq 25) = \frac{3}{10}.$$

   If we treat only one direction (positive or negative), we speak of a *one-sided* approach. But if we treat both directions simultaneously, we speak of a *two-sided* approach.

General procedure. The general technique for this kind of "permutation test" is first to decide on the statistic one is going to study. Here it is $\overline{Y} - \overline{X}$. Next, find the exact distribution of this statistic over the possible samples that could have been formed. Then refer the observed outcome to this distribution of all possibilities, and thus get a probability statement.

Listing a few extreme results. Sometimes the exact results can be obtained even when the number of possible samples is large, because it may be feasible to pick out and list those sample partitions that produce differences of at least a given size. In the easiest case, where the difference in means is the largest possible difference, the calculation is simple. But sometimes, even when the number of possible samples is enormous, it is easy to list the few instances in which the actually observed difference is exceeded.

Example 2. *Rosier futures.* How much better does the future look? Hadley Cantril surveyed samples of people in many countries, asking about what concerned them, how satisfied they were, and how satisfied they anticipated being five years later. He rated their satisfactions on a scale from 0 to 10. People from all countries thought the future would be better. Table 1–3 compares the anticipated improvements in ratings in 3 industrialized countries with those anticipated in 6 less industrialized countries. Use the permutation test to assess the difference in means of anticipated improvement scores in the two kinds of countries.

Table 1–3

Extent of anticipated improvements in two kinds of countries

Industrialized		Nonindustrialized	
United States	1.2	Yugoslavia	1.7
West Germany	0.9	Philippines	1.8
Japan	1.0	Nigeria	2.6
—		Brazil	2.7
Total	3.1	Poland	1.1
		India	1.4
		Total	11.3

Solution. The largest differences in means occur when the sums of the measurements for the 3 industrialized countries are smallest. We can list the few samples that give especially small sums of three measurements. The smallest measurements among the 9 are 0.9, 1.0, 1.1, and 1.2. These lead to the following smallest sums of sets of 3 measurements.

0.9	1.0	1.1	1.2	Sum	
X	X	X		3.0	*the smallest sum*
X	X		X	3.1	*the observed sum*
X		X	X	3.2	
	X	X	X	3.3	

There are $\binom{9}{3} = 84$ possible sums, and only 2 of them are as small as, or smaller than, the sum observed. Consequently, we can say that, by chance alone, for these 9 measurements

$$P(\text{sum} \le 3.1) = \frac{2}{84}.$$

Since the probability of getting sums less than or equal to 3.1 is small, we conclude that the industrialized countries do not have quite as high anticipations as the nonindustrialized ones for improvements in the future.

16   Normal and non-normal data

EXERCISES FOR SECTION 1–5

1. Two samples of size 2 give measurements as follows:

    X-sample: 2.2, 2.4;   Y-sample: 2.3, 5.6.

    Assuming identical populations, use the permutation approach to find how unusually large the difference in means is.

2. In Example 1, assuming identical populations, find

    a) $P(|\bar{Y} - \bar{X}| \geq 15)$,   b) $P(\bar{Y} - \bar{X} \geq 15)$.

3. Solve Example 2 (Rosier futures), given that the scores for the United States and India are interchanged. (Make a partial listing.)

4. Suppose, in Example 1, that we had obtained the following samples: X-sample, 108, 120, 150; Y-sample, 96, 126. Compute $\bar{Y} - \bar{X}$, and use the permutation approach to report (a) the one-sided descriptive level of significance, (b) the two-sided descriptive level of significance.

5. In Example 1, we used the permutation approach with Table 1–2 to study the statistic $\bar{Y} - \bar{X}$. Use Table 1–2 to study the statistic

    $$\text{smallest } Y - \text{smallest } X.$$

    Assuming identical populations, find

    (a) $P(\text{smallest } Y - \text{smallest } X \geq 24)$,   (b) $P(|\text{smallest } Y - \text{smallest } X| \geq 24)$.

6. Use Table 1–2 to study the statistic

    $$\text{largest } Y - \text{largest } X$$

    Assuming identical populations, find

    a) $P(\text{largest } Y - \text{largest } X \geq 30)$,   b) $P(|\text{largest } Y - \text{largest } X| \geq 30)$.

7. The *range* of values of a finite set of numbers is

    $$\text{its largest value } - \text{its smallest value.}$$

    Thus, in Example 1, the range of the X-sample is $120 - 96 = 24$. Use Table 1–2 to study the statistic

    $$\frac{\text{range of } Y}{\text{range of } X}.$$

    Find

    $$P\left(\frac{\text{range of } Y}{\text{range of } X} \geq 1\right)$$

    when the 10 selections are equally likely.

8. *Densities of planets.* The 5 planets known to the ancient world may be divided into two groups.

    Group 1: Planets farther from the sun than planet Earth (Mars, Jupiter, and Saturn).

    Group 2: Planets closer to the sun than planet Earth (Mercury and Venus).

    Taking the density of Earth as 1 unit, astronomers have made the following measurements of planetary densities.

	Y-sample			X-sample	
Planet:	Mercury	Venus	Earth	Mars	Jupiter	Saturn
Density:	0.68	0.94	1	0.71	0.24	0.12

Consider the 3 densities in Group 1 as an X-sample of measurements and the 2 densities in Group 2 as a Y-sample. Use the permutation approach to report a one-sided descriptive level of significance for the statistic $\bar{Y} - \bar{X}$.

9. If the X-sample has m measurements and the Y-sample n measurements, how small can the descriptive level of significance for the permutation test be?

1-6 PERMUTATION TEST WITH LARGE SAMPLES; SIMULATION

The student will at once note that, had the sizes of the samples in Section 1-5 been considerably larger, say two samples of size 10, the number of possible partitions into two samples would be large (184,756), and the listing of all possibilities or even the more extreme ones would be expensive, perhaps prohibitively so. In such situations, we may content ourselves with the results of a large random sample.

Given two samples of 10, we can readily draw many random partitions (say 1000) with a high-speed computer, or we can draw at least 100 by hand and make the appropriate calculations to get an approximate frequency distribution of the results. Then we can get an approximate value for the probability of a result at least as extreme as the one observed.

Example 1. *Marijuana.* The following data (Table 1-4) show the change in performance of naive subjects following certain treatments. The first set of data

Table 1-4

Data on test performance following smoking

X Ordinary cigarette	Y Marijuana cigarette
-3	5
10	-17
-3	-7
3	-3
4	-7
-3	-9
2	-6
-1	1
-1	-3
Total 8	-46

− means poorer performance.

corresponds to changes in intellectual performance following the smoking of an ordinary cigarette, the second to such changes following the smoking of a marijuana cigarette. Use the permutation test to treat the data from the subjects as if they were two independent samples from an arbitrary distribution, and find out how extreme the difference in means is.

Solution. Since we have two sets of 9 observations, the number of possible sets of samples is too large for easy listing. Using a high-speed computer, we drew 1000 random partitions from the 18 measurements and computed the differences in sums instead of means, since the sample sizes are equal. These differences in sums have the frequency distribution shown in Table 1–5. (Because the samples are of equal size, the theoretical distribution must be symmetric.)

Table 1–5

Empirical distribution of $\sum Y - \sum X$ in 1000 random partitions of the 18 measurements of Table 1–4

$\sum Y - \sum X$	Frequency	$\sum Y - \sum X$	Frequency	$\sum Y - \sum X$	Frequency
		− 24	13	20	26
− 74	1	− 22	17	22	19
− 66	1	− 20	23	24	17
− 62	2	− 18	27	26	22
− 60	1	− 16	21	28	18
− 58	4	− 14	21	30	25
− 56	3	− 12	19	32	13
− 54	3	− 10	36	34	19
− 52	5	− 8	32	36	13
− 50	7	− 6	36	38	8
− 48	8	− 4	28	40	15
− 46	8	− 2	25	42	9
− 44	11	0	21	44	9
− 42	12	2	33	46	5
− 40	8	4	35	48	8
− 38	9	6	32	50	5
− 36	17	8	34	52	4
− 34	12	10	30	54	3
− 32	15	12	23	56	3
− 30	13	14	29	58	1
− 28	14	16	24	62	1
− 26	18	18	26		
					1000

The actual difference in sums is $-46 - (8) = -54$. The number of simulated samples at -54 or below, or at $+54$ or above, is 23. So, by chance alone, we have

$$P(|\text{difference}| \geq 54) \approx 0.023,$$

and

$$P(\text{difference} \leq -54) \approx 0.012.$$

The latter report is appropriate if one wants to decide whether marijuana lowers the long-run average performance, as it did in these particular data.

EXERCISES FOR SECTION 1–6

1. Use Table 1–5 to report
   a) the two-sided descriptive level of significance if $\sum Y - \sum X = -52$,
   b) the one-sided descriptive level of significance if $\sum Y - \sum X = -46$.

2. Suppose that the same 18 numbers of Example 1 had been distributed so that $\sum X = 10$, $\sum Y = -40$. Use Table 1–5 to report the one-sided descriptive level of significance for $\bar{Y} - \bar{X}$.

3. Suppose, in Example 1, that we learn that the scores in the fourth row of Table 1–4, namely 3 and -3, were incorrectly recorded and should be interchanged. What change would we make in our report? (Use both the one-sided and the two-sided approach.)

4. *A class project.* Let each member of the class make 10 random partitions of the 18 measurements in Example 1. (Begin at a different entry in the table of random numbers for each partition.) Combine the results of the entire class to obtain by simulation a table of frequency distributions of $\sum Y - \sum X$ similar to Table 1–5.

5. Prove that when the X- and Y-samples are the same size, the distribution of $\bar{Y} - \bar{X}$ is symmetrical.

6. How might you use Table 1–5 to get, for $t > 0$, a better estimate of $P(\sum Y - \sum X > t)$ than the observed proportion of differences larger than t?

7. Explain how to use the permutation test to decide whether the variances of the populations from which two samples are drawn are equal, or unusually large or small in their ratio.

1–7 NORMAL APPROXIMATION FOR PERMUTATION TEST FOR DIFFERENCE OF MEANS

When the sample sizes are *large*, we can use a normal approximation to obtain the approximate descriptive significance level of the difference in means by developing the following theory.

Let us denote the total of all the measurements in the two samples drawn without replacement by S, so that

$$S = \sum X + \sum Y. \tag{1}$$

If the X- and Y-samples of sizes n_x and n_y, respectively, are randomly drawn from the $N\ (=n_x + n_y)$ measurements, then

$$E(\overline{Y} - \overline{X}) = \mu_{\overline{Y}} - \mu_{\overline{X}} = 0. \tag{2}$$

Although this result is obvious (from symmetry), a formal proof gives us a useful relation. First, let us express the difference of means in terms of S and $\sum X$ alone by replacing $\sum Y$ by $S - \sum X$ as follows:

$$\overline{Y} - \overline{X} = \frac{\sum Y}{n_y} - \frac{\sum X}{n_x} \tag{3}$$

$$= \frac{S - \sum X}{n_y} - \frac{\sum X}{n_x}.$$

After some simplification this gives

$$\overline{Y} - \overline{X} = \frac{S}{n_y} - \overline{X}\left(\frac{N}{n_y}\right). \tag{4}$$

Since $E(\overline{X}) = S/N$ (see Appendix VI–1), by taking expected values in equation (4), we get

$$E(\overline{Y} - \overline{X}) = 0, \tag{5}$$

as claimed.

Next we get the variance of $\overline{Y} - \overline{X}$ from the final expression in equation (4). Recall (Appendix VI–2) that the variance of the sample mean of a sample of size n, drawn without replacement from a finite population of size N with variance σ^2, is

$$\sigma_{\overline{X}}^2 = \frac{\sigma^2}{n}\left(\frac{N - n}{N - 1}\right). \tag{6}$$

Furthermore, if both $n_x = n$ and $n_y = N - n$ are not too small, say both greater than 3 or 4 (provided the individual measurements are not wildly distributed), we can as a practical matter expect approximate normality for the distribution of $\overline{X}$. It may interest the student to know that proving the approximate normality of $\overline{X}$ for sampling without replacement and for sampling with replacement are separate mathematical ventures. That is, these central limit theorems require different proofs because in the former case the observations are correlated, whereas in the latter they are not. It can be proved that the theorems hold in both cases.

Equation (4) shows, once S is fixed, that $\overline{Y} - \overline{X}$ is a linear function of $\overline{X}$. If $\overline{X}$ is approximately normal, then so is $\overline{Y} - \overline{X}$. This means that $\overline{Y} - \overline{X}$ under random partitions of the original observations is approximately normal, with mean 0 and, from equation (4), variance

$$\text{Var}\ (\overline{Y} - \overline{X}) = \frac{N^2}{n_y^2}\sigma_{\overline{X}}^2 = \frac{N^2\sigma^2}{n_y^2 n_x}\left(\frac{N - n_x}{N - 1}\right). \tag{7}$$

The step from equation (4) to equation (7) uses equation (2) of Appendix I–5. Since $N = n_x + n_y$, equation (7) reduces to

$$\text{Var}\,(\bar{Y} - \bar{X}) = \frac{N^2\sigma^2}{n_x n_y(N - 1)} = \frac{N}{N - 1}\left(\frac{1}{n_x} + \frac{1}{n_y}\right)\sigma^2. \tag{8}$$

Example 1. Use the data in Example 1 of Section 1–6 and the normal approximation to find how extreme the difference in means is.

Solution. Since the sample sizes are equal, $n_x = n_y = N/2$ and equation (8) reduces to

$$\text{Var}\,(\bar{Y} - \bar{X}) = 4\sigma^2/(N - 1). \tag{9}$$

We have $N = 18$, and σ^2, the variance of the original 18 measurements, is found to be 34.77. Since the difference of sums was -54, the observed difference of means is $\bar{y} - \bar{x} = -6.00$. (Note that we use lower-case letters to denote the *values* of the random variables, as discussed in Appendix I–1.) Substituting into equation (9) gives

$$\text{Var}\,(\bar{Y} - \bar{X}) = 4(34.77)/17 = 8.18,$$

and

$$\sigma_{\bar{Y} - \bar{X}} = \sqrt{8.18} = 2.86.$$

We next compute the standard normal deviate

$$\frac{\bar{y} - \bar{x} - 0}{\sigma_{\bar{Y} - \bar{X}}} = \frac{-6.00}{2.86} = -2.10.$$

Hence the observed result is more than 2 standard deviations away from 0, the mean when the samples are randomly selected. Looking this result up in the normal table, we find

$$P(|\bar{Y} - \bar{X}| \geq 6.00) = 0.0358.$$

This probability can be compared with that obtained from our simulation, namely 0.023. Either result tells us that the probability is small of getting by chance an observed difference at least this large.

Note. The original 18 measurements are samples from populations. Population variances are usually designated by σ^2, and sample variances are usually designated by s^2, with suitable subscripts. But from the point of view of the permutation calculation, these 18 measurements form a *population*. The calculation is made, *given* these 18 measurements. And so it is appropriate to use σ^2 to designate their variance in this calculation.

The idea of location. When we speak of *location* of a distribution, we are using a vague term to refer to the placement of a probability density function. Two specific measures of location that are not vague terms are *population mean* and *population median.* But there is no end of measures of location. For example, the position of the 10% point or any other percentage point of a distribution could be a measure of location.

We have discussed differences in means in the present chapter, and such discussions provide intuitive ideas about differences in location. But the methods we employ apply as well to differences in medians, or to distributions one of which is by and large to the left of the other. For example, it may be that a random observation from distribution *A* is likely to fall to the left of one drawn from distribution *B*. Generally, but not always, methods intended to assess one kind of difference in location also give similar inferences about other kinds. That is one reason why sturdy methods of statistics can and should be used.

Let us turn to an important but delicate point about means. By taking a tiny bit of probability from a distribution with a mean and moving it far to the right, one can make the mean of the distribution as large as he pleases. This implies that means are sensitive to the shape of the far tails of distributions. We are often more interested in where the central body of the distribution sits, and we may not be as concerned about the far tails as the population mean would imply. And so when we discuss means and differences of means or shifts or slippage (which just mean moving a distribution bodily to the right or left), it is well not to concentrate excessively on means, but to recognize that we are generally treating differences in location.

EXERCISES FOR SECTION 1–7

1. Suppose that a repetition of the marijuana experiment in Example 1, Section 1–6, with 9 pairs of subjects gives these data:

$$\sum X = 9, \qquad \sum Y = -41, \qquad \text{and} \qquad \sigma^2 = 30.42$$

   (Assume identical distributions.) Use the normal approximation to find the one-sided and two-sided descriptive levels of significance for $\bar{Y} - \bar{X}$.

2. Carry out the algebraic simplification leading from equation (3) to equation (4).

3. Prove that equation (7) follows from equation (4).

4. In equation (8), verify that the expression in the center transforms into the expression on the right.

5. In equations (6), (7), and (8) the expression σ^2 appears. Explain exactly what population of measurements this expression refers to. Why are we able to compute it exactly here, when in most problems we have only an estimate of σ^2?

6. Refer to Exercise 5. Although you could use the methods of Section 1–6 to assess the difference of the *medians* of two samples, you might have difficulty developing a normal approximation, as in the present section. Why? More generally explain why this section applies especially to means while other sections on the permutation test can be used for comparing a variety of statistics in the two-independent-sample situation.

7. In an article on all-time greatness in baseball, the *Baseball Digest* for June 1971 gave information about the best ten batters each club had ever had. Use the normal approximation to the permutation test to compare the mean slugging "percentage" for the Yankees and the Athletics.

[*Suggestion:* Round off the percentage to two decimals before computing σ^2.]

Yankees		Athletics	
Ruth	0.711	Foxx	0.640
Gehrig	0.632	Simmons	0.584
DiMaggio	0.579	R. Johnson	0.523
Mantle	0.557	Cochrane	0.490
Keller	0.518	Zernial	0.489
Maris	0.515	Hauser	0.478
Meusel	0.500	Higgins	0.478
Skowron	0.496	C. Walker	0.469
Heinrich	0.491	Baker	0.467
Dickey	0.486	E. Miller	0.464

CHAPTER 2

Applications of the Binomial and Multinomial Distributions

2–1 THE BINOMIAL DISTRIBUTION

The most frequently used family of discrete probability distributions is the binomial. Binomial distributions give us a probabilistic model for the outcomes of a large class of experiments illustrated by the following example.

Example 1. *Camera spring.* To be satisfactory, the tension in a camera spring should lie in a given interval. The manufacturer is having trouble because 10% of the springs are defective. Of the next 5 springs tested, how many will be defective?

Solution. The number may be 0, 1, 2, 3, 4, 5. The probability of 0 defectives is $(0.9)^5 \approx 0.590$, of 5 defectives is $(0.1)^5 \approx 0.000$. The other probabilities are: for 1 defective, 0.328; for 2 defectives, 0.073; for 3 defectives, 0.008; and for 4 defectives, 0.000. These results can be obtained by applying formula (1) below.

In this book, it is assumed that the reader has had experience with the binomial distribution. Hence we shall merely recall the facts and then illustrate them with the camera spring example.

A binomial experiment has four characteristics:

1. There is a *fixed number* of trials, *n*. In Example 1, the tests of springs were the trials, and $n = 5$.

2. Outcomes of the trials are *independent* of one another. A defect in one spring does not change the probability that another spring is defective.

3. Each trial falls into one of two categories, often called "*success*" or "*failure*". "Success" and "failure" are unemotional labels used for convenience in discussing the two categories of outcomes in a single trial. In Example 1, the outcomes were "defective" or "nondefective". The choice of the outcome to be called "success" is totally arbitrary. To drive this fact home, we shall call a defective a "success" in the camera spring example.

24

4. For each trial, the probability of a success is the same *constant*, *p*. Each new camera spring has probability $p = 0.1$ of being defective.

If the random variable *X* denotes the number of successes in a binomial experiment, then *X* has a binomial distribution, and we have the following:

a) The probability function of *X*, which depends on *n* and *p*, is given by

$$f(x) = P(X = x) = b(x; n, p) = \binom{n}{x} p^x (1 - p)^{n-x}, \qquad x = 0, 1, \ldots, n, \quad (1)$$

where

$$\binom{n}{x} = \frac{n!}{x!(n - x)!}.$$

For the binomial, $f(x) = 0$ except at the integers $x = 0, 1, \ldots, n$. Let *X* be the number of defective springs. The probability of exactly 2 defective springs is $[5!/2!3!](0.1)^2(0.9)^3 = 0.0729$.

b) The mean of *X* is $\mu_X = np$. Since $n = 5$, $p = 0.1$, the mean or "expected number" of defective camera springs is $\mu_X = 5(0.1) = 0.5$.

c) The variance of *X* is $\sigma_X^2 = np(1 - p) = npq$, where $q = 1 - p$. For the camera spring example, $\sigma_X^2 = 5(0.1)(0.9) = 0.45$, and the standard deviation is $\sigma_X \approx 0.68$.

The set of ordered pairs $(x, b(x; n, p))$ gives the discrete probability distribution of *X*, for which extensive tables are available. Our Table A–5 at the back of the book gives cumulative tables, and Table A–6 gives special tables for $p = 0.5$. If *n* is large, we can approximate the distribution of *X* by methods to be discussed in Section 2–2.

2–2 THE NORMAL APPROXIMATION FOR THE BINOMIAL

In a binomial experiment, if *X* is the number of successes, *p* the probability of a success, $q = 1 - p$, and *the number of trials, n, is large*, then

$$P(X \geq x) \approx P(Z > z),$$

where *Z* is the standard normal random variable, and

$$z = \frac{x - \frac{1}{2} - np}{\sqrt{npq}}.$$

(See Appendix III–3.)

The foregoing statement, which uses the Central Limit Theorem for the binomial distribution, involves some important ideas. The random variable *X* is discrete, but the random variable *Z* is continuous. How do we approximate the discrete binomial distribution by using the continuous standard normal distribution? Well, first we transform X into the *standardized* discrete random variable (see Exercise 4)

$$\frac{X - \mu_X}{\sigma_X} = \frac{X - np}{\sqrt{npq}},$$

which has mean 0 and standard deviation 1, just like the standard normal distribution. The Central Limit Theorem assures us that, for large n, and p not too near 0 or 1, the standardized random variable is distributed approximately like Z. To correct for the discreteness of X, we use the following reasoning.

Approximating a discrete distribution by a continuous one. In general, consider a continuous random variable, V, and a discrete random variable, T, taking a set of successive integral values, as shown in Fig. 2–1(a, b). We shall suppose that the distribution of T can be approximated by that of V. For the continuous random variable V, $P(V \geq 3)$ is given by the *area* of the shaded part of Fig. 2–1(a); for the *discrete* random variable T, $P(T \geq 3)$ is given by the sum of the ordinates shown in Fig. 2–1(b). Let us replace each ordinate $P(t)$ by a rectangle centered at t and having a width of one unit and height equal to the ordinate $P(t)$, as shown in Fig. 2–1(c). Then the area of each rectangle has the same numerical measure as the ordinate that it replaced. For example, the area of the rectangle from $3 - \frac{1}{2}$ to $3 + \frac{1}{2}$ has the same numerical measure as $P(3)$, the ordinate at 3 in Fig. 2–1(b).

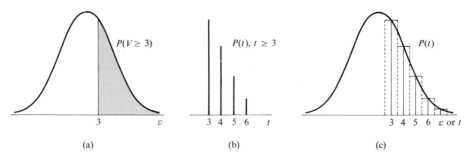

(a)   (b)   (c)

Fig. 2–1 The details of the approximation.

Now $P(T \geq 3)$ is given by the area of all rectangles to the right of $t = 2\frac{1}{2}$ in Fig. 2–1(c). If we were to take only *areas* to the right of $t = 3$ in that figure, we should lose one half of $P(3)$. As an approximation to the sum of the rectangular areas to the right of $t = 2\frac{1}{2}$, we use the area under the continuous curve to the right of $2\frac{1}{2}$. We now apply this idea to the binomial distribution.

Returning to the binomial random variable X and the standard normal random variable Z, in order to get $P(X \geq x)$, we correct for continuity by using the value $x - \frac{1}{2}$, computing the deviate

$$z = \frac{x - \frac{1}{2} - np}{\sqrt{npq}},$$

and then using the normal tables.

Similarly, to find $P(X \leq x)$, we compute

$$z = \frac{x + \frac{1}{2} - np}{\sqrt{npq}},$$

and then refer to the normal tables.

Example 1. A coin is tossed 100 times. If X is the number of heads, find $P(X \geq 60)$.

Solution. Since n is large, we use the normal approximation. The required deviate for entering the normal table is

$$z = \frac{60 - 0.5 - 100(0.5)}{\sqrt{(100)(0.5)(0.5)}} = \frac{9.5}{5} = 1.9.$$

Then

$$P(X \geq 60) \approx P(Z > 1.9)$$
$$= 0.5000 - 0.4713 = 0.0287.$$

Confidence limits for a binomial probability p based on large samples. Based on a sample size of n with X successes, the usual estimate of the probability of success, p, is X/n. Such an estimate is called a *point estimate* because when evaluated it gives a single number. We use the notation $\bar{p} = X/n$. The mean and variance of $\bar{p}$ are, respectively, p and $p(1 - p)/n$. We usually write pq/n for $p(1 - p)/n$.

Sometimes we want an interval estimate for p, that is, an interval likely to include the true value of p. We use the fact that

$$\frac{X - np}{\sqrt{npq}} = \frac{(X/n) - p}{\sqrt{pq/n}} = \frac{\bar{p} - p}{\sqrt{pq/n}}$$

is approximately a standard normal random variable. (See Appendix VII–1.)

When p is not too near 0 or 1, a convenient way to get a fair approximation to 95% confidence limits for *large* samples is to recall that

$$P\left(\left| \frac{\bar{p} - p}{\sqrt{pq/n}} \right| \leq 1.96 \right) \approx 0.95.$$

The statement inside the parentheses can be rewritten as

$$\bar{p} - 1.96\sqrt{pq/n} < p < \bar{p} + 1.96\sqrt{pq/n}. \tag{1}$$

(See Appendix VII–1 and Exercise 5.) Since we don't know p, we have trouble substituting for $\sqrt{pq/n}$ in inequalities (1). So we use $\sqrt{\bar{p}\bar{q}/n}$ and thus get a good estimate. Approximately, the probability is 0.95 that

$$\bar{p} - 1.96\sqrt{\bar{p}\bar{q}/n} < p < \bar{p} + 1.96\sqrt{\bar{p}\bar{q}/n}. \tag{2}$$

It is convenient to think of this as the statement "p is in the interval between $\bar{p} + 1.96\sqrt{\bar{p}\bar{q}/n}$ and $\bar{p} - 1.96\sqrt{\bar{p}\bar{q}/n}$".

The upper and lower limits for p are called the *upper and lower confidence limits*; and $[\bar{p} - 1.96\sqrt{\bar{p}\bar{q}/n}, \bar{p} + 1.96\sqrt{\bar{p}\bar{q}/n}]$ is the *confidence interval*. The *confidence coefficient* is 0.95, which is the probability that p is in the interval as long as we regard $\bar{p}$ as a random variable.

In specific examples, when we replace $\bar{p}$ and n by their numerical values, either the true p lies in the interval or it does not; and so the statements corresponding to (2) will be true or false. About 95% of these statements will be true.

Example 2. Of a random sample of 900 people responding to a questionnaire, 80% favored a proposed amendment to the Constitution. Set 95% confidence limits on the proportion of the population who would respond and favor the amendment if a census were taken.

Solution. $\bar{p} = 0.80$, $\sqrt{\bar{p}\bar{q}/n} = \sqrt{(0.8)(0.2)/900} \approx 0.013$.   The 95% confidence limits are $0.800 \pm 1.96(0.013) = 0.774, 0.826$.

To set different confidence limits on p, we need only replace the multiplier 1.96 in inequality (2). In general, if the confidence coefficient is $1 - \alpha$, then the general confidence limits on p are given by

$$\bar{p} - z_{\frac{1}{2}\alpha}\sqrt{\bar{p}\bar{q}/n} < p < \bar{p} + z_{\frac{1}{2}\alpha}\sqrt{\bar{p}\bar{q}/n}, \tag{3}$$

where $z_{\frac{1}{2}\alpha}$ is the standard normal deviate having probability $\frac{1}{2}\alpha$ to the right of it:

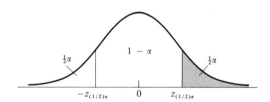

For example, for confidence 0.90, $\alpha = 0.010$, $z_{\frac{1}{2}(0.10)} = z_{0.05} = 1.65$, from Table A–1.

EXERCISES FOR SECTION 2–2

1. A basketball player sinks 60% of his foul shots on the average. During the season, he gets 100 tries. What is the approximate probability that he sinks at least 50 of them? 50 or less?

2. Given 1000 tosses of a coin, approximate the probability of exactly 500 heads.

3. If $n = 2m$ coins are tossed, approximate the probability of getting exactly m heads.

4. If X is a random variable with mean μ and standard deviation σ, show that $(X - \mu)/\sigma$ is a random variable with mean 0 and standard deviation 1. (See Appendix I–5.)

5. If X is a normal random variable with mean μ and variance σ^2, show that

   a) $P\left(\left|\dfrac{X-\mu}{\sigma}\right| \leq 1.96\right) \approx 0.95;$

   b) $P(X - 1.96\sigma \leq \mu \leq X + 1.96\sigma) \approx 0.95.$

6. In Example 2, set 99% confidence limits on the proportion of the population who would respond and favor the amendment.

7. *Bowling.* In a league season, a good bowler bowled 400 frames and got 120 strikes. Set 95% confidence limits on p, the probability that the bowler gets a strike. (Assume independence between frames. Strikes beyond the tenth frame in a game have not been counted.)

8. A random sample of 50 families from Cambridge, Massachusetts, showed 10 families with incomes over $10,000 during 1970. Set a 90% confidence interval on p, the percentage of families with incomes over $10,000.

9. *Even integers.* Suppose that a discrete distribution were confined to the even integers, as would happen, for example, if we studied $Y = 2X$ where X is the number of successes in n binomial trials. Carry out the discussion corresponding to that given for Fig. 2-1(a, b, c) if the continuous variable U were being used to fit Y. Instead of the usual "$\frac{1}{2}$ correction", what would the correction be? To compute the approximation $P(Y \geq y)$, what formula should we use for the deviate z?

2-3 THE SIGN TEST: MORE WORK FOR THE BINOMIAL

The binomial distribution is the basis for the sign test, which is a simple and effective method of testing statistical hypotheses. We shall first illustrate this test with an example.

Example 1. *UCP excretion* (J. T. Galambos and R. B. Cornell). In medical diagnosis, the level of UCP (urinary coproporphyrin) excretion is important. Doctors Galambos and Cornell measured, over a number of days, the UCP excretions of individuals during two 12-hour (day and night) periods. The average day and night excretions for 8 healthy individuals were as follows:

Case number:	1	2	3	4	5	6	7	8
Day average, X:	35.3	65.9	73.4	70.6	56.3	73.4	39.3	36.9
Night average, Y:	39.0	58.8	70.6	58.7	53.1	72.6	42.2	63.1

Is there evidence of a difference between amounts of day and night excretions?

Solution. The data may be tabulated as follows, with D standing for the day measurement and N for night.

| | | Sign of |
D	N	D − N
35.3	39.0	−
65.9	58.8	+
73.4	70.6	+
70.6	58.7	+
56.3	53.1	+
73.4	72.6	+
39.3	42.2	−
36.9	63.1	−

The difference $D − N$ is plus 5 times out of 8. How shall we use this result?

The big idea. Roughly speaking, a large majority of pluses strongly suggests a higher daytime average, and a large majority of minuses a higher nighttime average. The following reasoning leads to more precise conclusions.

For identical populations (i.e., those with no difference between the distributions of day and night excretions), we may regard the investigation as a binomial experiment with $n = 8$ and $p = \frac{1}{2}$, where $p = P(D − N > 0)$. To put it another way, if the D- and N-populations are identical, then $P(\text{plus}) = P(\text{minus}) = \frac{1}{2}$, or $P(D − N > 0) = \frac{1}{2}$. Hence changes in the numbers of plus and minus signs from one experiment to another are due to chance in the random sampling. Since each sign of $D − N$ is obtained independently of the others, indeed from a different individual, we have satisfied the criteria for a binomial experiment.

Therefore, letting $X = $ number of $+$'s, we have from the binomial tables:

$$P(X \geq 5) = \sum_{x=5}^{8} b(x; 8, \tfrac{1}{2}) = 0.363.$$

This means that identical populations would give us 5 or more pluses more than 36% of the time. Hence, if we reject the hypothesis of identical populations only when the probability is small, say 5% or less, we would accept the hypothesis of nearly identical populations and reject the idea of substantial differences between day and night averages. In performing the test, we have taken a one-sided position, as if we had some reason to expect more pluses than minuses. If the original investigators did not have such a reason, then we should use a two-sided test and compute $P(5 \text{ or more pluses or minuses}) = 0.726$. Again the evidence against identical populations is negligible because in both instances the probability is substantially greater than zero.

In summary, the sign test for the comparison of two procedures using a sample of matched pairs involves the following steps.

1. Match n pairs.

2. Where possible, randomly assign the two procedures to the members of each matched pair (in our example, day and night can't be moved).

3. Get the sign of each difference of matched-pair measurements, discarding any zero differences and using as n the count of nonzero differences.

4. Set a level of probability below which we reject the proposed hypothesis.

5. Use the binomial tables for the given n and $p = \frac{1}{2}$ to compute the probability of a result at least as extreme as that observed (the descriptive level of significance). If the descriptive level of significance is less than or equal to the chosen probability level, reject the proposed statistical hypothesis; otherwise, accept it.

Note: Ties. When we use the sign test in practice, ties sometimes do occur. These ties make 0's instead of $+$'s or $-$'s. In the sign test, we exclude such zero differences from consideration, and apply the test only to the nonzero differences, as stated in step 3. The reason is that zero differences contribute no information for deciding whether positive or negative differences are more likely. They do, of course, contribute to an estimate of the probability that a tie occurs.

Sometimes binomial problems merely produce a collection of signs or categories, and there is no intervening difference of measurements.

The sign test is not restricted to tests of $p = \frac{1}{2}$. For example, one can test whether $p = \frac{1}{4}$, if that is of interest. Also, when measurements are available, one can test whether the difference exceeds a given amount, say 10 units. For such a test in the example, we would assign a $+$ if $D - N - 10 > 0$, and we would assign a $-$ if $D - N - 10 < 0$. Since our purpose here is not so much to present the sign test as to review the binomial, we shall not give further examples in this section.

In Chapter 5, we shall deal with a refinement of the sign test that offers some advantages in testing statistical hypotheses.

EXERCISES FOR SECTION 2–3

Use Table A–5 or Table A–6.

1. An investigator applies the sign test to 10 matched pairs. He finds that 6 differences give plus and 4 give minus. Should the investigator reject the hypothesis of identical location of distributions at the two-sided 25% level?

2. Applying the sign test to 25 matched pairs, an investigator finds that 18 differences are plus and 7 are minus. Find the descriptive two-sided level of significance.

3. A dermatologist wishes to test the effectiveness of a new lotion, Y, in preventing sunburn. He uses six subjects. In treating one shoulder of each subject with lotion Y and the other shoulder with lotion X, he follows the standard procedure. After iden-

tical exposures, the degree of sunburn is measured on a scale, and the following matched pairs of measurements x_i, y_i are obtained.

X:	34	72	122	14	52	75
Y:	72	109	74	32	92	60

Do these results justify rejecting equality of effectiveness of the lotions at the one-sided 25% probability level?

4. A manufacturer wishes to decide whether method A for making light bulbs should replace method B. He conducts 15 independent investigations, in each of which bulbs are made by method A and by method B. If method A produces a better bulb in 10 of the 15 investigations, does he reject the hypothesis of equality of methods A and B at the one-sided 20% level? At the one-sided 10% level?

2–4 SIGNIFICANCE LEVEL AND POWER

In the example of the sign test, we used properties of the distribution of the binomial random variable X, the number of plus signs in the differences $D - N$, under the following assumption

$$p = P(\text{success}) = P(D - N > 0) = \tfrac{1}{2}.$$

This assumption that pluses and minuses in the $D - N$ differences are equally likely is called a "null hypothesis". There are many possible null hypotheses, among which $p = \tfrac{1}{2}$ is only one. Other null hypotheses might assume that p has some other specific value, say $p = \tfrac{1}{4}$, or that p is in an interval such as $0 \le p \le \tfrac{1}{4}$. The case $p = \tfrac{1}{2}$ occurs so often that we supply for it a special table, Table A–6.

When one uses a statistical test, he naturally wants to know how good the test is at *correctly signaling departures from the "null" conditions*. The technical name for such ability is *power*. In the next two sections, we illustrate this idea by discussing the power of the sign test. Later, we shall introduce and study the power of other tests.

Example 1. *Sign test.* Suppose that we use the sign test of Section 2–3 on a sample of size $n = 10$ and that we decide to reject the hypothesis that $p = P(\text{success}) = \tfrac{1}{2}$ if we observe 0, 1, 2, 8, 9, or 10 successes. The special assumption $p = \tfrac{1}{2}$ is the *null hypothesis*. (The word "null" here suggests "no difference".) What is the probability of rejection when p actually is $\tfrac{1}{2}$?

Solution. From one of the binomial tables, Table A–5 or Table A–6, we find that, if we use the foregoing *rejection rule* with $n = 10$ and $p = \tfrac{1}{2}$, the probability of our rejecting the null hypothesis when it is true is

$$P(0, 1, 2, 8, 9, \text{ or } 10 \text{ successes} | p = \tfrac{1}{2}) = 0.11.$$

(For the vertical bar read "given".) This probability of rejection is called the *significance level* of the test. The numbers 0, 1, 2, 8, 9, and 10 form the *rejection region* and the numbers 3, 4, 5, 6, 7 the *acceptance region*.

> **Definition.** *Significance level.* In a test, the probability of rejecting the null
> hypothesis when it is true is called the *significance level* of the given test.

A small significance level means that there is only a slight chance of rejecting a
true null hypothesis. In Example 1, a true null hypothesis would be rejected about
11% of the time, on the average.

The null hypothesis may be a set of p-values rather than a single value, as in
Example 1. Then each p-value in the set has its own "significance level". In such
cases, we choose the most unfavorable—that is, the largest—among the possible
significance levels, as "the" significance level. (Sometimes in continuous problems
there is no greatest significance level, as when the set of p's in the null hypothesis
is defined by $p < \frac{1}{2}$ and the alternatives by $p \geq \frac{1}{2}$. Then we make do with the
level for $p = \frac{1}{2}$.)

Since the significance level is the probability of signaling a departure from null
conditions, we shall see that *it is also a very special value of power*, even though the
signal is incorrect.

Returning to Example 1, we note that all values of p except $p = \frac{1}{2}$ were excluded from the null hypotheses. These excluded values of p are *alternative hypotheses*. Under the rejection rule of Example 1, if the actual value of p is 0.8, then
from binomial tables we get

$$P(\text{reject}) = P(0, 1, 2, 8, 9, \text{ or } 10 \text{ successes}|p = 0.8)$$
$$= 0.678.$$

*Each possible hypothesis gives a probability of rejection, and this probability is the
power of the test for the particular hypothesis.* Examples of hypotheses or sets of
hypotheses for binomials are $p = 0.8$, $p = 0.6$, or all the p's in an interval, say
$0.50 \leq p \leq 1$.

> **Definition.** *Power of a test.* In a test, the probability of rejecting the null
> hypothesis when a given hypothesis is true is called *the power of the test for the
> given hypothesis.*

A large power, near 1, for an alternative hypothesis means that there is a very good
chance of rejecting a false null hypothesis or, what amounts to the same thing, of
accepting a true alternative hypothesis. Note that each alternative has its own
power; the set of ordered pairs (hypothesis, power) is a function called the *power
function.* Note that we also include the pair (null hypothesis, significance level) in
the power function, or we may include a set of pairs when the null hypothesis has
a set of values.

To sum up, in testing statistical hypotheses, we want to avoid the following
two errors.

1. *The error of rejecting the null hypothesis when it is true.* A small significance level indicates that the test offers little chance of making this error.

2. *The error of accepting the null hypothesis when it is false.* A large power for alternatives indicates that the chances of accepting a false null hypothesis are small.

The ideal test would have significance level equal to 0, and power equal to 1. In life, we must not always expect the best of both worlds.

Figure 2–2(a, b, c) represents ideal power curves for testing various hypotheses. In these representations, it is rather arbitrary which set is called the null hypothesis and which the alternative.

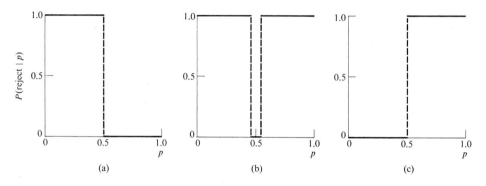

(a)   (b)   (c)

Fig. 2–2 (a) Ideal power curve for testing the null hypothesis $p \geq \frac{1}{2}$ versus $p < \frac{1}{2}$. (b) Ideal power curve for testing the null hypothesis p near $\frac{1}{2}$ versus p not near $\frac{1}{2}$. (c) Ideal power curve for testing the null hypothesis $p \leq \frac{1}{2}$ versus $p > \frac{1}{2}$.

EXERCISES FOR SECTION 2–4

1. Under the rejection rule of Example 1, find the power of the test for the alternative hypothesis $p = 0.9$.

2. In applying the sign test to a sample of 12, we decide to reject the null hypothesis $p = P(\text{success}) = 0.5$ if we observe 0, 1, 2, 10, 11, or 12 successes. Find the significance level of this test.

3. What is the power of the test in Exercise 2 for the alternative hypothesis $p = 0.7$?

4. In Exercise 2, let us decide to reject the null hypothesis $p = 0.5$ whenever 8 or more successes are observed. Find (a) the significance level of the test, (b) the power of the test for the alternative hypothesis $p = 0.7$.

5. We apply the sign test to a sample of 20 matched pairs. If the null hypothesis is $p = P(\text{success}) = 0.5$, and if X denotes the number of successes, complete the following table for each of the given rejection regions.

Rejection region	Significance level	Power for $p = 0.8$

a) $X \leq 5$ or $X \geq 15$
b) $\qquad X \geq 15$
c) $\qquad X \geq 14$
d) $\qquad X \geq 13$

6. In applying the sign test to a sample of 15, take as the null hypothesis $p = P(\text{success}) = 0.5$. Use the binomial tables to get a rejection region for a one-sided test having a significance level about 15% and, for the alternative hypothesis $p = 0.2$, power about 0.94.

7. Two fertilizers, A and B, are used on 18 pairs of adjacent, randomly selected plots of cabbage. The yields are measured in pounds, and for each pair of plots, the sign of the difference A-yield minus B-yield is recorded. Use as the null hypothesis $P(\text{success}) = P(+) = 0.5$. Then apply the sign test, using the rejection region $X \geq 13$, where X is the number of pluses. What is the significance level of this test?

8. What is the power of the test in Exercise 7 for the alternative hypothesis $P(+) = 0.7$?

9. In Exercise 7, how might we improve the power of the test for the alternative hypothesis $P(+) = 0.7$?

10. Suppose that a test rejects only when n successes occur in n trials. If the null hypothesis is $p = \frac{1}{2}$, and the alternative is a value of p larger than $\frac{1}{2}$, obtain a formula for n to ensure that the ratio of the power for p to the significance level is at least R. Find the least value of n when $p = 0.8$ and $R = 4$.

2–5 POWER FUNCTIONS FOR THE SIGN TEST

Table 2–1 gives the basic data for sign tests with $n = 10$ and rejection criteria as indicated. The corresponding power curves are shown in Fig. 2–3. The sets of values of p that lead to rejection correspond to the regions in Fig. 2–2(a, b, c) where $P(\text{reject}|p) = 1$.

The tests corresponding to Fig. 2–3(a, c) are *one-sided* because, in each test, rejection occurs if the observation goes too far in *one direction*. The test corresponding to Fig. 2–3(b) is *two-sided* because we reject if the observation goes too far in *either direction*.

Note that the ideal power functions in Fig. 2–2 are imitated in a somewhat slumped or round-shouldered way by the curves in Fig. 2–3. To come closer to the ideal curves requires larger sample sizes. To show how changing sample sizes can move the power curves toward the ideals of Fig. 2–2, we show in Fig. 2–4 power functions for $n = 4$, 10, and 25, all having about the same significance level. As n grows large, the power curve can be made to look more like the ideal curve. We expect this because of the law of large numbers. (See *PWSA*, p. 231.†)

---

†The letters *PWSA* refer to *Probability with Statistical Applications,* 2nd edition, by F. Mosteller, R. E. K. Rourke, and G. B. Thomas, Jr., Addison-Wesley, 1970. Such references are only for those who feel that they need a review of the ideas.

If p is not in the null set, then as n grows large, $\bar{p}$ is almost certain to approach p and thus fall into the rejection region.

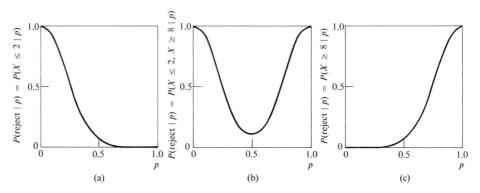

Fig. 2–3 (a) Power for sign test, $n = 10$, rejection region $X \leq 2$. (b) Power for sign test, $n = 10$, rejection region $X \leq 2$, $X \geq 8$. (c) Power for sign test, $n = 10$, rejection region $X \geq 8$.

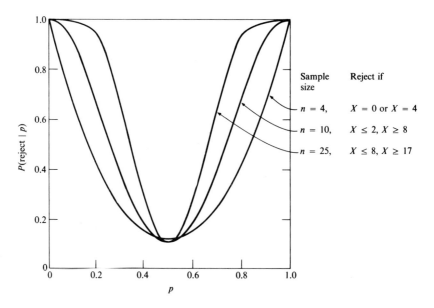

Sample size	Reject if
$n = 4$,	$X = 0$ or $X = 4$
$n = 10$,	$X \leq 2$, $X \geq 8$
$n = 25$,	$X \leq 8$, $X \geq 17$

Fig. 2–4 Power curves for sign test for three sample sizes having about the same significance level, 0.11.

Table 2–1

Binomial probabilities for $n = 10$ and various p's
(used for the power curves for sign test in Fig. 2–3)

p	Fig. 2–3(a) $P(X \leq 2)$	Fig. 2–3(b) $P(X \leq 2$ or $X \geq 8)$	Fig. 2–3(c) $P(X \geq 8)$
0.00	1	1	0
0.01	1	1	0
0.05	0.988	0.988	0
0.10	0.930	0.930	0
0.20	0.678	0.678	0
0.30	0.383	0.385	0.002
0.40	0.167	0.179	0.012
0.50	0.055	0.110	0.055
0.60	0.012	0.179	0.167
0.70	0.002	0.385	0.383
0.80	0	0.678	0.678
0.90	0	0.930	0.930
0.95	0	0.988	0.988
0.99	0	1	1
1.00	0	1	1

EXERCISES FOR SECTION 2–5

The sign test is used for a sample of 10. Use Table 2–1 to answer Exercises 1, 2, and 3.

1. If the null hypothesis is $p = 0.5$ and the rejection region is $X \leq 2$, find (a) the significance level of the test, (b) the power of the test for the alternative hypothesis $p = 0.2$.

2. If the null hypothesis is 0.5 and the rejection region is $X \leq 2$ or $X \geq 8$, find (a) the significance level of the test, (b) the power of the test for the alternative hypothesis $p = 0.9$.

3. If the null hypothesis is $p \leq 0.5$ and the rejection region is $X \geq 8$, find the significance level of the test and the power of the test for $p = 0.8$.

4. Use Fig. 2–3(b) to estimate the power of the sign test for $n = 10$ and rejection region $X \leq 2$ or $X \geq 8$, if (a) $p = 0.25$, (b) $p = \frac{2}{3}$.

5. Use the graphs in Fig. 2–4 to estimate, for the given rejection regions, the power for the hypothesis $p = 0.75$ when (a) $n = 4$, (b) $n = 10$, (c) $n = 25$.

6. Refer to Example 1, Section 2–4. Use Fig. 2–4 to find the improvement in power for the hypothesis $p = 0.8$ when the sample size is increased to 25 with rejection rule $X \leq 8$ or $X \geq 17$.

2-6 AN INFORMAL AND OFTEN USED APPROACH TO TESTING

Although the description given in Sections 2-3, 2-4, and 2-5 outlines the formal steps in a significance test, we do not often find ourselves wanting to make such formal tests. Instead we ordinarily use the testing approach as a method of reporting how surprising the result is or how far it seems to be removed from a standard value or null hypothesis.

If we compute the number of standard deviations by which the result departs from the null hypothesis, that number tells us a great deal about what we need to know. Even rounded results like 1, 2, 3, or 4 standard deviations are valuable and often adequate for reporting. Sometimes instead of rounding, it is worth looking up the result in a table of probabilities to get the descriptive level of significance. Finer calculations may be of use when the data are to be used further, as when they are to be combined with other data.

We turn now to problems that are not quite as simple as those treated so far, problems where there is a question about just how the data should be treated. In this context, we need to look at data in more than one way. An example mentioned earlier will illustrate the ideas.

Example 1. *Effect of marijuana on naive subjects.* In a now famous experiment (Andrew T. Weil, Norman E. Zinburg, Judith M. Nelsen, *Science*, Vol. 162, December 13, 1968, pp. 1234–1242), the investigators studied in 9 adult subjects the effect of smoking marijuana. To test effects on intellectual functioning, one part of the work dealt with each subject's scores in a digit symbol substitution test under two conditions: (a) 15 minutes after smoking a placebo cigarette that had no marijuana and (b) 15 minutes after smoking a marijuana cigarette. The subjects were new to marijuana and didn't know whether the drug was present in a cigarette. These data, shown in Table 2–2, are scores measured from the subject's base level. Analyze the results to see whether there is reason to suppose that performance has been lowered by the marijuana.

Solution

a) *The permutation test* (*simulation*). In Example 1, Section 1–6, we ignored the matching of scores and treated the data as two independent samples of 9 scores each. We got a one-sided descriptive level of 0.012.

b) *The t-test.* (See Appendix VII–4 and Exercise 1.) In this approach, we have to assume that differences are approximately normally distributed (often differences are more nearly normally distributed than are the original measurements). The null hypothesis is that the mean difference $\mu = 0$. Under this hypothesis, the details of the t-test are as shown in Table 2–2. We get $t = 1.87$ with 8 degrees of freedom. From Table A–3 we find that

$$P(t \geq 1.87) \approx 0.05,$$

which is the one-sided descriptive significance level. It means that, for a sample

Table 2–2

Performance on digit symbol substitution test 15 minutes after smoking cigarette (naive users)

Subject	Placebo score, P	High dose score, H	$D = P - H$ d_i	d_i^2
1	-3	$+5$	-8	64
2	$+10$	-17	27	729
3	-3	-7	4	16
4	$+3$	-3	6	36
5	$+4$	-7	11	121
6	-3	-9	6	36
7	$+2$	-6	8	64
8	-1	$+1$	-2	4
9	-1	-3	2	4

$$\sum d_i = 54 \qquad \sum d_i^2 = 1074$$

$$\bar{d} = \frac{54}{9} = 6$$

$$s_D^2 = \frac{\sum d_i^2 - n\bar{d}^2}{n(n-1)} = \frac{1074 - 324}{9(8)} = \frac{750}{72} = 10.4$$

$$s_{\bar{D}} = \sqrt{10.4} \approx 3.22$$

$$t = \frac{\bar{d} - \mu_0}{s_{\bar{D}}} = \frac{54/9 - 0}{3.22} = \frac{6}{3.22} = 1.87$$

of 9, we have fair evidence that the subjects performed better after the placebo than after the marijuana. However, we are a bit worried about Subject 2's scores. His difference is so outstanding that we wonder if the total result is misleading because of him. The sign test offers some protection against very large differences because it gives all measurements equal weight.

c) *The sign test.* In this approach we need not assume normality of the distribution of differences. Seven of nine subjects performed better with the placebo. Under the null hypothesis, we have a binomial experiment with $P(+) = \frac{1}{2}$. The one-sided descriptive significance level is $P(X \geq 7)$, where X is the number of pluses. We can use Table A–5 or A–6 or calculate directly and get

$$P(X \geq 7) = P(X = 7) + P(X = 8) + P(X = 9)$$

$$= \binom{9}{7}\left(\frac{1}{2}\right)^9 + \binom{9}{8}\left(\frac{1}{2}\right)^9 + \binom{9}{9}\left(\frac{1}{2}\right)^9$$

$$= \frac{36 + 9 + 1}{512} \approx 0.09.$$

This result is of the same order of magnitude as that given by the *t*-test, and we are comforted that Subject 2's observation seems not to have led us much astray.

The *t*-test let the 27 of Subject 2 count at its full numerical value; the sign test gave it a 1 like all the other +'s. Later (Chapter 5) we shall return to this example and give Subject 2's score more weight than the others but not so much as is implied by the 27.

The permutation test gave us a more significant result than did either the *t*-test or the sign test, but it did not use the matching. Often the matching gives a more significant result, but it did not do so here.

The skeptic, seeing us try out several different tests on the same data, may accuse us of hunting for the result we prefer. Far from it. We are trying to make sure that we are in a position to give an informative report. If the significance levels are widely different, we want to report that fact.

EXERCISES FOR SECTION 2–6

1. Refer to Example 1, Solution (a). A formula for the estimated variance of $\bar{D}$, the mean of the sample differences, is

$$s_{\bar{D}}^2 = \frac{\sum(d_i - \bar{d})^2}{n(n-1)} .$$

(See Appendix VII–4.) Prove that

$$\sum(d_i - \bar{d})^2 = \sum d_i^2 - n\bar{d}^2 = \sum d_i^2 - \frac{(\sum d_i)^2}{n}.$$

2. *Effect of marijuana on naive subjects.* As part of the experiment described in Example 1, the investigators obtained scores 90 minutes after the subjects had smoked cigarettes, as follows:

Subject:	1	2	3	4	5	6	7	8	9
Placebo score, *P*:	−7	−1	−10	x	6	3	3	4	6
High dose score, *H*:	8	−5	−1	x	−8	−12	−4	−3	−10

Use the *t*-test and the sign test to analyze these results. Is there reason to suppose that performance has been lowered by the marijuana? (Subject 4 had incomplete data.)

3. *Methods of memorizing.* Two methods of memorizing difficult material are tried to see which gives better retention. Pairs of students are matched for both IQ and academic performance. They then receive instruction on the same material; but one member of a pair uses method *A* to learn it while the other member uses method *B*. The students are then tested for recall, and the following scores are obtained.

Pair:	1	2	3	4	5	6	7	8	9	10	11	12
A-score:	71	82	59	78	92	85	81	79	77	54	61	83
B-score:	64	72	61	75	88	84	88	82	70	49	64	81

Use the *t*-test and the sign test to analyze these data. Should we reject, at the 5% level, the hypothesis that the two learning methods are equally effective?

4. *Airline punctuality*. The punctuality of 12 airlines, based on flights between North American cities at least 200 miles apart, was measured by the percentage of flights arriving within 15 minutes of the scheduled time. The records for July 1969 (midsummer) and January 1970 (midwinter) were as follows:

Airline	% July 1969	% January 1970
A	77	81
B	65	75
C	76	55
D	72	68
E	66	62
F	72	68
G	63	45
H	66	51
I	58	63
J	60	74
K	72	71
L	66	61

Do these data support the hypothesis that the listed airlines are more punctual in July than in January? Use the *t*-test and the sign test. (Assume that the operations of each airline are independent of the operations of the others.) Guess whose advertisement gives these data.

2-7 THE MULTINOMIAL DISTRIBUTION

Consider an experiment that has all the characteristics of the binomial, with one exception: There are *more than two* possible outcomes, or categories. Thus there are n trials, each classified into one of the c categories. Such an experiment gives rise to the multinomial distribution.

Example 1. *Public opinion poll.* In a public opinion poll, the responses in the population to a certain question are "yes", "no", or "no opinion", with the following probabilities:

$P(\text{yes}) = p_1 = 0.5;$ $P(\text{no}) = p_2 = 0.3;$ $P(\text{no opinion}) = p_3 = 0.2.$

Ten ballots are randomly drawn from the thousands obtained. What is the probability that exactly 5 ballots show yes, 3 show no, and 2 show no opinion?

Solution. We are drawing a small sample without replacement from a very large population. Since the sampling ratio is small, we regard the dependence of one draw on the next as negligible, and we use the multinomial model.

The probability of drawing 5 ballots showing yes, 3 no, and 2 no opinion, *in that order*, is

$$p_1^5 p_2^3 p_3^2.$$

The number of rearrangements of 10 objects of which 5 are alike of one kind, 3 of another, and 2 of another is given by the multinomial coefficient

$$\frac{10!}{5!\,3!\,2!} = 2520,$$

which is the number of ways of drawing the specific sample.

Since all these ways are equally likely, we get

$$P(5 \text{ yes, 3 no, 2 no opinion}) = \frac{10!}{5!\,3!\,2!}\, p_1^5 p_2^3 p_3^2 = 0.08505.$$

Thus, in this example, there is less than 1 chance in 10 that the sample of ten will exactly reproduce the population proportions. The reason is: If X_1 is the number of "yes", X_2 the number of "no", and X_3 the number of "no opinion" responses, then the outcome of the experiment can be represented by the number triple (X_1, X_2, X_3), where $X_1 + X_2 + X_3 = 10$. In our example, $X_1 = 5$, $X_2 = 3$, $X_3 = 2$. The outcomes offer many possible triples: $(10, 0, 0), \ldots$, $(5, 3, 2), \ldots, (1, 0, 9), \ldots, (0, 0, 10)$, 66 in all (see *PWSA*, p. 63). Although all triples are not equally likely, they spread the probability around.

The generalization to the case with c outcomes, or categories, follows readily.

Definitions. *Multinomial experiment; multinomial distribution.* A multinomial experiment has n independent trials, each of which results in one of c categories numbered $1, 2, \ldots, c$, with probabilities $p_1, p_2, \ldots, p_c$, respectively. Let $P(x_1, x_2, \ldots, x_c)$ denote the probability that x_i trials fall into the ith category, where $i = 1, 2, \ldots, c$; then the set of probabilities

$$P(x_1, x_2, \ldots, x_c), \tag{1}$$

where the x_i's are integers,

$$0 \le x_i \le n \quad \text{and} \quad \sum_{i=1}^{c} x_i = n,$$

form a multinomial distribution.

It may be worth saying explicitly that X_i is the random variable giving the number of trials falling into category i, and that expression (1) is a shorthand notation for

$$P(X_1 = x_1, X_2 = x_2, \ldots, X_c = x_c). \tag{2}$$

Theorem. *Multinomial distribution.* In a multinomial distribution,

$$P(x_1, x_2, \cdots, x_c) = \frac{n!}{x_1! \, x_2! \ldots x_c!} \, p_1^{x_1} p_2^{x_2} \cdots p_c^{x_c}, \qquad (3)$$

where the x_i's are integers,

$$0 \le x_i \le n \quad \text{and} \quad \sum_{i=1}^{c} x_i = n.$$

Proof. Exercise 8 requires that you generalize the approach given in Example 1 and thus supply the proof.

The probabilities in (3) include exactly all the terms in the multinomial expansion of

$$(p_1 + p_2 + \ldots + p_c)^n.$$

Therefore the sum of terms of the form (3) over all $(x_1, x_2, \ldots, x_c)$ such that $\sum x_i = n$ is 1 because $\sum p_i = 1$.

The counts $x_1, x_2, \ldots, x_c$ are respectively special values of the multinomial random variables $X_1, X_2, \ldots, X_c$. Let us focus attention on the random variable X_1, the count for category 1. The value of X_1, namely x_1, varies from one sample of n trials to another. What are the mean and the variance of X_1? If we pool all the categories except category 1, we have in effect only two categories, and we have reduced the multinomial to a binomial without bringing about any change in the behavior of X_1. We have let p_1 be the probability for category 1, and we have let $1 - p_1 = (p_2 + p_3 + \cdots + p_c)$ be the probability of "not category 1". But now we know the mean and variance of X_1:

$$\mu_{X_1} = \mu_1 = np_1 \quad \text{and} \quad \text{Var}(X_1) = \sigma_1^2 = np_1(1 - p_1). \qquad (4)$$

The same device can be applied for any category i, and so the mean and variance of X_i are

$$\mu_{X_i} = \mu_i = np_i \quad \text{and} \quad \text{Var}(X_i) = \sigma_i^2 = np_i(1 - p_i). \qquad (5)$$

To study the degree to which X_i and X_j vary together, we need to know the covariance between X_i and X_j. (See Appendix IV–1 and IV–3 for definition of covariance.) For the binomial, this procedure amounts to correlating X with $n - X$; this correlation, ρ, is -1, for as X gets larger, $n - X$ gets exactly that much smaller, and vice versa. The binomial covariance, then, is

$$\rho\sigma_1\sigma_2 = \rho\sigma_X\sigma_{n-X} = \rho\sigma_X^2 = (-1)np(1 - p) = -npq. \qquad (6)$$

To get $\text{Cov}(X_i, X_j)$, we first tabulate the category outcomes with their probabilities *for one fixed trial.* The possible outcomes from the viewpoint of categories i, j are: Category i does or doesn't occur; category j does or doesn't occur.

	Y_i \\ Y_j	In category j 1	Not in category j 0	
In category i	1	0	p_i	p_i
Not in category i	0	p_j	$1 - p_i - p_j$	$1 - p_i$
		p_j	$1 - p_j$	1

We assign a 1 if the trial falls into category i, a 0 if it doesn't, and we make similar assignments for category j. We want the covariance of the scores in this one chosen trial, because these 1 or 0 scores, when added over all the trials, give the values of X_i and X_j, which are the total counts for categories i and j, respectively, in n trials.

Let Y_i and Y_j be respectively the 1 or 0 scores for categories i and j in our chosen trial. The covariance of Y_i and Y_j (see Appendix IV–2) is

$$\text{Cov}\,(Y_i,\ Y_j) = E(Y_iY_j) - E(Y_i)E(Y_j).$$

The table shows that

$$E(Y_iY_j) = 0, \qquad E(Y_i) = 1(p_i) = p_i, \qquad E(Y_j) = 1(p_j) = p_j.$$

Therefore,

$$\text{Cov}\,(Y_i,\ Y_j) = 0 - p_ip_j = -p_ip_j. \tag{7}$$

In view of this result for a single trial, the student will not be surprised (see Exercise 12) that, for n trials,

$$\text{Cov}\,(X_i,\ X_j) = -np_ip_j. \tag{8}$$

With this result in hand, we are ready to compute the correlation between X_i and X_j.

Example 2. *Correlation between counts.* For the public opinion poll of Example 1, find the correlation between the count of the "yes" ballots, X_1, and that of the "no" ballots, X_2.

Solution. By definition (see Appendix IV–3), the correlation between two random variables X and Y is

$$\rho = \frac{\text{Cov}\,(X,\ Y)}{\sigma_X\sigma_Y}.$$

Applying this formula to X_1 and X_2, we get, in general,

$$\rho_{X_1, X_2} = \frac{\text{Cov}(X_1, X_2)}{\sigma_{X_1}\sigma_{X_2}}$$

$$= \frac{-np_1 p_2}{\sqrt{np_1(1 - p_1)np_2(1 - p_2)}} \tag{9}$$

$$= -\sqrt{\frac{p_1 p_2}{(1 - p_1)(1 - p_2)}}.$$

For the particular example, we get

$$\rho_{X_1, X_2} = -\sqrt{\frac{(0.5)(0.3)}{(0.5)(0.7)}} = -0.65.$$

If we have many yeses, we have relatively few noes. Note that the formula for the correlation of X_1 and X_2 does not depend on n.

EXERCISES FOR SECTION 2–7

1. Work the problem in Example 1, given that $p_1 = 0.7$, $p_2 = 0.2$, and $p_3 = 0.1$. Also compute $P(7 \text{ yes}, 2 \text{ no}, 1 \text{ no opinion})$.

2. In Exercise 1, if X is the number of yeses and Y is the number of noes drawn in a sample of 10, compute $E(X)$, Var (X), $E(Y)$, Var (Y), Cov (X, Y), and $\rho_{X, Y}$.

3. A poll shows that for a certain electoral district the voters are 40% Republican, 30% Democrat, and 30% Independent. If a random sample of 8 voters is drawn, find the probability of getting 4 Republicans and 4 Democrats.

4. For the population of Exercise 3, let R denote the number of Republicans drawn in the sample of 8, D the number of Democrats, and I the number of Independents. Compute $\rho_{R, D}$ and $\rho_{D, I}$.

5. Of some thousands of students taking a test, 10% get A's, 20% get B's, 30% get C's, and 40% get D's. Write expressions for the probability that a random sample of 10 students will show (a) one A, two B's, three C's, and four D's; (b) four A's, three B's, two C's, and one D. (c) Compute the ratio of the probabilities of (a) and (b).

6. In Exercise 5, let A denote the number of A-students, B the number of B-students, and so on. Compute, for the sample of 10, $\rho_{A, B}$ and $\rho_{A, D}$.

7. A true die is tossed n times, and the face with i dots shows on top X_i times, where $i = 1, 2, 3, 4, 5, 6$ and $\sum X_i = n$. Write an expression for the probability of the outcome $(x_1, x_2, \ldots, x_6)$. Find the correlation between X_1 and X_2.

8. Prove the theorem of this section.

9. Use equation (9) to prove $\rho_{X_1 X_2}^2 \leq 1$.

10. Apply equation (8) to the binomial case and thus obtain equation (6).

11. An experiment consists of n independent trials. On trial i, the random variables U_i, V_i occur, and they have means, variances, and covariances. Let

$$U = \sum_{i=1}^{n} U_i, \qquad V = \sum_{i=1}^{n} V_i,$$

and prove

$$\text{Cov}\,(U,\,V) = \sum_{i=1}^{n} \text{Cov}\,(U_i,\,V_i).$$

[*Hint:* The independence of U_i and V_j, $i \neq j$, ensures that $E(U_iV_j) = E(U_i)E(V_j)$.]

12. Use the result of Exercise 11 to justify the step from equation (7) to equation (8).

2–8 APPLICATIONS AND TABLES OF THE MULTINOMIAL DISTRIBUTION

Let us illustrate the multinomial with an example.

Example 1. *Automobile accidents* (Joseph Ferreira, Jr.). In a given state, records of accidents were available for six successive calendar years for many drivers. Of these, 595 had exactly two reported accidents. If the equally likely multinomial applies (that is, $n = 2$, $c = 6$, $p_i = \frac{1}{6}$, $i = 1, 2, \ldots, 6$, where i corresponds to the ith year), how many drivers would have had both their accidents in the same year?

Solution. Think of 6 categories, each with associated probability $\frac{1}{6}$. We can compute the probability that both accidents happen in the first year and then multiply by 6. For two first-year accidents, we get, from equation (3) of Section 2–7,

$$P(2, 0, 0, 0, 0, 0) = \frac{2!}{2!\,(0!)^5} \left(\frac{1}{6}\right)^2 \left[\left(\frac{1}{6}\right)^0\right]^5 = \frac{1}{36}.$$

Multiplying by 6 gives $\frac{1}{6}$, and so the estimated number of drivers having both accidents in the same year is $\frac{595}{6} \approx 99$. The number observed was 110, which is slightly more than one standard deviation away.

Let us next illustrate the use of the three-category multinomial tables in the back of the book, Table A–7.

Example 2. *Five accidents* (Joseph Ferreira, Jr.). Ferreira gives data showing that 13 drivers had 5 reportable accidents over the six-year period. Divide the 6 years into 3 two-year periods (first two, second two, and third two), and use Table A–7 to compute the probability that such a driver did not have 3 or more accidents in the same two-year period. (Assume that the multinomial with 3 equally likely periods applies.)

Solution. The required outcomes are (2, 2, 1), (2, 1, 2), and (1, 2, 2). Since $n = 5$ and $p_1 = p_2 = p_3 = \frac{1}{3}$, the probability of each outcome is given as 0.1235. The three together have probability 0.3705. Ferreira finds that 4 drivers out of 13 belonged to one of these outcomes, giving the empirical value as 0.31.

EXERCISES FOR SECTION 2–8

Use Table A–7 for three-category multinomials.

1. In Exercise 3 of Section 2–7, a random sample of 10 voters is drawn. Find the probability of getting an overall Republican majority (not plurality) in the sample. Check by using binomial tables.

2. In Example 2, use the multinomial to find the probability that such a driver did not have 4 or more accidents in the same two-year period.

3. A sample of 10 items is randomly drawn from a trinomial distribution with probabilities 0.2, 0.3, 0.5. (a) Find the mean number of items for each of the three categories. (b) What is the probability that the number of items drawn in each category is at most 1 more than the mean for that category?

4. The accident records for a large factory show that each of 12 workers had 4 accidents over a three-year period. Consider the 3 years as 3 one-year periods, and assume that the distribution of accidents is an equally likely multinomial. Find (a) the probability that a worker had 3 or more of his 4 accidents in a single one-year period; (b) the expected number of workers having 3 or more accidents in a single one-year period. Check the work algebraically.

5. In a keg of mixed nails, similar except for the finish, 60% of the nails have one kind of finish, 30% another kind, and 10% a third kind of finish. If a random sample of 10 is drawn from the keg, find the probability that the sample contains (a) exactly two kinds of nails, (b) only one kind of nail, and (c) all three kinds of nails.

6. *Imitating the probabilities.* A sample of 10 is drawn from each of two multinomials, one with probabilities (0.1, 0.2, 0.7) and the other with probabilities (0.3, 0.3, 0.4). Which distribution is more likely to give sample proportions identical with its true proportions? Discuss.

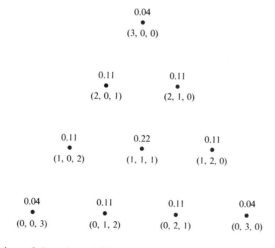

Fig. 2–5 Distribution of the trinomial for $n = 3$, with equally likely categories.

7. Figure 2–5 shows for $n = 3$ the distribution of probabilities for the multinomial with 3 equally likely categories. At what point is the greatest probability found? As we move out from the centroid of the triangular array, how are the point probabilities affected?

8. Use Table A–7 to make, for $n = 5$, a diagram corresponding to Fig. 2–5. Find the total probability at the three points that are closest to the centroid of the triangular array. As points get farther from the centroid, how are the associated probabilities affected? Where do the probabilities appear to peak? Draw an analogy between the representation of a trinomial and the usual graph of a binomial probability distribution.

9. In a sample of n drawn from a trinomial distribution, the observed counts are (x_1, x_2, x_3). Use the data to estimate $p_1, p_2,$ and p_3.

★10. (Continuation) In Exercise 9, suppose that you know that $p_2 = p_3$. How would you then use the data to estimate the p's?

★11. Suppose that you have four equally likely categories and $n = 3$. Describe the kind of array you would use instead of that in Fig. 2–5.

2–9 UNBIASED AND MAXIMUM LIKELIHOOD ESTIMATION

Unbiased estimates. To estimate a binomial probability p from the number of successes X in a sample of n, we have used the sample proportion $\bar{p} = X/n$. This estimate has the expected value

$$E(\bar{p}) = E(X/n) = E(X)/n = np/n = p.$$

Because this equation holds for each p and n, we say that $\bar{p}$ is an *unbiased* estimate of p. To make more explicit in our notation that we are finding the mean of $\bar{p}$ *for a given p and n*, let us use $E(\bar{p}|n, p)$. Restating the result of the above equation gives

$$E(\bar{p}|n, p) = p, \qquad \text{for all } n \geq 1.$$

Since we get the same value for all n, we could write the condition as $E(\bar{p}|p) = p$. In general, using the newly introduced notation, we define an unbiased estimate as follows:

Definition. *Unbiased estimate.* A statistic T is an unbiased estimate of a population parameter θ if $E(T|\theta) = \theta$, for all admissible values of the parameters.

Thus, for category i of a multinomial, an unbiased estimate of p_i is X_i/n, where the random variable X_i is the count in the ith cell. These estimates of p and p_i are special cases of the use of the sample mean, $\bar{X}$, as an unbiased estimate of the population mean μ (see Exercise 7).

Maximum likelihood estimates. Another approach to estimation in discrete problems asks us to maximize the probability of getting the observed sample in the particular order in which the items in the sample appeared. Thus, for a binomial with n trials and X successes with $P(\text{success}) = p$, we ask: What value of p maximizes the probability of getting the sample that we actually got? One basis for this approach is that we are more likely to get a frequently occurring sample than a rare one. Consequently, we tie our estimate to *the value of p that would make the observed sample most likely to occur.* We call this estimate of p its *maximum likelihood estimate,* and we denote it by $\hat{p}$ (usually read "p-hat").

Example 1. In a binomial sample of three trials, 2 successes and 1 failure occur. Find $\hat{p}$, the maximum likelihood estimate of p.

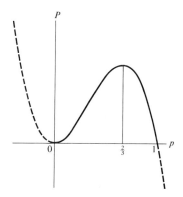

Solution 1. The probability of getting 2 successes and 1 failure in a particular order is $P = p^2(1 - p)^1$. What value of p maximizes P? If we graph P as a function of p, we see that the maximum of p is near $p = \frac{2}{3}$. This result is adequate for this particular problem, but we want a general formula. Let's try to find what values of p, if any, give higher probabilities P than $p = \frac{2}{3}$ gives. Is there a value of p that satisfies the inequality

$$p^2(1 - p) > (\tfrac{2}{3})^2(1 - \tfrac{2}{3}), \qquad \text{where} \quad 0 \le p \le 1? \tag{1}$$

To analyze the problem, let $p = \frac{2}{3} + u$, where $-\frac{2}{3} \le u \le \frac{1}{3}$. Then, after substitution, the inequality (1) becomes

$$-u^2 - u^3 > 0,$$

or

$$-u^2(1 + u) > 0, \qquad \text{where} \quad -\tfrac{2}{3} \le u \le \tfrac{1}{3}. \tag{2}$$

Let $g(u) = -u^2(1 + u)$, the left side of (2). By inspection, we see that

$$g(u) = 0, \qquad \text{if } u = 0;$$

and

$$g(u) < 0, \qquad \text{if } u \ne 0 \text{ and } -\tfrac{2}{3} \le u \le \tfrac{1}{3}.$$

Thus, for $-\frac{2}{3} \le u \le \frac{1}{3}$, no value of u satisfies the inequality (2). Hence, $p = \frac{2}{3}$ maximizes the probability $p^2(1 - p)$, and so $\hat{p} = \frac{2}{3}$. Note that the maximum likelihood estimate $\hat{p}$ is our usual estimate $\bar{p} = X/n$ for this particular case with $n = 3$ and $X = 2$. These facts are illustrated by the graph of $g(u)$.

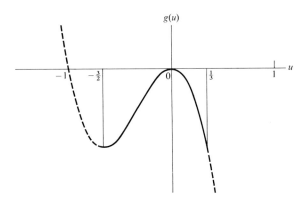

★ **Solution 2.** (Calculus) In order to maximize $P = p^2(1 - p)$, we take the derivative of P with respect to p:

$$\frac{dP}{dp} = 2p(1 - p) - p^2.$$

If $dP/dp = 0$, then

$$2p(1 - p) - p^2 = 0$$

or

$$p(2 - 3p) = 0.$$

The critical values are $p = 0$ and $p = \frac{2}{3}$. The value $p = 0$ does not maximize $p^2(1 - p)$, but $p = \frac{2}{3}$ does, as the graph of P shows. One could check this fact with the second derivative.

Solution 1 is essentially the same as solution 2, but the calculus approach is quicker and readily generalizes. We shall therefore use calculus to prove the following theorem.

Theorem 1. $\hat{p} = \bar{p}$. The maximum likelihood estimate of p for a binomial distribution with X successes in n trials is $\bar{p} = X/n$.

★ *Proof.* First, consider $0 < X < n$. Then for X equal to a certain value x, we have

$$P = p^x(1 - p)^{n-x}, \tag{3}$$

and

$$\frac{dP}{dp} = xp^{x-1}(1 - p)^{n-x} - (n - x)(1 - p)^{n-x-1}(p^x)$$

$$= p^{x-1}(1 - p)^{n-x-1}(x - np).$$

The values of p that make the first derivative zero are 0, 1, and x/n. One way of sorting out these values is to use the second derivative. Alternatively, we can see by inspecting formula (3) that P vanishes for $p = 0$ and $p = 1$ and stays positive for $0 < p < 1$. Hence P has a maximum between $p = 0$ and $p = 1$, because P is a polynomial and therefore continuous. This maximum is at x/n, which is the value of the desired estimate, $\hat{p} = x/n$.

Second, we still have to deal with the cases $X = 0$ and $X = n$. For $X = 0$, the probability is $(1 - p)^n$. As p takes values closer and closer to 0, the value of $1 - p$ gets larger and larger. Since the lower limit of p is 0, the maximizing value of p is 0, which equals X/n for this case.

In Exercise 1, you are asked to prove that, when $X = n$, the maximizing value of p is 1. Since $X/n = n/n = 1$, this completes the proof that the maximum likelihood estimate of p is $\hat{p} = X/n$.

We present here the corresponding theorem for the multinomial distribution without proving it.

Theorem 2. $\hat{p}_i = \bar{p}_i$. If a sample of size n, drawn from a multinomial with probabilities $(p_1, p_2, \ldots, p_c)$, gives counts $(x_1, x_2, \ldots, x_c)$ for the c categories, then the value of the maximum likelihood estimate of the set of p's is given by

$$\hat{p}_i = x_i/n, \qquad i = 1, 2, \ldots, c.$$

Note. It is usual to think of an estimate as a formula that is a function of a random variable. For example, an estimate of μ is $\hat{\mu} = \bar{X}$, or of p_i, $\hat{p}_i = X_i/n$; and the value of the estimate is what is obtained when values of the random variables are substituted. For example, $\hat{\mu} = \bar{x}$, or $\hat{p}_i = x_i/n$. But it is inconvenient to print everything twice, and authors often do not find it worthwhile to make a distinction between an estimate and its value. This can lead to mild confusion. For example, if we say the estimate $\hat{\mu} = \bar{X}$ is an unbiased estimate of the population mean, μ, we understand that $E(\bar{X}) = \mu$, and all is well. But to say that $\hat{\mu} = \bar{x}$ is an unbiased estimate of μ is not true, since $\bar{x}$ is some number like 3.0, and it either is or is not μ; indeed, $E(\bar{x}) = \bar{x}$. It is wise to be careful about these distinctions at first, but the practical literature in statistics ordinarily does not make a distinction between X_i and x_i. The reader is often expected to know when to consider x_i a random variable and when to consider it a numerical value realized in an experiment. When properties of estimates are being discussed, they are usually being thought of as random variables. Mathematics is full of such elliptical devices.

A helpful theorem that we may often use but shall not prove follows.

Theorem 3. *Estimating functions.* If $\hat{\theta}_1, \hat{\theta}_2, \ldots, \hat{\theta}_c$ are maximum likelihood estimates of the parameters $\theta_1, \theta_2, \ldots, \theta_c$, then the maximum likelihood estimate of a function of the θ's, $g(\theta_1, \theta_2, \ldots, \theta_c)$, is the same function of the maximum likelihood estimates, $g(\hat{\theta}_1, \hat{\theta}_2, \ldots, \hat{\theta}_c)$.

This important result makes it easy to use maximum likelihood estimates to estimate complicated functions.

Example 2. If $\hat{p}_1$, $\hat{p}_2$, $\hat{p}_3$ are the maximum likelihood estimates of p_1, p_2, p_3 for the trinomial, find the maximum likelihood estimate of the probability of 3 outcomes in category 1, 4 in category 2, and 2 in category 3 in 9 trials of an experiment based on the same trinomial.

Solution. The probability of the outcome is

$$\frac{9!}{3!\,4!\,2!}\, p_1^3 p_2^4 p_3^2$$

and so the theorem tells us that its maximum likelihood estimate is

$$\frac{9!}{3!\,4!\,2!}\, \hat{p}_1^3 \hat{p}_2^4 \hat{p}_3^2.$$

Note. It would go well beyond the scope of this book to explain the many advantages of maximum likelihood estimators. Under rather general conditions, at least in large samples, the maximum likelihood approach will find good estimators if there are any. The reason $\bar{p}$ and $\bar{p}_i$ are good estimators stems little from their unbiasedness but, instead, largely from their basis in maximum likelihood. Exercises 2, 3, and 8 illustrate this point. A major weakness of the maximum likelihood approach is that one must know the form of the distribution. Since we often have only a slight grasp on the form, maximum likelihood estimation may not be available.

EXERCISES FOR SECTION 2-9

The notation in the exercises is that used in the section.

1. For the binomial distribution, prove that, when $X = n$, the value of the maximum likelihood estimate of p is 1.

2. Each binomial trial has probability p (> 0) of success, $1 - p$ of failure. If the first success comes on trial n, where $n = 1, 2, 3, \ldots$, the probability distribution of n is
$$P(n) = p(1 - p)^{n-1}.$$
Show that the maximum likelihood estimate of p is $1/n$.

3. Using the distribution of Exercise 2, show that the estimate that gives
$$\hat{p} = 1, \quad \text{for } n = 1, \quad \text{and}$$
$$\hat{p} = 0, \quad \text{for } n > 1,$$
is an unbiased estimate of p (its average value is p: $E(\hat{p}|p) = p$). Do you see any objections to this estimate?

4. (Calculus) You know that a trinomial has probabilities $(1 - 2p, p, p,)$, and the observed counts are (x_1, x_2, x_3). Find the maximum likelihood estimate. Does it agree with the estimate $\hat{p}_1 = x_1/n$?

5. (Calculus) Prove Theorem 1 for $0 < X < n$ by taking advantage of the fact that the maximizing value of P and of log P are the same. And so, before differentiating, take the logarithm of $p^x(1 - p)^{n-x}$.

6. $\bar{X}$ versus $\bar{p}$. Given n multinomial trials, each resulting in category i or not, we define n random variables B_j, where $j = 1, 2, \ldots, n$.

$$B_j = \begin{cases} 1 \text{ if trial } j \text{ results in category } i, \\ 0 \text{ otherwise.} \end{cases}$$

Then

$$X_i = \sum_{j=1}^{n} B_j.$$

Use this idea to explain the textual remark that the usual estimates of the binomial p and the multinomial p_i are special cases of the unbiasedness of $\bar{X}$, the sample mean, as an estimate of the population mean μ.

7. If the value of the maximum likelihood estimate of a binomial probability p is 0.2, what is the maximum likelihood estimate of $p^2(1 - p)$?

★8. Prove that the unbiased estimate given in Exercise 3 is the only unbiased estimate of p.

CHAPTER 3

Ranking Methods for Two Independent Samples

3–1 A PROBLEM AND A METHOD

An investigator has *two small, independent, random samples of measurements.* He does not know any details of the shape of the distributions from which the measurements were drawn, and he wishes to analyze the data of the samples. Do the samples come from the same distribution or from two different distributions? If the distributions are different, how do they differ? Answers to such questions may facilitate decisions in important practical investigations. For example: Is one medical treatment better than another? Does one method of annealing glass produce tougher tumblers than another? Does one group of people perform better than another?

To answer such questions, we may sometimes use the measurements directly as they come from a study, as we do in the *t*-test and the permutation test of Chapter 1. At other times, we may not wish to use measurements as they stand because, for example, they may exaggerate or underrate the importance of certain effects. Or we may want a quick method of testing. On such occasions, an analysis based on *ranking methods* may offer a simple and effective procedure.

To get the *rank* of a measurement, given two samples of measurements, proceed as follows: (1) Combine the two samples, and list their measurements in order of ascending magnitude, from left to right. (2) Beginning at the left, assign the number 1 to the first measurement, 2 to the second, 3 to the third, . . ., and *n* to the *n*th. Then *the rank of each measurement is the number that has been assigned to it.*

Example 1. *Densities of planets.* Taking the density of planet Earth as 1 unit, astronomers have made the following measurements of planetary densities (see Exercise 8, Section 1–5):

Planet:	Mercury	Venus	Earth	Mars	Jupiter	Saturn
Density:	0.68	0.94	1	0.71	0.24	0.12
	Group 2			Group 1		

Regard the densities in Group 1 as an *A*-sample of measurements and the densities in Group 2 as a *B*-sample. Then rank the measurements of the two samples.

Solution. Arrange the 5 measurements in order of ascending magnitude, and assign the ranking numbers thus:

0.12	0.24	0.68	0.71	0.94
1	2	3	4	5
A	*A*	*B*	*A*	*B*

Measurement 0.24 has rank 2, measurement 0.94 has rank 5, and so on. The *rank sum* of the *A*-sample is $1 + 2 + 4 = 7$; the rank sum of the *B*-sample is $3 + 5 = 8$.

In this chapter we shall use ranking methods to present and develop a test for analyzing two independent samples of measurements drawn from continuous distributions. The continuity guarantees no tied measurements. (When one is dealing with real data, ties sometimes occur. We shall discuss ties in Section 4–4.) This test has been chosen for a number of reasons:

1. It is in common use and offers the reader an introduction to practical statistics.

2. It enables us to work out some important probabilities for *small* samples by simple counting methods. Recall that the *t*-distribution approach for small sample tests (used in Section 2–6) assumes that the two populations from which the samples are drawn have approximately equal variances and roughly the shape of normal distributions. No such assumptions are needed for the test discussed in this chapter.

3. It informally introduces, or reviews, the ideas and nomenclature of statistical tests of hypotheses.

4. It uses, in a new setting, some basic ideas—the mean, the variance, and the normal approximation to a discrete distribution.

5. Its development can be regarded as a prototype for other tests. Once we have explored it carefully, we can study other tests more easily and quickly because the fundamental ideas will have been introduced and exemplified. The ideas of other rank tests are extremely close to these; and so are the ideas of all tests.

In the next section, we introduce this test, often called the Wilcoxon-Mann-Whitney test. It is a statistical test that many authors have discovered independently. Wilcoxon seems to have been the first in modern times, but Mann and Whitney did much to develop and popularize the test. To avoid confusion with another Wilcoxon test, we shall refer to the Wilcoxon-Mann-Whitney test as the *Mann-Whitney* test.

EXERCISES FOR SECTION 3–1

1. In Example 1 of this section, what is the total sum of all the ranks? If the two samples together have N measurements, what is the total sum of all the ranks?

2. If an A-sample has m measurements and a B-sample has n measurements, find (a) the total sum of all the ranks; (b) the minimum possible rank sum of A; (c) the maximum possible rank sum of B.

3. (Continuation) In Exercise 2, what relation holds among the three answers?

3–2 THE MANN-WHITNEY TWO-SAMPLE TEST

Suppose that A and B denote the samples having, respectively, m and n measurements, where $m \geq n$. The general plan of the Mann-Whitney test is as follows: (1) Get the rank sum of sample B. (2) Assuming that the measurements come from distributions that are identical, find the probability distribution of the possible rank sums of B. (3) Consider the probability distribution of rank sums under alternative assumptions about the distributions, such as the assumption that they are displaced from one another. Such considerations give us an idea of the strength or power of the test. (4) Use the results of the foregoing three steps to make decisions about the distributions from which the samples were drawn.

Note. We have chosen the rank sum of B as our Mann-Whitney statistic because it is a little easier to compute than other statistics that might be used.

Example 1. *Densities of planets.* (Refer to Example 1, Section 3–1.) Is it reasonable to decide that the densities in the B-sample come from a distribution generally to the right of that for the A-sample? Discuss.

Solution. In Example 1 of Section 3–1, we ranked the five measurements of densities and found the B rank sum equal to 8. If we assume that the five measurements are independently drawn from the same distribution, then the ranks of the two B's and of the three A's are completely random. Under this assumption, it is easy to compute probabilities. Here is the reason: Since 5 objects (2 of one kind and 3 of another kind) can be arranged in $\binom{5}{2}$ or 10 ways (see *PWSA*, p. 60), there are 10 possible outcomes for the sampling, as shown in Table 3–1.

Table 3–1
Distribution of the sum of the B ranks: 3 A's, 2 B's

1	2	3	4	5	B's rank sum	1	2	3	4	5	B's rank sum
A	A	A	B	B	9	A	B	A	B	A	6
A	A	B	A	B	8	B	A	A	B	A	5
A	B	A	A	B	7	A	B	B	A	A	5
B	A	A	A	B	6	B	A	B	A	A	4
A	A	B	B	A	7	B	B	A	A	A	3

Under the assumption that the samples from A and B are independent random samples from the same distribution, each of the foregoing 10 arrangements is equally likely. Since the 5 measurements come from the same distribution, they could be thought of as a single sample of 5 measurements, two of which are to be randomly designated as coming from population B. Hence we can construct the probability table shown as Table 3–2 by associating with each possible rank sum of B the probability that it occurs. We shall use T to denote the random variable "the rank sum of B" and t to denote "a value of T". Note that the distribution of T is symmetrical about its middle value. This symmetry always occurs for T when the observations are independently drawn from the same distribution.

Table 3–2
Probability distribution of T, the rank sum of B: 3 A's and 2 B's

$P(t)$	$\frac{1}{10}$	$\frac{1}{10}$	$\frac{2}{10}$	$\frac{2}{10}$	$\frac{2}{10}$	$\frac{1}{10}$	$\frac{1}{10}$
t	3	4	5	6	7	8	9

Figure 3–1 gives a *probability graph* corresponding to the probability table.

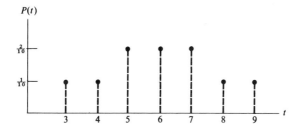

Fig. 3–1 Probability graph for distribution of T, the rank sum of B: 3 A's and 2 B's.

Suppose that we make a decision on the basis of the following rule: *If the rank sum of B is 8 or 9, then decide that the B-measurements come from a distribution generally to the right of the A-distribution; otherwise, decide that the distributions are identical.* From Table 3–2 we see that, for identical distributions, the rule leads to a correct decision 80% of the time and to an incorrect decision 20% of the time. In other words, if the distributions are identical, the probability is 0.80 that our test will assert that they are identical.

In our example we have decided to reject the hypothesis of a random distribution of densities.

Discussion

1. *Why ranks?* Why do we use the ranks of the measurements instead of the measurements themselves?

If we have had little experience with a distribution, we may prefer not to make strong assumptions about its shape. If we expect occasional "wild" observations, that is, unusually large or small ones, we may prefer not to toss them out, but to discount them automatically by using ranks. Also, for a preliminary analysis, the ranks offer a quick and simple test. Finally, in some situations only the ranks of the observations are available.

2. If the two samples are drawn not from the same distribution but from two where the B-measurements are generally larger than the A-measurements, then the probability table will differ from Table 3–2. Higher rank sums of B will be more likely. Indeed, if the distributions do not overlap (Fig. 3–2), the probability that the rank sum of B equals 9 is 1. Such nonoverlapping is unusual in statistical work. In more common situations, illustrated in Fig. 3–3, overlapping occurs. The figure shows two normal distributions with a common standard deviation, but their means are one standard deviation apart. We shall explain in Section 3–6 how to compute the probabilities for such an example, but their values are given approximately by Table 3–3.

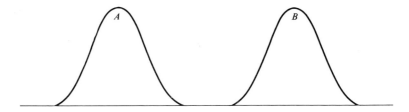

Fig. 3–2 Nonoverlapping distributions.

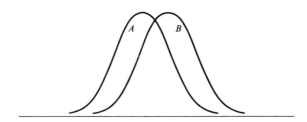

Fig. 3–3 Overlapping distributions.

Table 3–3
Distribution for T, the rank sum of B: 3 A's and 2 B's, drawn from populations represented by Fig. 3–3

$P(t)$	0.02	0.02	0.06	0.13	0.20	0.20	0.37
t	3	4	5	6	7	8	9

As an alternative assumption, suppose that the distributions are as shown in Fig. 3–3, giving probabilities as shown in Table 3–3. Then, *on the basis of the foregoing rejection rule* (reject for $T \geq 8$), we would correctly decide 57% of the time that the *B*-measurements had a larger mean than the *A*-measurements, and we would be mistaken 43% of the time. With only five measurements, we cannot expect wonders; the large probability of error stems partly from the small samples and partly from the overlap of the distributions.

Sometimes we need to enumerate a few rank sums rather than the entire set of rank sums. Then a simple layout facilitates and systematizes the work.

Example 2. *Listing.* If a sample of m A's and one of n B's, where $m = 4$ and $n = 3$, have been independently drawn from the same distribution, find the probability that the rank sum of B's is at least 14.

Solution. Examine Table 3–4. The top row shows that the maximum B rank sum, 18, is obtained when the 3 B's get the 3 highest possible ranks, namely 5, 6, and 7. The second row shows that the next highest B rank sum, 17, corresponds to B ranks of 4, 6, and 7; and so on. The table shows 11 samples with a B rank sum of 14 or more. The total number of samples is $7!/3!\,4! = 35$. Therefore, $P(\text{rank sum of } B \geq 14) = 11/35$.

Table 3–4

Listing for large rank sums of B: $m = 4$, $n = 3$

Outcome	1	2	3	4	5	6	7	B-sum
1					B	B	B	18
2				B		B	B	17
3			B			B	B	16
4		B				B	B	15
5	B					B	B	14
6				B	B		B	16
7			B		B		B	15
8		B			B		B	14
9			B	B			B	14
10				B	B	B		15
11			B		B	B		14

In this section, we merely wish to indicate the general idea of the test.

60   Ranking methods for two independent samples

EXERCISES FOR SECTION 3–2

1. In Example 1 suppose that we had decided to reject the hypothesis of identical distributions when and only when the sum of the ranks was 7 or more. What would the probability of a correct decision be if the distributions were identical? What would the probability be if the distributions had the form shown in Fig. 3–3?

2. What set of rejection scores would minimize the sum of the probabilities of wrong decisions in the two situations—identical distributions versus those of Fig. 3–3—in our example? (This might be a reasonable criterion if we believed the two situations were equally likely.)

3. Suppose that 2 measurements are of type A and 2 are of type B and that the A's and B's are drawn from identical continuous distributions. Find the distribution of the sum of the ranks for observations of type B.

4. If there are 3 observations of type A and 3 of type B, how many possible outcomes (rankings) are there?

5. If there are m observations of type A and n of type B, how many possible outcomes (rankings) are there?

6. Suppose that there are 5 observations of type A and 4 of type B, all from the same distribution. Find (a) the maximum rank sum of B; (b) the number of rank sums of B that are ≥ 27; (c) P(rank sum of $B \geq 27$).

7. Suppose that there are 6 measurements of type A and 4 of type B, all from the same distribution. Find (a) the maximum value of T, the rank sum of B; (b) the number of rank sums of B for which $T \geq 31$; (c) $P(T \geq 31)$.

3–3 TABLES FOR THE MANN-WHITNEY TEST

Consider m measurements in sample A and n measurements in sample B, where $m \geq n$. We have seen in Example 1, Section 3–2, that the number of possible ranking orders is

$$\binom{m + n}{n}.$$

Since this number gets very large, even for modest values of m and n, the listing by hand of possible outcomes soon becomes impractical. Fortunately, tables come to our rescue. These tables, constructed for small values of m and n, tell us how rare a large or small rank sum is, *assuming that the distributions for the two samples are alike*.

Table 3–5 illustrates such complete distributions. As before, we denote the rank sum of B by T, and we use t to denote a value of the random variable T.

Example 1. If $m = 4$, $n = 3$, find $P(T \geq 14)$ and $P(T \leq 10)$.

Solution. For $m = 4$ and $n = 3$, Table 3–5 shows that $P(T \geq 14) = 0.314$ and that $P(T \leq 10) = 0.314$. The integers beside the probabilities give the number *at or above* which the probability lies; and the integers at the *far* left and right side of the whole table give the number *at or below which the probability lies*.

Table 3-5

Exact cumulative distribution for the Mann-Whitney test. The three-digit entries are probabilities multiplied by 1000. The integers at the extreme left and right are values of t to be used in finding $P(T \le t)$ on that line. The integers beside the probabilities are values of t for which the probability is $P(T \ge t)$.

$n = 3; m = 8$	$m = 7$	$m = 6$	$m = 5$	$m = 4$	$m = 3$	
6	30 006	27 008	24 012	21 018	18 029	15 050	6
7	29 012	26 017	23 024	20 036	17 057	14 100	7
8	28 024	25 033	22 048	19 071	16 114	13 200	8
9	27 042	24 058	21 083	18 125	15 200	12 350	9
10	26 067	23 092	20 131	17 196	14 314	11 500	10
11	25 097	22 133	19 190	16 286	13 429		11
12	24 139	21 192	18 274	15 393			12
13	23 188	20 258	17 357	14 500			13
14	22 248	19 333	16 452				14
15	21 315	18 417					15
16	20 388	17 500					16
17	19 461						17

$n = 4; m = 8$	$m = 7$	$m = 6$	$m = 5$	$m = 4$	
10	42 002	38 003	34 005	30 008	26 014	10
11	41 004	37 006	33 010	29 016	25 029	11
12	40 008	36 012	32 019	28 032	24 057	12
13	39 014	35 021	31 033	27 056	23 100	13
14	38 024	34 036	30 057	26 095	22 171	14
15	37 036	33 055	29 086	25 143	21 243	15
16	36 055	32 082	28 129	24 206	20 343	16
17	35 077	31 115	27 176	23 278	19 443	17
18	34 107	30 158	26 238	22 365		18
19	33 141	29 206	25 305	21 452		19
20	32 184	28 264	24 381			20
21	31 230	27 324	23 457			21
22	30 285	26 394				22
23	29 341	25 464				23
24	28 404					24
25	27 467					25

These two results stem from the symmetry of the probability graph of T when the samples come from the same distribution. (Refer to Table 3–2 and Fig. 3–1.) Observe that $P(T \geq 7) = P(T \leq 4)$ and that $P(T \geq 6) = P(T \leq 5)$, and so on. In constructing Table 3–5, we use this symmetry to associate two values of T with each probability tabulated and to reduce the size of the table by about half.

Example 2.   For $m = n = 3$, Table 3–5 tells us that $P(T \geq 15) = 0.05$. Hence, if we decide to reject the hypothesis of identical distributions when the rank sum of B's is 15 or more, then the probability of our being wrong is 0.05 when the distributions are actually identical. A study of Table 3–5 shows that $n = 3$ and $m = 3$ give the smallest total sample size, 6, that provides such a small probability of our deciding that the distributions are different when they are in fact alike.

In Example 2, suppose that we decide to reject the hypothesis of identical distributions if $T \geq 15$. Implicit in this decision is the assumption that the B-distribution might be slipped to the *right* of the A-distribution (see Fig. 3–3), thus tending to yield large values of T, the rank sum of B. So our test is one-sided. Similarly, if we assume that the B-distribution might lie to the *left* of the A-distribution, a one-sided test might reject the hypothesis of identical distributions if $T \leq 6$. (See Table 3–5.)

On the other hand, if we assume that the B-distribution may lie *either to the right or to the left* of the A-distribution, then it would be appropriate to reject the hypothesis of identical distributions for either large or small values of T. For example, we might decide to reject for $T \geq 15$ *or* for $T \leq 6$. Our test is then two-sided (see Section 2–5).

Significance levels.   (Table A–9 at the back of the book) In Example 1 of Section 3–2, where $m = 3$ and $n = 2$, we decided, when $T \geq 8$, to reject the hypothesis that the A's and B's come from the same distribution. This special assumption of identical distributions is the null hypothesis. Table 3–2 shows that, when we use the foregoing rejection rule, the probability is 0.2 that we reject the null hypothesis when it is true. Thus, by the definition in Section 2–4, the significance level of the given test is 0.2; given a true null hypothesis, this test rejects it about 20% of the time.

Table A–9 at the back of the book gives in summary form probabilities near the levels commonly used in significance tests. It gives the counts just above and just below the one-sided 0.005, 0.025, 0.05, and 0.10 levels of significance, together with their associated probabilities.

Example 3.   If $m = n = 8$, what rank sum is just beyond the one-sided 0.05 level.

Solution.   Table A–9 gives $T = 85$ as the smallest total just beyond the one-sided 0.05 level, and the actual probability is 0.0415 rather than 0.05. On the other hand, for $T = 84$ the probability is 0.0524.

Example 4. *Does Brand X prevent infection?* An investigator wishes to test the effectiveness of a certain germicidal soap in preventing infections in the day-to-day cuts, bruises, and abrasions of school children. He uses the children in 10 dormitories of a boarding school to provide two independent samples. Sample *A* consists of the children in 5 of the dormitories; sample *B* consists of those in the remaining 5 dormitories. Children in sample *A* are provided with soap that resembles the germicidal soap in appearance but does not have any of the alleged germicidal ingredients (a placebo); children in sample *B* are provided with the germicidal soap. Records are kept, and the proportions of infections are computed as follows:

Infection rates for 10 dormitories (F. Carlborg)

Sample *A* (placebo)	Sample *B* (germicidal soap)
0.333	0.240
0.125	0.425
0.229	0.268
0.129	0.536
0.061	0.083

Question: What can be said about the effectiveness of the germicidal soap?

Solution. To see whether the germicidal soap is more effective than the placebo soap, we might ask if T is smaller than what one might expect to get by pure chance. But something has gone awry, and we can plainly see that the *B*'s are running a bit larger than the *A*'s. We had better ask if the *B*'s are running unreasonably high. The germicidal soap appears to be something less than effective. To apply our test, we rank the proportions:

0.061	0.083	0.125	0.129	0.229	0.240	0.268	0.333	0.425	0.536
1	2	3	4	5	6	7	8	9	10
A	*B*	*A*	*A*	*A*	*B*	*B*	*A*	*B*	*B*

T = rank sum of *B*'s = $2 + 6 + 7 + 9 + 10 = 34$.

For $m = 5$ and $n = 5$, Table A–9 gives

$$P(T \geq 34) = 0.1111.$$

We have performed a one-sided test of the hypothesis that the *B*-distribution is to the *right* (more infections) of the *A*-distribution. The result leads to the following conclusions.

1. Under the null hypothesis, the probability of getting a rank sum of 34 or more is 0.1111.

2. If we reject the null hypothesis in such cases, we'll be wrong just over 11% of the time when the null hypothesis is true.

3. The significance level of this test is 11.11%.

The practical report is this: Although the results point toward more infections for germicidal soap than for placebo soap, they do not reach the levels ordinarily used for such a conclusion. The results go in a direction opposite from that expected. We would ordinarily use a one-sided test to see if the B scores had lower ranks. This is one of the many cases where the actual findings send us scurrying to see whether the results are significant in a direction opposite to that we set out to test for. Certainly the developers of the germicidal soap hoped to reduce the number of infections. Since the result did not reach significance, they considered it possible that they had had an unlucky break in the random selection of dormitories, and they planned a further, larger test.

EXERCISES FOR SECTION 3–3

1. If $m = 6$ and $n = 4$, find $P(T \geq 26)$ and $P(T \leq 18)$. Use Table 3–5.

2. If $m = 7$ and $n = 3$, find $P(T \geq 20)$ and $P(T \leq 13)$. Use Table 3–5.

3. Given $m = 7$ and $n = 4$, find the probability of an incorrect decision if the distributions are alike and we decide to reject the hypothesis of identical distributions when (a) $T \geq 36$, (b) $T \geq 29$, (c) $T \geq 25$.

4. In each of the following, use Table 3–5 to evaluate u, l, and p whenever they occur.

   a) $m = 5$, $n = 3$, $P(T \geq u) = 0.196 = P(T \leq l)$
   b) $m = 4$, $n = 4$, $P(T \leq l) = p = P(T \geq 23)$
   c) $m = 8$, $n = 4$, $P(T \leq 20) = p = P(T \geq u)$

5. If $m = 7$ and $n = 5$, find the smallest value of u that makes $P(T \geq u) \leq 0.05$. Use Table A–9.

6. If $m = n = 6$, find the largest value of l that makes $P(T \leq l) \leq 0.025$. Use Table A–9.

7. When $m = 8$ and $n = 3$, we decide to reject the hypothesis of identical distributions for $T \geq 25$. What is the probability of our making an incorrect decision when the distributions are identical.

8. Refer to Example 4. Suppose that the investigator discovers that clerical errors were made in tabulating the infection rates and that the correct listing is:

Sample A (placebo):	0.129	0.333	0.536	0.240	0.268
Sample B (germicidal soap):	0.425	0.125	0.083	0.061	0.229

   Draw conclusions about the effectiveness of germicidal soap from these data, using the Mann-Whitney test.

9. Refer to Example 2, Section 1–5 (Rosier futures). Use the Mann-Whitney test on the data of this example to get the descriptive significance level for the null hypothesis that distributions for both kinds of countries are identical. Compare your result with that obtained in Example 2, Section 1–5.

10. Following a first heart attack, a patient receives one of two kinds of treatment, and the number of years of survival is noted for each patient.

Years of survival under treatment A: 4.3   5.0   7.5   9.4   12.2
Years of survival under treatment B: 4.5   6.0   15.0

(a) What conclusion should be drawn if we reject the hypothesis of identical distributions for $T \geq 14$? If the distributions are alike, (b) what is P(incorrect decision), and (c) what rejection plan would make P(incorrect) ≤ 0.05?

11. *The race.* The *World Almanac* for 1970 lists the winning times for the annual Harvard-Yale varsity eights rowing races from 1952 to 1969. The times (in minutes and seconds) for the first five years (1952–1956) are

$$22:49 \qquad 20:09 \qquad 21:58 \qquad 20:05 \qquad 19:26$$

and the times for the last five years (1965–1969) are

$$19:42 \qquad 19:44 \qquad 22:43 \qquad 20:21 \qquad 19:37$$

Using the Mann-Whitney test, should we accept, at the 5% level, the null hypothesis of identical distributions for the two sets of times?

3–4 FORMULAS FOR $n = 1$, $n = 2$

Although we do not give tables for them, we may need probabilities when the smaller sample is of size 1 or 2.

Case $n = 1$. Since a single measurement may have any one of $N = m + n$ equally likely ranks, the probability that $T \leq t$ when $n = 1$ is

$$P(T \leq t \mid n = 1) = \frac{t}{N}, \qquad N = m + n = m + 1. \qquad (1a)$$

Similarly, the probability that $T \geq t$ is

$$P(T \geq t \mid n = 1) = 1 - \frac{t - 1}{N}, \qquad N = m + 1. \qquad (1b)$$

(See Exercises 4 and 5.)

Case $n = 2$. To get the probability that $T \leq t$ when $t \leq N + 1$, we shall consider a special case and generalize the results. Examine Table 3–6, which exhibits rank sums of B when $n = 2$ and $N = 5$. For this case, the B-sample consists of two measurements, arbitrarily denoted by b_1 and b_2, without regard to their sizes. Measurement b_1 may have any rank from 1 to 5, and so may b_2. Of course, both measurements may not have the same rank. Table 3–6 shows the rank sum for each possible pair of ranks; for example, if b_1 has rank 3 and b_2 has rank 4, the table shows that the corresponding rank sum is 7.

Table 3–6

Possible rank sums for B-sample
when $n = 2$ and $N = 5$

		Rank of b_2				
		1	2	3	4	5
	1	X	3	4	5	6
	2	3	X	5	6	7
Rank of b_1	3	4	5	X	7	8
	4	5	6	7	X	9
	5	6	7	8	9	X

For this special case, $3 \leq t \leq 9$, and the middle value of t is $\bar{t} = (3 + 9)/2 = 6$; in general, $3 \leq t \leq 2N - 1$ and $\bar{t} = (3 + 2N - 1)/2 = N + 1$. Table 3–6 shows $5^2 - 5 = 20$ admissible outcomes; in general, there are $N^2 - N = N(N - 1)$ admissible outcomes.

Now begin at the upper left corner of Table 3–6 and consider the diagonal rows of cells that slope upward to the right. The first diagonal row has one cell, marked X; the second has two cells, marked 3, 3; the third has three cells, marked 4, X, 4; and so on. Thus each admissible value of t occurs in diagonal row number $t - 1$. When $3 \leq t \leq 6$, observe that $T \leq t$ corresponds to the outcomes in the $t - 1$ upper left diagonal rows. The same result holds when $3 \leq t \leq N + 1$ and $T \leq t$. *In general*: The number of admissible outcomes for $T \leq t$ is

$$1 + 2 + 3 + \cdots + (t - 1) = \frac{(t - 1)t}{2}$$

less the number of cells marked X in the $t - 1$ diagonal rows.

Since there is one X-cell on every other diagonal row, the number of these X-cells for a given t is $[t/2]$, which means "the greatest integer contained in $t/2$". (Thus $[3/2] = 1$, $[4/2] = 2$, $[5/2] = 2$, and so on.)

Therefore, if $t \leq N + 1$, the number of admissible outcomes is

$$(t - 1)(t)/2 - [t/2].$$

Since all outcomes are equally likely, we get

$$P(T \leq t \mid n = 2, t \leq N + 1) = \frac{(t - 1)t/2 - [t/2]}{N(N - 1)}. \tag{2a}$$

By similar reasoning, we can show that, for $t \geq N + 1$, we get

$$P(T \geq t \mid n = 2, t \geq N + 1)$$
$$= \frac{(2N - t + 1)(2N - t + 2)/2 - [(2N - t + 2)/2]}{N(N - 1)}, \tag{2b}$$

where, as before, the square bracket means "the greatest integer contained in". The proof of formula (2b) is left as an exercise.

Example 1. Given $m = 3$ and $n = 2$, use formulas (2a) and (2b) to find $P(T \geq 8)$ and $P(T \leq 5)$. Check your results with Table 3–2.

Solution. Since $n = 2$ and $8 \geq N + 1 = 6$, formula (2b) gives

$$P(T \geq 8) = \frac{(3)(4)/2 - [4/2]}{(5)(4)} = \frac{2}{10}.$$

Since $5 \leq N + 1 = 6$, formula (2a) gives

$$P(T \leq 5) = \frac{(4)(5)/2 - [5/2]}{(5)(4)} = \frac{4}{10}.$$

These results check those obtained from Table 3–2.

EXERCISES FOR SECTION 3–4

1. Use formulas (1a) and (1b) to find $P(T \leq 4 \mid n = 1)$ and $P(T \geq 8 \mid n = 1)$ when $m = 10$.

2. Use formulas (2a) and (2b) to find $P(T \geq 25 \mid n = 2)$ and $P(T \leq 17 \mid n = 2)$ when $m = 18$.

3. *Hummingbird speeds.* Walter Scheithauer timed the flight of Blue Sylph hummingbirds over a measured course in a large aviary. During a run of 10 flights, the flight times in seconds for flights 1 and 2 were, respectively, 3.6 and 4.2; and the times for flights 9 and 10 were, respectively, 5.1 and 3.9. If we decide to accept the hypothesis that the last two flights are slower than the first two when $T \geq 6$, what is the significance level of the test? Use formula (2b), and check the result by making tables like Table 3–1 and Table 3–2.

4. Prove formula (1a).

5. Use formula (1a) to derive formula (1b).

6. Prove formula (2b). [*Hint:* Begin numbering diagonal rows at the lower right corner of Table 3–6. Show that a given rank sum t, where $t \geq N + 1$, occurs in diagonal row number $(2N - 1) - (t - 1) + 1 = 2N - t + 1$.]

7. For t odd and for t even find $P(T = t)$ for $n = 2$.

8. (Continuation) Check the results of Exercise 7 by counting for the cases $m = 5$ and $m = 6$. (Make a table like Table 3–6.)

9. (Continuation) If N is even, check the results in Exercise 7 by summing the probabilities to get unity.

10. *Mass of planets (World Almanac, 1970).* The following table gives the masses of the planets in terms of the mass of Earth, which is taken as 1 unit. The A planets are farther from the sun than Earth, and the B's are nearer.

Mercury	Venus	Earth	Mars	Jupiter	Saturn	Uranus	Neptune	Pluto
0.05	0.81	1	0.11	318	95	15	17	0.18

$$\underbrace{\hspace{3cm}}_{B} \qquad \underbrace{\hspace{7cm}}_{A}$$

Does the Mann-Whitney accept, at the 10% level, the hypothesis of identical distributions for groups A and B?

3–5 POWER OF THE MANN-WHITNEY TEST

In discussing Example 1 of Section 3–2, we noted the possibility of making assumptions other than the null hypothesis. Figure 3–2 represents one such assumption, and Fig. 3–3 represents another. As before, these other possible assumptions are called *alternative hypotheses*. Thus Fig. 3–3 exhibits one alternative hypothesis, and Table 3–3 gives its corresponding T-distribution. We have seen that, for this alternative hypothesis and rejection region $T \geq 8$, the probability of rejecting the null hypothesis is 0.57, which is also the probability of accepting this alternative hypothesis when it is true.

For a given test, *each alternative hypothesis has a table of probabilities* like that shown as Table 3–3.

In our example, we may not be happy about the results: The significance level is rather high, 0.20, and the power is rather low, 0.57. How might we improve matters? Would it improve matters to change the rejection number? Table 3–7 compares results for different decisions regarding the rejection number. As we decrease the significance level, we get tests of decreasing power for the given alternative hypothesis. To get a test that gives good protection against the two kinds of errors mentioned in Section 2–4, we need larger samples.

Table 3–7

Comparison of results for different decisions regarding the rejection number (Mann-Whitney test)

Reject null hypothesis for	Significance level	Power of alternative hypothesis
$T \geq 6$	0.6	0.90
$T \geq 7$	0.4	0.77
$T \geq 8$	0.2	0.57
$T = 9$	0.1	0.37

3–6 TABLES FOR COMPUTING POWER FOR NORMALLY DISTRIBUTED ALTERNATIVES FOR MANN-WHITNEY TEST

Power is usually hard to compute even for very specific situations, and often one must simulate the process to get answers. That is how we got Table 3–3. We took many samples of 3 from distribution A, and many samples of 2 from distribution B in Fig. 3–3. We then computed the proportion of times we got each value for T, the rank sum of the B-sample. High-speed computers make this relatively easy.

What we did, as an illustration, was to consider the two normal distributions, both with standard deviation $\sigma = 1$. The A-distribution was taken to have mean 0 and the B-distribution mean d. We drew 3 random normal deviates from the A-distribution, to form the A sample. We drew 2 more random normal deviates from the A—yes, A— distribution but added d to each of them to form the B sample. Then we ranked the 5 measurements and computed the B rank sum. We did this 1000 times for each d. We let d run from -2.0 to $+5$ by intervals of $\frac{1}{2}$. The resulting frequency distributions appear in Table 3–8.

Table 3–8

Frequency distributions of the B rank sum, T, for 1000 simulations for various differences, d, between means: $m = 3, n = 2$ (Mann-Whitney test)

T = rank sum of B	d: -2	-1.5	-1	-0.5	0	0.5	1	1.5
3	712	547	370	211	96	40	16	5
4	156	190	191	153	100	47	16	8
5	94	164	215	236	197	122	58	21
6	22	62	119	184	197	193	134	83
7	13	26	73	133	208	222	204	147
8	0	5	21	37	108	173	203	188
9	3	6	11	46	94	203	369	548
Total for $T = 8$ or 9	3	11	32	83	202	376	572	736

T = rank sum of B	2	2.5	3	3.5	4	4.5	5
3	0	0	0	0	0	0	0
4	2	0	0	0	0	0	0
5	14	4	0	0	0	0	0
6	28	9	1	0	0	0	0
7	90	43	16	1	0	0	0
8	164	121	65	23	12	3	0
9	702	823	918	976	988	997	1000
Total for $T = 8$ or 9	866	944	983	999	1000	1000	1000

Note that the entries in Table 3–8 may be modified to give estimates of probabilities. We just divide each entry by 1000. Thus, for 712 read 0.712, and for 94 read 0.094, and so on. It is common practice to obtain estimates of probabilities by simulation because corresponding theoretical calculations are often impossible to carry out.

Table 3–8 shows clearly that as d, the distance between the population means, grows, the distribution of rank sums, T, piles up at higher values. When $d \geq 3$, practically all the B-samples have a rank sum of 8 or 9. Note also that when d is negative, the probability of $T \geq 8$ becomes practically 0.

When $d = 0$, we know the theory, and from Table 3–2 we get the exact probability: $P(T \geq 8) = 0.2$. The simulation gives 202 samples, or 0.202, which is extremely close.

The power of the test that rejects the null hypothesis when $T \geq 8$ can be approximated by computing, for each d in Table 3–8, the proportion of samples having $T \geq 8$. Since each column sums to 1000, this computation is especially easy. We have plotted the resulting power curve in Fig. 3–4.

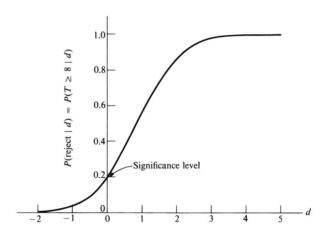

Fig. 3–4 Power of Mann-Whitney test when samples of size 3 and 2 are drawn from normal populations with $\sigma = 1$, means 0 and d, and rejection criterion $T \geq 8$.

One would expect that when m and n are both larger than 3 and 2, respectively, the power curve will change and become sharper when the significance level is held constant. For tests having significance levels almost identical at 0.2, Fig. 3–5 shows the power curves for $(m = 3, n = 2)$, $(m = 5, n = 3)$, and $(m = 7, n = 4)$. Note that the larger samples do increase the power for $d > 0$.

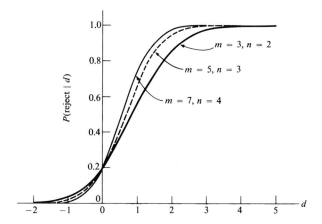

Fig. 3–5 Comparison of power curves for ($m = 3$, $n = 2$), ($m = 5$, $n = 3$), ($m = 7$, $n = 4$) for test having approximately equal significance levels 0.2. Mann-Whitney test.

EXERCISES FOR SECTION 3–6

Decide whether each of the statements in Exercises 1 through 4 is true or false.

1. In a given one-sided Mann-Whitney test, if we increase the rejection number, both the power and the significance level of the test are certain to decrease.

2. Figure 3–4 shows that as the distance between the population means grows, the power of the given test increases.

3. Figure 3–4 shows that as d increases, the significance level increases.

4. Figure 3–5 shows that as the sample size increases, the power of a test with significance level 0.2 increases when $d > 0$.

5. Suppose that an A-sample of size 3 and a B-sample of size 2 are drawn from different normal distributions, both with standard deviation 1. Under the alternative hypothesis that the mean of the B-distribution is 2 units greater than the mean of the A-distribution, we get the following table for the T-distribution.

t:	3	4	5	6	7	8	9
$P(t)$:	0.000	0.002	0.014	0.028	0.090	0.164	0.702

Find the power of a test that rejects the null hypothesis for (a) $T \geq 6$; (b) $T \geq 7$; (c) $T \geq 8$; (d) $T = 9$.

6. Repeat Exercise 5 under the alternative hypothesis that the mean of the B-distribution is 1.5 units greater than the mean of the A-distribution. (Use Table 3–8 to construct the T-distribution of the B-sample.)

In Exercises 7 through 10, suppose that $m = 1$, $n = 1$, where m is the number of A observations and n the number of B's. Suppose that the null hypothesis is rejected if

$T = 2$, based on the rank of the B observation. Suppose that the A observation comes from the uniform distribution on the interval $[0, 1]$, and that the B observation comes from a uniform distribution on the interval a to $a + 1$.

7. What value does a have when the null hypothesis is true?

8. What is the significance level of the test?

9. Suppose that $a = \frac{2}{3}$. What is the power of the test?

10. Graph the power as a goes from -1 to $+2$.

Normal Approximation
for the Mann-Whitney Test

4–1 MEAN AND VARIANCE OF T, THE RANK SUM OF THE B-SAMPLE

Although we can actually enumerate the distribution of the rank sum T for modest values of N, and tables can take us a step further, for large values of N we need an approximation. Even high-speed computers cannot handle the enormous number of arrangements implied by $\binom{N}{n}$ for large N. Since we shall use an approximation that requires us to know the mean and variance of the distribution of T, we now proceed to find them.

We have seen that T, the rank sum of B's, is a simple and useful test statistic. In the examples discussed in Section 3–2, T is a random variable with values $t = 3, 4, \ldots, 8, 9$. Moreover, from the data of these examples, we can compute the mean and the variance of T. Using the distribution in Table 3–2 and the definition of the mean, μ, we get

$$\mu_T = \sum_{t=3}^{9} tP(t) = 3(\tfrac{1}{10}) + 4(\tfrac{1}{10}) + 5(\tfrac{2}{10}) + 6(\tfrac{2}{10}) + 7(\tfrac{2}{10}) + 8(\tfrac{1}{10}) + 9(\tfrac{1}{10}) = 6.$$

This result is obvious from the symmetry of the distribution, as shown in Fig. 3–1. It is also obvious because the possible ranks of a measurement are $1, 2, \ldots, 5$, with average 3; and since the average value of a sum is the sum of the average values, the mean rank sum of our two measurements is $3 + 3 = 6$.

The variance of T is, by definition,

$$\sigma_T^2 = \sum_{t=3}^{9}(t - \mu_T)^2 P(t) = 9(\tfrac{1}{10}) + 4(\tfrac{1}{10}) + 1(\tfrac{2}{10}) + 0(\tfrac{2}{10})$$
$$+ 1(\tfrac{2}{10}) + 4(\tfrac{1}{10}) + 9(\tfrac{1}{10})$$
$$= 3.$$

Hence the standard deviation is

$$\sigma_T = \sqrt{3} \approx 1.7.$$

Table 3–2 shows that 60% of the T-distribution is within one standard deviation of the mean, μ_T, and that all of the T-distribution is within two standard deviations of μ_T.

An alternative way of getting the variance of T is to note that we are drawing a sample of two ranks, without replacement, from a population of the 5 ranks 1, 2, 3, 4, 5. The variance of the sum of the measurements of a sample of n objects, taken from a population of N objects with variance σ_I^2, is (Appendix VI–2 and Exercise 8)

$$\sigma_T^2 = n\sigma_I^2 \left(\frac{N - n}{N - 1} \right). \tag{1}$$

Here I stands for the random variable being measured, and in this example the values of I are integers 1, 2, 3, 4, 5.

For our population of 5 equally likely ranks, we have $N = 5$, $n = 2$, and

$$\sigma_I^2 = E(I^2) - (\mu_I)^2 = E(I^2) - 3^2.$$

To evaluate $E(I^2)$, recall from algebra that

$$1^2 + 2^2 + 3^2 + \cdots + n^2 = \frac{n(n + 1)(2n + 1)}{6}.$$

It follows that

$$E(I^2) = \frac{5(6)(11)}{6} \cdot \frac{1}{5} = 11,$$

and

$$\sigma_I^2 = 11 - 9 = 2.$$

Hence, from formula (1),

$$\sigma_T^2 = 2(2) \left(\frac{5 - 2}{5 - 1} \right) = 3,$$

which checks with the result obtained by direct attack. The following theorem generalizes these results.

Theorem. *Mean and variance of rank sum.* If T is the rank sum of n B's, randomly distributed among m A's, then

$$\mu_T = \frac{n(m + n + 1)}{2}, \tag{2}$$

and

$$\sigma_T^2 = \frac{mn(m + n + 1)}{12}. \tag{3}$$

Proof. The ranks form an arithmetic progression, and so the average of the $(m + n)$ ranks is the average of the first and last ranks, $[1 + (m + n)]/2$. Since the average of the sum of n ranks is the sum of their averages, we have

$$\mu_T = \frac{n(m + n + 1)}{2},$$

as required.

We can get the variance of *T*, the rank sum of *B*'s, by using formula (1) with reference to the population of integers consisting of the $(m + n)$ ranks $1, 2, 3, \ldots, N$, where $N = m + n$. Then

$$\mu_I = \frac{N + 1}{2},$$

and

$$\sigma_I^2 = E(I^2) - (\mu_I)^2 = \frac{N(N + 1)(2N + 1)}{6} \cdot \frac{1}{N} - \frac{(N + 1)^2}{4} \tag{4}$$

$$= \frac{(N + 1)(N - 1)}{12} = \frac{N^2 - 1}{12}.$$

(This is the standard formula for the variance of *N* successive integers; it is well worth remembering.) Therefore, by formula (1), we have

$$\sigma_T^2 = n \left(\frac{N^2 - 1}{12} \right) \left(\frac{N - n}{N - 1} \right)$$

$$= \frac{n(N + 1)(N - n)}{12}$$

$$= \frac{mn(m + n + 1)}{12}.$$

EXERCISES FOR SECTION 4–1

1. Check that formulas (2) and (3) work for Example 1, Section 3–2, where $m = 3$ and $n = 2$.

2. Show by direct calculation, that is, use

$$\sigma_I^2 = \sum_{i=1}^{N} (i - \mu)^2 P(i),$$

that formula (4) gives the correct variance σ_I^2 for $m + n = 5$ ranks.

3. Use the exact distribution obtained for Exercise 3 of Section 3–2, along with the definitions of mean and variance, to compute μ_T and σ_T^2. Compare the results with those given by formulas (2) and (3) of this section.

4. Use formulas (2) and (3) to compute μ_T and σ_T^2 for (a) $m = 7$, $n = 6$; (b) $m = 6$, $n = 6$.

5. Compute the mean and variance of T for the data of Exercise 10, Section 3–3.

6. The proofs of formulas (2) and (3) do not depend on the relation $m \geq n$. Use this fact to obtain formulas for the mean and the variance of the rank sum of the A-measurements when all measurements are drawn from the same population.

7. *Hair and chemistry.* Parker and Holford measured concentrations of chemical elements in human hair. The following table gives the natural logarithms of the concentrations, in parts per million, of iodine and copper found in 10 samples of human hair.

Iodine, A: 2.945, 3.642, 1.631, 2.339, 4.488, 2.589, 2.355, 2.594, 2.174, 2.104
Copper, B: 2.729, 1.948, 3.666, 2.442, 1.360, 3.152, 3.931, 3.640, 3.869, 2.626

Find T, the rank sum of the copper measurements in the pooled sample of 20. Also find the mean of T and the variance of T under the null hypothesis that the distributions of concentrations for the two chemicals are identical. What do you conclude?

8. From formula (2) of Appendix VI–2 and the fact that $\sigma = \sigma_I$ and $T = n\bar{X}$, derive formula (1) of this section.

9. Verify the first part of the parenthetical remark following equation (4).

4–2 NORMAL APPROXIMATION FOR THE MANN-WHITNEY TEST

Given that the A- and B-samples come from the same population, let us consider the shape of the distribution of T for a B-sample of n measurements, where n is fixed and $n \leq m$.

If $n = 1$, we get a uniform distribution on the integers $1, 2, 3, \ldots, N = m + 1$:

$$P(T = t) = 1/N, \qquad 1 \leq t \leq N = m + 1. \tag{1}$$

If $n = 2$, the facts become evident when we inspect Table 3–6. Note the following.

1. $T =$ sum of ranks of the two B's, which is the sum of two slightly correlated random variables. (Sampling without replacement leads to slightly correlated, or approximately independent, random variables when N is large.)

2. The probabilities associated with values of T increase to a peak as the corresponding diagonal rows lengthen, and then they decrease as the diagonal rows shorten. The empty X-cells prevent a perfect pattern. Table 3–6 illustrates the fact that in general the distribution of T for $n = 2$ is approximately triangular.

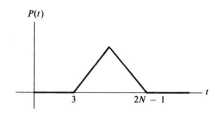

For n as large as 3, characteristics of the shape of the normal distribution appear, as can be seen by inspecting Table 3–5. In particular, we always have symmetry. For $n = 4$, we get very good normal approximations.

Let us consider $m = 4$ and $n = 4$. Table 4–1 shows $P(T \geq t)$ computed exactly and also from the normal approximation. Fig. 4–1 shows a graph of the distribution of T and its approximating normal curve. Both the figure and the table show that the normal fits very well.

Table 4–1

Comparison of exact probabilities $P(T \geq t)$ with normal approximations for $m = 4$, $n = 4$

t	26	25	24	23	22	21	20	19
Exact	0.014	0.029	0.057	0.100	0.171	0.243	0.343	0.443
Approximation	0.015	0.030	0.056	0.097	0.156	0.235	0.333	0.443

Tables for the Mann-Whitney test do not need to be impressively extensive for this reason: *It can be proved that, when the A- and B-distributions are identical and when m and n both become large, the random variable T is approximately normally distributed with mean μ_T and standard deviation σ_T.* As we have seen in Section 4–1, the mean and variance of the T-distribution are

$$\mu_T = \frac{n(m + n + 1)}{2}, \qquad \sigma_T^2 = \frac{mn(m + n + 1)}{12}.$$

This means that the corresponding standardized random variable, that is, the corresponding random variable with zero mean and unit variance, is

$$Z_T = \frac{T - \mu_T}{\sigma_T}.$$

This Z_T is approximately distributed according to the *standard* normal distribution.

Since T takes only integral values, the random variable Z_T has a *discrete* distribution. Consequently, in order to use the normal approximation for Z_T, we may profitably apply the continuity correction, as discussed in Section 2–4. Thus, to get $P(T \geq 3)$, we correct for continuity by using the value $t = 2\frac{1}{2}$, computing

$$z = \frac{3 - \frac{1}{2} - \mu_T}{\sigma_T},$$

and using the fact that

$$P(T \geq 2\tfrac{1}{2}) \approx P\left(Z_T \geq \frac{2\tfrac{1}{2} - \mu_T}{\sigma_T}\right).$$

Tables for the standard normal distribution finish the job.

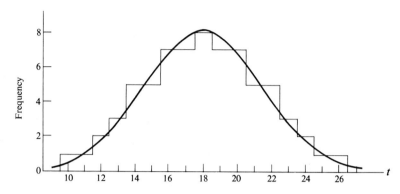

Fig. 4–1 Normal approximation to distribution of T: $m = 4$, $n = 4$.

The $\frac{1}{2}$ that is subtracted from t corrects for the discreteness of T and Z_T. In general, for t an integer, to find $P(T \geq t)$, we compute the standard normal deviate

$$z = \frac{t - \frac{1}{2} - \mu_T}{\sigma_T} \tag{2}$$

and refer z to the tables of the standard normal distribution. Similarly, to find $P(T \leq t)$, we compute

$$z = \frac{t + \frac{1}{2} - \mu_T}{\sigma_T}$$

and then refer z to the standard normal tables. The procedure, parallel to that used in Section 2–2 for the binomial, is illustrated in the following example.

Example 1. For the case discussed in Section 3–2, where $m = 3$ and $n = 2$, find $P(T \geq 8)$.

Solution. Assuming that the distribution of T is approximately normal, we use the foregoing results and get

$$z = \frac{8 - \frac{1}{2} - 6}{\sqrt{3}} \approx 0.866.$$

Thus, when $T \geq 8$, we have $z \geq 0.866$. From tables of the normal distribution, the probability to the right of 0.866 is about 0.193. For such small values of m and n, this approximation is surprisingly close to the correct value 0.2, as shown in Table 3–2.

Example 2. *Do famous men die after reaching their birthmonths?* One might want very much to live until his birthday. Is it possible that some people can live a little longer if they have a goal? The following data were obtained by David P. Phillips, who investigated the relationship of the months in which 1251 famous men died to their birthmonths (i = number of months after the birthmonth).

i:	-6	-5	-4	-3	-2	-1	0	$+1$	$+2$	$+3$	$+4$	$+5$
Number of deaths:	90	100	87	96	101		86	119	118	121	114	113	106

Use these data and the Mann-Whitney test to decide whether or not famous men usually die after their birthmonths.

Solution. Note that the given data offer several possible interpretations. It seems natural to let $i > 0$ indicate "after birthmonth" and $i < 0$ indicate "before birthmonth". But how shall we deal with the birthmonth itself, where $i = 0$? We might interpret "before birthmonth" and "after birthmonth", respectively, by (1) $i \leq 0$, $i > 0$; or by (2) $i < 0$, $i \geq 0$; or by (3) $i < 0$, $i > 0$, in which we omit the $i = 0$ record.

For each of these interpretations, the data suggest that famous men die after their birthmonths. In order to get precise information about the strength of the evidence, we shall apply the Mann-Whitney test, using interpretation (1). The other interpretations are left as exercises.

Let the A-sample include the numbers of deaths in the months before the birthmonth ($i \leq 0$), and let the B-sample include the numbers of deaths in months after the birthmonth ($i > 0$). Rank the numbers, and indicate the ranks belonging to sample B.

86	87	90	96	100	101	106	113	114	118	119	121
1	2	3	4	5	6	7	8	9	10	11	12
						B	B	B	B		B

We have $m = 7$, $n = 5$, and $T = 7 + 8 + 9 + 10 + 12 = 46$. In order to use the normal approximation, we get from formulas (2) and (3) of Section 4–1:

$$\mu_T = (5)(13)/2 = 32.5,$$

$$\sigma_T^2 = (35)(13)/12 = 37.9.$$

The corresponding standard normal deviate for $T = 46$ is

$$\frac{46 - \frac{1}{2} - 32.5}{\sqrt{37.9}} \approx \frac{13.0}{6.16} \approx 2.11.$$

Using the normal approximation and the normal tables, we find

$$P(T \geq 46) \approx P(Z \geq 2.11) \approx 0.017.$$

It follows that we have strong evidence that famous men die after their birthmonths, under the foregoing interpretation.

As a check, we can use Table A–9 for $m = 7$ and $n = 5$. We get $P(T \geq 46) <$ 0.024, which agrees quite well with the normal approximation result.

The reader may wish to know that in further work Phillips has taken account of the distribution of births and deaths by time of year. The effect persists. One suggested explanation: Famous men find their birthdays worth living for. A related phenomenon of living for celebrations is suggested by the fact that Adams and Jefferson both died on the fiftieth anniversary of the signing of the Declaration of Independence. Phillips has also gathered special data for deaths before and after important religious holidays which suggest similar results for the general population. Will there be too few deaths before and an excessive number after January 1, 2000?

EXERCISES FOR SECTION 4–2

For each of the following, use the normal approximation to find the probability indicated and compare the result with the value given in Table 3–5.

1. $P(T \geq 20)$ when $m = 7$ and $n = 3$.

2. $P(T \geq 29)$ when $m = 8$ and $n = 4$.

3. Use the normal approximation to find $P(T \geq 38$ or $T \leq 14)$ when $m = 8$ and $n = 4$. Compare the result with Table 3–5.

4. Use the normal approximation to compute the probability that $T \geq 27$ for $m = 5$ and $n = 4$. Compare the result with the solution to Exercise 6 of Section 3–2.

5. Solve the problem in Example 2, using interpretation (2).

6. Solve the problem in Example 2, using interpretation (3).

7. Solve the problem in Example 2, using the following interpretation. Let "before birthmonth" mean "during the birthmonth itself or in one of the 5 preceding months," and let "after birthmonth" mean "during one of the 6 months following the birthmonth". (This interpretation changes the 90 at -6 into 90 at $+6$ in the data.)

8. In Exercise 7 of Section 4–1 (Hair and chemistry), would you accept, at the 10% level of significance, the hypothesis that distributions of iodine and copper in human hair are identical?

9. Use the normal approximation to check the result of Example 4, Section 3–3 (Germicidal soap).

10. Explain why *both* m and n must become large to ensure the approach to normality. [*Hint:* Consider $n = 1$, with m large.]

11. Explain why the distribution of T must be symmetrical when the distributions of measurements of A's and B's are identical.

4–3 A CLOSER LOOK AT THE MANN-WHITNEY STATISTIC

How do we measure the difference between the two distributions in problems where we use the Mann-Whitney test? To this point, we have used the difference d between the means when working with normal distributions having equal variances. We now give another way to measure the difference between two distributions: Suppose that we draw one observation from the A-distribution and one from the B. *The probability that the B-value exceeds the A-value relates to the "overlap" between the distributions.* If we repeatedly make such paired drawings, then we can compute the observed proportion of times the B-value is larger, and thus we have an estimate of this parameter that measures the "difference" between the two distributions. Two extremes occur: (1) When the distributions are identical or symmetrical about the same point, this probability is $\frac{1}{2}$. (2) As the B-distribution moves far to the left or right of the A, the probability that the B-value is the larger approaches 0 or 1, respectively.

How would this work out in the case of our standard example, where the A-sample has $m = 3$ measurements and the B-sample has $n = 2$? We look at each A, B pair and see which is the larger. For our standard example, the number of times B's exceed A's is given in Table 4–2.

Table 4–2

Distribution of number of times B values exceed A values when null hypothesis holds: $m = 3$, $n = 2$

1	2	3	4	5	No. of times $B > A$	B-rank sum	1	2	3	4	5	No. of times $B > A$	B-rank sum
A	A	A	B	B	6	9	A	B	A	B	A	3	6
A	A	B	A	B	5	8	B	A	A	B	A	2	5
A	B	A	A	B	4	7	A	B	B	A	A	2	5
B	A	A	A	B	3	6	B	A	B	A	A	1	4
A	A	B	B	A	4	7	B	B	A	A	A	0	3

These numbers range from 0 to 6. More important, when we compare these counts with the Mann-Whitney sum of ranks, we find that *each sum of ranks is just three more than the number of times B exceeds A.* This three is the smallest sum of B-ranks, $(1 + 2)$. When this three is subtracted from the B-rank sum, the remainder can be used to estimate the probability that a random B observation exceeds a random A observation.

If we let

$U = $ the observed number of times B's exceed A's,

we can establish a general relation between U's and rank sums.

Theorem. Given n B's with ranks $r_1, r_2, r_3, \ldots, r_n$, where $r_1 < r_2 < r_3 < \cdots < r_n$, and m A's, where $m \geq n$, then

$$U = T - n(n + 1)/2. \tag{1}$$

Proof. First consider $n = 2$ and a general value of m. Then for the two B's, with ranks r_1 and r_2, we have $r_1 < r_2$. The number of A's to the left of r_1 is $r_1 - 1$; and the number of A's to the left of r_2 is $r_2 - 2$, because there are $r_2 - 1$ slots to the left of r_2 and one of these slots is filled with the leftmost B. Therefore, we have

$$T = r_1 + r_2 \quad \text{and} \quad U = r_1 + r_2 - 3.$$

By the same reasoning, if $n = 3$, we have three B's with ranks $r_1, r_2,$ and r_3, where $r_1 < r_2 < r_3$. As before, the number of A's to the left of r_1 is $r_1 - 1$, and the number of A's to the left of r_2 is $r_2 - 2$. The number of A's to the left of r_3 is $r_3 - 3$, because there are $r_3 - 1$ slots to the left of r_3 and two of these are filled with the two leftmost B's. Therefore,

$$T = r_1 + r_2 + r_3 \quad \text{and} \quad U = r_1 + r_2 + r_3 - 6.$$

Similarly, for n B's, we can show that

$$T = \sum_{i=1}^{n} r_i \quad \text{and} \quad U = \sum_{i=1}^{n}(r_i - i) = \sum r_i - \sum i.$$

Hence, since $\sum i = n(n + 1)/2$, we get

$$U = T - n(n + 1)/2,$$

as required.

When we have N measurements, m A's and n B's, where $m \geq n$, the foregoing theorem gives us an important fact:

If the observed B rank sum is T, then the observed U is $T - n(n + 1)/2$.

This fact enables us to compute an estimate of the probability that a B-measurement exceeds an A-measurement. We proceed as follows:

Given m A-measurements $A_1, A_2, \ldots, A_m,$ and n B-measurements $B_1, B_2, \ldots, B_n,$ from the original populations we randomly draw an A_i, where $i = 1, 2, \ldots, m,$ and a B_j, where $j = 1, 2, \ldots, n$. Then we define a random variable X_{ij} thus:

$$X_{ij} = 1 \quad \text{if} \quad B_j > A_i;$$

$$X_{ij} = 0 \quad \text{if} \quad B_j < A_i.$$

Since we are sure that $B_j \neq A_i$ because of continuity, it follows that

$$\sum_{i,j} X_{ij} = U,$$

the observed number of times that B's exceed A's.

Using the fact that there are mn possible AB-pairs, we get

$$V = \text{estimate of } P(B\text{-measurement} > A\text{-measurement})$$

$$= \frac{\sum X_{ij}}{mn} = \frac{U}{mn}$$

$$= \frac{T - n(n + 1)/2}{mn}. \tag{2}$$

Note that the expected value of X_{ij} is $P(B_j > A_i) = p$. Therefore,

$$E(U) = E(\sum X_{ij}) = mnp,$$

and

$$E(V) = E(U)/mn = p.$$

In words, V is an unbiased estimate of p.

Under the null hypothesis, the random variable V has mean $\frac{1}{2}$; under the alternative hypotheses, V has a mean running from 0 to 1.

Example 1. For the special case discussed in Section 3–2, where $m = 3$ and $n = 2$, let $d = 1$ be the difference in the means between A and B, two normal distributions with standard deviation 1. Find $E(V)$, the mean value of the random variable V.

Solution 1. From formula (2), we get

$$E(V) = \mu_V = \frac{\mu_T - n(n + 1)/2}{mn}.$$

We can compute μ_T approximately from the data in Table 3–8, which we obtained by simulation, as discussed in Section 3–6. For $d = 1$, we get

$$\mu_T = E(T) = 3(0.016) + 4(0.016) + 5(0.058) + 6(0.134) + 7(0.204)$$

$$+ 8(0.203) + 9(0.369) = 7.579.$$

The foregoing formula gives

$$\mu_V = \frac{7.579 - 2(3)/2}{6} = 4.579/6 = 0.763.$$

Solution 2. For this particular problem, we can solve theoretically, as follows:

Let Y be a measurement drawn from a normal distribution with $\mu = 1$ and $\sigma = 1$, and let W be a measurement independently drawn from a standard normal distribution with $\mu = 0$ and $\sigma = 1$. Then the random variable $D = Y - W$ is normally distributed with $\mu_D = 1$ and $\sigma_D^2 = 1^2 + 1^2 = 2$. (See Appendix V–4.)

Since $E(V) = P(D > 0)$, we can compute $E(V)$ from the tables of the standard normal distribution. The standardized random normal variable D is

$$Z = \frac{D - \mu_D}{\sigma_D},$$

and the standard normal deviate corresponding to $D = Y - W = 0$ is

$$\frac{0 - 1}{\sqrt{2}} = -0.707.$$

Therefore, $D > 0$ when $Z > -0.707$, and $P(D > 0)$ is the probability to the right of -0.707 in the standard normal distribution. From Table A–1, we find

$$E(V) = P(D > 0) = P(Z > -0.707) = 0.760,$$

which is close to the result 0.763 obtained by simulation from Table 3–8. The value 0.760 obtained from the standard normal distribution is the correct value; the value 0.763 obtained by simulation is an approximation. It estimates the probability that a random measurement from population B exceeds that of one drawn from population A.

EXERCISES FOR SECTION 4–3

1. Repeat Example 1, Solution 1, given that $d = 2$.
2. Repeat Example 1, Solution 2, given that $d = 2$.
3. Given $m = 2$ and $n = 2$, construct a table like Table 4–2.
4. *Birthmonths.* For Example 2, Section 4–2, find an estimate of V, the probability that a B-measurement exceeds an A-measurement.
5. In the theorem in Section 4–1, we proved that $\mu_T = n(m + n + 1)/2$. Use this fact along with the formula in Example 1, Solution 1, to obtain $E(V)$. Explain the result.
6. *Hair and chemistry.* Use the data of Exercise 7, Section 4–1, to obtain an estimate of $P(B\text{-measurement} > A\text{-measurement})$.
7. *Germicidal soap.* Use the data in Example 4, Section 3–3, to obtain an estimate of $P(B\text{-measurement} > A\text{-measurement})$.

4–4 HOW TO RANK TIES

Contrary to the assumptions made earlier, ties occur with real data, and we may still wish to use the Mann-Whitney statistic, T. Ties occur because of discreteness in the original scale of measurement, because of discreteness in the probability distribution, and because of rounding or inability of a judge to rank the observations.

Consider a modification of Example 1 in Section 3–2. Suppose that we have two samples, one of size $m = 3$ and one of size $n = 2$, five measurements in all; and suppose that in the pooled set of measurements the smallest three are tied. If we want to use T, the Mann-Whitney statistic, we need to decide what to do with the ties. The easiest procedure is to assign to each tied item the average rank available to the tied measurements. Here the ranks available are 1, 2, 3, and their average is 2. If items holding ranks 6 and 7 are tied in some problem, we assign "rank" $6\frac{1}{2}$ to both. If the three items that would have received 5, 6, and 7 are tied, all receive "rank" 6. The average tied "rank" is always either an integer or one-half of an integer. (See Exercise 4.)

Example 1. *A-sample size, m $= 3$; B-sample size, n $= 2$.* Suppose that the lowest 3 measurements are tied in a set of 5 observations having 3 *A*-measurements and 2 *B*-measurements. Treating tied ranks as having the average rank of those tied, and assuming that the null hypothesis is true, find the distribution of the sum of the ranks in the smaller sample.

Solution. The 10 possible positions of the sample of size 2 from the *B*-measurements, the summed ranks, and the distribution of the rank sum, T, are shown in Table 4–3. The calculation of the variance follows Table 4–3.

Table 4–3
Samples of size $n = 2$ from the five ranks

2	2	2	4	5	T		2	2	2	4	5	T
			B	B	9				B		B	6
		B		B	7		B				B	6
	B			B	7			B	B			4
B				B	7		B			B		4
		B	B		6		B	B				4

Distribution of Rank Sum, T

t	4	5	6	7	8	9
$P(t)$	$\frac{3}{10}$		$\frac{3}{10}$	$\frac{3}{10}$		$\frac{1}{10}$

Since

$$\mu_{\text{sum}} = \tfrac{3}{10}(4) + \tfrac{3}{10}(6) + \tfrac{3}{10}(7) + \tfrac{1}{10}(9) = 6,$$

$$\sigma^2_{\text{sum}} = \tfrac{3}{10}(4 - 6)^2 + \tfrac{3}{10}(6 - 6)^2 + \tfrac{3}{10}(7 - 6)^2 + \tfrac{1}{10}(9 - 6)^2 = \tfrac{24}{10} = 2\tfrac{2}{5}.$$

The new distribution differs from that of Section 3–2, where there were no ties. The original distribution in Section 3–2 was

t	3	4	5	6	7	8	9
$P(t)$	$\frac{1}{10}$	$\frac{1}{10}$	$\frac{2}{10}$	$\frac{2}{10}$	$\frac{2}{10}$	$\frac{1}{10}$	$\frac{1}{10}$

Note that the new distribution of Table 4–3 has lost the symmetry of the original, and that it has only 4 values with positive probabilities instead of the original 7. The new distribution has the same mean as the original distribution in Section 3–2, but the variance is smaller.

The effect of ties on the mean and the variance in Example 1 is typical of what happens in the general case. If some of the measurements in the pooled set are tied, it can be proved that

$$\mu_T = n(N + 1)/2,$$

as in the untied case, and that

$$\sigma_T^2 = \frac{mn(N + 1)}{12} - \left(\frac{mn}{12N(N - 1)} \cdot \sum_{i=1}^{r} K_i \right), \tag{1}$$

where

$N = m + n,$

$r = $ the number of sets of tied measurements,

$\tau_i = $ the number of tied ranks in the ith set of ties,

$K_i = (\tau_i - 1)\tau_i(\tau_i + 1).$

If there are no ties, $\sum K_i = 0$ and formula (1) reverts to formula (3) in Section 4–1, as might be expected. Note that the expression in the large parentheses of formula (1) measures the decrease in the variance as a result of the ties.

Example 2. Use formula (1) to find the variance for the data in Example 1.

Solution. We have $m = 3, n = 2, N = 5, r = 1,$ and $K_1 = (2)(3)(4) = 24.$ Therefore,

$$\sigma_T^2 = \frac{(6)(6)}{12} - \left(\frac{6}{12(5)(4)} \cdot 24 \right) = 3 - \frac{6}{10} = 2.4,$$

as before.

Example 3. *Earthquakes* (C. Emiliani, C. G. A. Harrison, and M. Swanson). In eight-hour intervals just before an atomic underground explosion and in eight-hour intervals 90 days later, the numbers of earthquakes at the Nevada Test Site were recorded. The results are given in Table 4–4.

Table 4-4

Number of earthquakes in eight-hour intervals before and after atomic underground explosion in Nevada

8-hour interval	Before explosion (*A*-sample)	90 days after explosion (*B*-sample)
1	48	39
2	28	32
3	33	61
4	22	31
5	29	29
6	31	42
7	24	45
8	33	29
9	35	45
10	58	37
11	45	16
12	27	26
13	35	37

Find the mean and the variance of the appropriate Mann-Whitney statistic, T. [Assume that the measurements in the eight-hour intervals are independent.]

Solution. The measurements in ascending order are

16, 22, 24, 26, 27, 28, 29, 29, 29, 31, 31, 32, 33, 33,

35, 35, 37, 37, 39, 42, 45, 45, 45, 48, 58, 61.

Therefore, we have

$m = 13, n = 13, N = 26$;

$\tau_1 = 3, \tau_2 = 2, \tau_3 = 2, \tau_4 = 2, \tau_5 = 2, \tau_6 = 3$;

$K_1 = K_6 = 2 \times 3 \times 4 = 24, K_2 = K_3 = K_4 = K_5 = 1 \times 2 \times 3 = 6$,

and

$\Sigma K_i = 2(24) + 4(6) = 72.$

From formula (2) of Section 4-1,

$$\mu_T = 13(27)/2 = 175.5,$$

and from formula (1),

$$\sigma_T^2 = \frac{13(13)(27)}{12} - \left(\frac{13(13)}{12(26)(25)} \cdot 72 \right)$$

$$= 405.25 - 1.56 = 403.69.$$

Thus the adjustment for ties here is slight.

EXERCISES FOR SECTION 4-4

1. Repeat Example 1 of this section, given that the *largest two* measurements are tied.

2. Use formula (1) to check the variance found in Exercise 1.

3. Repeat Example 1 of this section, given that the smallest three measurements are tied and the largest two measurements are tied.

4. Prove that the average of n consecutive, positive integers is an integer when n is odd, and half of an integer when n is even.

5. In a sample of 15 measurements, 10 are from population A and 5 from population B. If the pooled measurements contain one set of 4 ties and one set of 3 ties, find the variance of T.

6. If $m = 5$ and $n = 2$, and the 7 pooled measurements have the "ranks" 1, 2, 4, 4, 4, 6, 7, find the variance of T.

7. If $m = 5$ and $n = 3$, and the pooled 8 measurements have the "ranks" 2, 2, 2, 4.5, 4.5, 7, 7, 7, find the variance of T.

8. A physics teacher has two classes taking the same course. On the final examination, the following marks actually occurred:

   Class A: 92   88   88   82   75   70   66   62
   Class B: 95   88   52   52   48.

   Use the normal approximation to answer the following questions. Should the teacher reject the null hypothesis of identical distributions at the 5% significance level? At the 10% significance level?

9. Refer to Example 3 (Earthquakes). Would you accept, at the 5% level, the hypothesis that the observations in this example come from identical distributions?

10. *Enzyme activity* (G. E. Gibbs and G. D. Griffin). The activity of an enzyme (beta glucuronidase) was measured for the sweat glands of a group of diseased patients and a group of control patients. The results were as follows:

    Diseased patients: 1.2, 2.9, 1.9, 1.5, 0.5, 1.5, 1.4, 2.1, 3.3, 1.4, 2.2, 1.6, 1.8

    Control patients: 2.7, 3.1, 2.2, 1.3, 4.2, 2.2, 2.3, 2.9, 2.2, 2.2, 2.7, 2.8, 1.7, 4.1, 1.9

    Use the normal approximation for the Mann-Whitney test to see if these measurements might reasonably have come from identical distributions.

CHAPTER 5

Wilcoxon's Signed Rank Test

5–1 THE WILCOXON STATISTIC, *W*

To compare the means, or other measures of location, of two populations on the basis of two samples, we frequently pair objects in the samples. Such pairing may cut down the *variability of differences* between observations from the two populations, and at the same time leave the *average difference* unchanged. For example, to reduce the effect of differences between materials, we may cut two pieces from one bolt of cloth, treat one piece in one way and the other piece in another way, and then compare the results. Similarly, pieces from other bolts are cut and treated. Bolts may differ a lot, but adjacent pieces from the same bolt are much alike. The matching plays an important part in the analysis.

Alternatively, the same subject may be treated in two different ways, either at the same time (as when one medication is applied to a rash on the left hand and another on the right) or at different times. In such cases, we speak of a subject as acting as his "own control". Experiments with twins in humans, or littermates in animals, attempt to reduce the variation in the differences in outcomes between two treatments by controlling more background variables than would be possible with independent samples. The idea is that both members of a matched pair tend to give similar responses, thus making the effect of the treatment stand out more clearly.

★**Theoretical discussion of matching.** The following theory shows when matched pairs reduce the variance of differences and thus enable us to get a more accurate estimate of $\mu_X - \mu_Y$. (See Appendix V–1.)

The variance of a difference $X - Y$ is

$$\sigma^2_{X-Y} = \sigma^2_X + \sigma^2_Y - 2\rho\sigma_X\sigma_Y, \tag{1}$$

where ρ is the correlation between X and Y in the population. When ρ is high and positive, as is usual in matched pairs, the variance of the difference is considerably reduced in comparison with the result when $\rho = 0$, as occurs in independent samples. Thus, to simplify the discussion, if $\sigma_X = \sigma_Y = \sigma$, then

$$\sigma^2_{X-Y} = 2\sigma^2(1 - \rho).$$

For *n independent* pairs formed from two samples of *n*, we have

$$\rho = 0 \quad \text{and} \quad \sigma_{\bar{X}}^2 = \sigma_{\bar{Y}}^2 = \sigma^2/n \quad \text{(Appendix VI-1)}.$$

Hence, from (1), the variance of the difference of sample means is

$$\sigma_{\bar{X}-\bar{Y}}^2 = 2\sigma^2/n;$$

and for *correlated* pairs, this variance is

$$\sigma_{\bar{X}-\bar{Y}}^2 = \frac{2\sigma^2}{n} - 2\rho\frac{\sigma^2}{n} = \frac{2\sigma^2}{n}(1-\rho) = \frac{2\sigma^2}{k},$$

where $k = n/(1 - \rho)$. These formulas show that, as far as the variance is concerned, the correlation has the effect of making the sample size for the correlated pairs equivalent to $n/(1 - \rho)$ independent measurements. Hence, if $\rho = \frac{1}{2}$, for example, then the variance is that produced by two samples, each of size $n/(\frac{1}{2}) = 2n$. The effect is that of twice as many independent pairs.

As a result of such a reduction in variance, confidence limits on $\mu_X - \mu_Y$, for example,

$$\bar{x} - \bar{y} \pm 1.96\sigma_{\bar{X}-\bar{Y}}$$

(see Appendix VII-5), give us a more accurate estimate of the difference in means than we would be able to get from independent samples or from statistical calculations that ignore the matching.

This is the fundamental idea. We shall not be directly using the raw difference between means, but the ideas of the matching carry over to the discussion of the Wilcoxon approach to testing differences.

In these matched pair situations, the Wilcoxon test, using signed ranks of differences, offers one way to assess the difference in the location of two populations. For example, suppose that we have *n* *X*-measurements and *n* *Y*-measurements, and that we want to test the null hypothesis $\mu_X - \mu_Y = 0$. If the 2*n* measurements come as *n* pairs so that each pair has one measurement from each population, we denote the *i*th pair by

$$x_i, y_i, \quad i = 1, 2, 3, \ldots, n.$$

Then the Wilcoxon statistic, *W*, is obtained as follows:

1. Rank the absolute differences of the original measurements, $|x_i - y_i|$.
2. To the rank of the *i*th absolute difference attach the sign of $x_i - y_i$, and denote this signed rank by R_i.
3. Obtain the sum of the signed ranks, *W*:

$$W = R_1 + R_2 + R_3 + \cdots + R_n.$$

A large positive sum suggests that $\mu_X - \mu_Y > 0$; a large negative sum suggests that $\mu_X - \mu_Y < 0$.

Other equivalent test statistics are sometimes used for convenience. One might use:

a) the sum of the positive signed ranks, W^+; or

b) the sum of the negative signed ranks, W^-; or

c) W_S, the smaller of $|W^+|$ and $|W^-|$.

In addition to W, we tabulate W_S in Table A–11 at the back of the book. Thus, if $W^+ = +12$ and $W^- = -5$, we tabulate $W_S = |-5| = 5$, because $|-5| < |+12|$.

Example 1. Refer to the experiment in Example 1, Section 1–6 (marijuana). Consider the data in Table 1–4 as provided by matched pairs of subjects, and find W. (Rank ties as described in Section 4–4.)

Solution. The data may be arranged as in the following table.

X	Y	$X - Y$	Rank of $|X - Y|$	Signed ranks, R_i
-3	5	-8	$6\frac{1}{2}$	$-6\frac{1}{2}$
10	-17	$+27$	9	9
-3	-7	$+4$	3	3
3	-3	$+6$	$4\frac{1}{2}$	$4\frac{1}{2}$
4	-7	$+11$	8	8
-3	-9	$+6$	$4\frac{1}{2}$	$4\frac{1}{2}$
2	-6	$+8$	$6\frac{1}{2}$	$6\frac{1}{2}$
-1	1	-2	$1\frac{1}{2}$	$-1\frac{1}{2}$
-1	-3	$+2$	$1\frac{1}{2}$	$1\frac{1}{2}$

$$W = 29$$

The indication is that X-scores are higher. To find out how significant the result is, we shall need to develop the theory of the distribution of W when values of X and Y come from the same distribution, as we do in the next section.

Obviously, the Wilcoxon statistic represents a move in between the sign test and the direct treatment of the raw observations through the t-statistic. The sign test disregards the size of the measurements: differences of 10 and 1000 each rate one plus sign. The Wilcoxon test takes modest account of size, but this test is not sensitive to great differences in the size of $X - Y$ unless the sample size is large. Thus the Wilcoxon prevents giving huge weights to wild observations.

EXERCISES FOR SECTION 5–1

1. For the following set of matched pairs $x_i - y_i$, compute the Wilcoxon statistic, W.

X:	45	· 68	70	72	78	83	95
Y:	35	52	71	80	70	85	99

2. For the data in Example 1, compute the statistic W_S.

3. A dermatologist wishes to test the effectiveness of a new lotion, Y, in preventing sunburn. He uses six subjects. One shoulder of each subject is treated with lotion Y and the other shoulder with lotion X, which is the standard treatment. After identical exposure, the degree of sunburn is measured on a scale. Given the following matched pairs of measurements x_i, y_i, find W and W_S.

X:	34	72	165	14	52	75
Y:	72	109	74	32	92	60

4. For the data of Exercise 3, Section 2–6 (methods of memorizing), find W.

5. Consider the data of Exercise 4, Section 2–6 (airline punctuality), as matched pairs. Compute W and W_S.

6. Given n matched pairs and no ties, find (a) the maximum value of W, (b) the minimum value of W.

5–2 THE DISTRIBUTION OF W UNDER THE NULL HYPOTHESIS

We should think of our assumptions as approximately true rather than exactly true. Under the assumption that the differences $x_i - y_i$ are randomly drawn from a continuous, symmetrical population centered at 0, we can compute some properties of the distribution of W. A *continuous population* has a continuous cumulative distribution, one with no jumps, and if a *symmetrical population* centered at 0 has a density function f, then $f(x) = f(-x)$. The continuity avoids ties, and the symmetry guarantees that the number of plus signs in a sample of n differences is binomially distributed with $P(+ \text{ sign}) = \frac{1}{2}$. Indeed, a helpful way of thinking about the sample is to imagine that it was drawn only from the right half of the distribution of differences $x_i - y_i$, and that we then assigned to each difference a positive or negative sign by tossing a fair coin. If in particular the null hypothesis assumes identical and independent distributions for X and Y, then these assumptions imply symmetry of the distribution of $X - Y$ and mean zero. If we want to relax the independence assumptions and still be sure of getting symmetry, one way is to consider the null situation where a coin chooses which member of a pair is to be the X measurement. Then even with dependence, $X - Y$ will be symmetrically distributed about 0. And so the symmetry condition is satisfied.

The discussion above about coins was intended to help us think about the null hypothesis. The fact that in an experiment we would actually choose by a coin flip or other randomization device which member of a matched pair gets treatment A is of course related to this thinking.

Given a sample of n pairs, the ranks of the absolute values run from 1 to n. We therefore have 2^n possible allocations of signs. The following example illustrates the distribution of W for $n = 3$.

Example 1. *Orange trees.* In an orange orchard, each plot has 8 trees. The plots form a rectangular array composed of 27 rows and 10 columns. A grower measures, for each plot, the mean yield over a 5-year period. The following mean yields were recorded.

	Row 6	Row 16	Row 26
X (leftmost plot, column 1):	111	100	97
Y (adjacent plot, column 2):	110	116	121

Under the null hypothesis of identical populations, find the descriptive level of significance to assess whether the plots in the leftmost column produce lower mean yields than those in column 2. (The grower thinks they do.)

Solution. The assumption of the null hypothesis is that differences in plot production stem solely from the different column locations and random variation. Using the Wilcoxon approach gives

X	Y	$D = X - Y$	Signed rank, R_i
111	110	1	$+1$
100	116	-16	-2
97	121	-24	-3
			$W = -4$

Table 5–1 exhibits the $2^3 = 8$ possible outcomes for values of W when $n = 3$.

Table 5–1

Possible signs for ranks 1	2	3	Possible sums of signed ranks, W
+	+	+	6
−	+	+	4
+	−	+	2
+	+	−	0
−	−	+	0
−	+	−	−2
+	−	−	−4
−	−	−	−6

Since these 8 outcomes are equally likely under the null hypothesis, we have the exact distribution for W for this special case. It follows that $P(W \leq -4) = \frac{2}{8} = 0.25$, which means that we would get a sum of -4 or less 25% of the time.

Note that we have used a one-sided test; it would be appropriate here if the grower based his opinion on something about the physical situation. If he just noticed that the output was less in these particular data, then a two-sided approach would be more appropriate. We have used rows spaced 10 rows apart because in side-by-side rows the X values might not be independent of one another.

EXERCISES FOR SECTION 5–2

1. Work the problem in Example 1, given the following data.

	Row		
	6	16	26
X (leftmost column):	110	100	97
Y (adjacent column):	111	116	121

2. For the data in Exercise 1, would you conclude that the results are significant at the one-sided 10% level. What must be done if the test is to accommodate lower significance levels?

3. Construct a table similar to Table 5–1 showing the $2^4 = 16$ possible outcomes for values of W when $n = 4$.

4. Use the table obtained in Exercise 3 to find

a) $P(W \geq 4)$,   b) $P(W \geq 8)$,
c) $P(W \leq -10)$,   d) $P(|W| \geq 8)$.

5. For $n = 6$, construct a table like Table 5–1, but list only those possible outcomes for which $W \geq 19$. Find $P(W \geq 19)$ and $P(W = 21)$.

5–3 TABLES FOR THE DISTRIBUTION OF W; THE NORMAL APPROXIMATION

For values of n greater than 4, the method used in Section 5–2 for getting the distribution of W becomes laborious. Tables such as Table A–11 at the back of the book provide the exact distribution of W for modest values of n and probability levels nearest 0.01, 0.025, 0.05, and 0.10. For larger values of n, the normal distribution once more saves the day. That is, as the sample size increases, large numbers of random variables are being added, and the distribution of W, the sum of the signed ranks, can be approximated by a normal distribution. To use this approximation, we need to know the mean and the variance of W.

> **Theorem.** *Mean and variance of W.* Given a sample of n from a continuous distribution symmetrical about 0, if W is the sum of the signed ranks, then
>
> $$\mu_W = 0 \quad \text{and} \quad \sigma_W^2 = n(n + 1)(2n + 1)/6. \qquad (1)$$

Proof. Under the null hypothesis of identical populations, if R_i is the signed rank of the ith measurement in absolute size, we have

$$P(R_i = +i) = P(R_i = -i) = \tfrac{1}{2}.$$

It follows that the mean value of each R_i is 0 because

$$E(R_i) = \tfrac{1}{2}(+i) + \tfrac{1}{2}(-i) = 0.$$

Since $W = R_1 + R_2 + \cdots + R_n$, and since the mean of a sum equals the sum of the means, we get

$$\mu_W = 0.$$

To get the variance of W, we note that the signed ranks R_i are independent random variables because the signs are independently assigned, as described in Section 5–2. In particular, $P(R_i = +i) = \tfrac{1}{2}$, even when any or all of the signs of the other ranks are known. Therefore, since we have independence, we can use the fact that the variance of the sum of independent random variables is the sum of their variances. (See Appendix V–2.) Thus

$$\sigma_W^2 = \sigma_{R_1}^2 + \sigma_{R_2}^2 + \cdots + \sigma_{R_n}^2.$$

Because the mean of R_i is 0, its variance is

$$\sigma_{R_i}^2 = E(R_i^2) - \mu_{R_i}^2 = [\tfrac{1}{2}(+i)^2 + \tfrac{1}{2}(-i)^2] - 0^2 = i^2.$$

Finally, we have

$$\sigma_W^2 = \Sigma\sigma_{R_i}^2 = \Sigma i^2 = 1^2 + 2^2 + \cdots + n^2 = n(n + 1)(2n + 1)/6.$$

This completes the proof that, under the null hypothesis, the mean and variance of W, the sum of the signed ranks, are given by formulas (1).

Let us check formulas (1) against the data in Table 5–1, where $n = 3$. By inspecting Table 5–1, we see that $\mu_W = 0$. By definition,

$$\sigma_W^2 = \Sigma(x_i - \mu)^2 f(x_i)$$
$$= \tfrac{1}{8}[6^2 + 4^2 + 2^2 + 0^2 + 0^2 + (-2)^2 + (-4)^2 + (-6)^2]$$
$$= 14.$$

By formula (1), we get for $n = 3$

$$\sigma_W^2 = (3)(4)(7)/6 = 14,$$

which agrees with the result obtained from the definition.

Example 1. *Marijuana.* In Example 1, Section 5–1, we found $W = 29$ for 9 matched pairs. Use the normal approximation for the distribution of W to find, under the null hypothesis, the approximate probability that $W \geq 29$. Compare the result with that given in Table A–11.

Solution. We have $\mu_W = 0$ and $\sigma_W^2 = 9(10)(19)/6 = 285$. We compute the adjusted deviate that we use to enter the standard normal table thus:

$$z = \frac{w - \frac{1}{2} - \mu_W}{\sigma_W} = \frac{28.5 - 0}{16.88} = 1.69.$$

From the normal tables,

$$P(Z \geq 1.69) = 0.046,$$

which is close to the tabulated value 0.049.

In Section 2–3 we treated the following example with the sign test.

Example 2. *UCP excretion* (J. T. Galambos and R. G. Cornell). In medical diagnosis, the level of UCP (urinary coproporphyrin) excretion is important. Doctors Galambos and Cornell measured, over a number of days, the UCP excretions of individuals during two 12-hour (day and night) periods. The average day and night excretions for 8 healthy individuals were as follows:

Case number:	1	2	3	4	5	6	7	8
Day average, X:	35.3	65.9	73.4	70.6	56.3	73.4	39.3	36.9
Night average, Y:	39.0	58.8	70.6	58.7	53.1	72.6	42.2	63.1

Under the null hypothesis of identical populations, are differences in day and night excretions significant at the 5% level?

Solution. The data may be tabulated as in Table 5–2.

Table 5–2

X	Y	$\|X - Y\|$	Rank of $\|X - Y\|$	Signed rank, R_i
35.3	39.0	3.7	5	-5
65.9	58.8	7.1	6	$+6$
73.4	70.6	2.8	2	$+2$
70.6	58.7	11.9	7	$+7$
56.3	53.1	3.2	4	$+4$
73.4	72.6	0.8	1	$+1$
39.3	42.2	2.9	3	-3
36.9	63.1	26.2	8	-8
				$W = 4$

We shall use the normal approximation to find $P(W \geq 4)$. Under the null hypothesis of identical populations, $\mu_W = 0$ and $\sigma_W^2 = 8(9)(17)/6 = 204$. Hence the adjusted deviate is

$$z = \frac{w - \frac{1}{2} - \mu_W}{\sigma_W} = \frac{3.5 - 0}{14.3} = 0.245.$$

From the normal table, we get: $P(Z \geq 0.245) = 0.403$.

Since identical populations would give us $W \geq 4$ over 40% of the time, we accept the null hypothesis. Accepting the null hypothesis is equivalent to rejecting the alternative hypothesis that $\mu_X - \mu_Y \neq 0$. If we use the 5% significance level, then having a difference that corresponds to the 40% level means we do not reject the null hypothesis at the 5% level.

The sign test gave us a probability level of 0.363, and therefore taking some account of the sizes of the differences did not change our conclusions.

The bottom value of $|X - Y|$, namely 26.2, appears to be in a class by itself. It is an "outlier". Note that the sign test in Section 2–3 gave no extra weight to the magnitude of this value. The Wilcoxon test gives it some weight—about 1.8 times that of the average absolute weight, 4.5—but nothing comparable to its relative size. The Wilcoxon test protects us from undue influence of bad outliers, as noted in Section 5–1.

Ties in the Wilcoxon. The Wilcoxon signed rank test can have two kinds of ties: absolute zeros and tied absolute nonzero values.

Zeros. Zeros occur when the paired values are identical so that their difference is zero. Recall that we are testing to see whether the distributions have different location parameters. A zero difference shows two things: discreteness and alikeness. Since zeros offer no discriminating power for the purpose of testing for a difference, one reasonable procedure is to set them aside (though not if one is estimating the average difference). And then the sample size for the test is reduced by the number of zeros.

Nonzero ties. Ties in the absolute differences $|X - Y|$ are assigned the average of the ranks available to the tied $|X - Y|$ values, as explained in Section 4–4. If there are many ties, one may wish to correct the variance. The corrected variance is

$$\sigma_W^2 = \tfrac{1}{6}n(n + 1)(2n + 1) - \sum(\tau_i + 1)\tau_i(\tau_i - 1)/12,$$

where the sum is over the tied sets. Here $\tau_1, \tau_2, \ldots, \tau_r$ are the numbers of ties in the set of absolute differences (see Table 5–3). As usual, if a set has only one value, $\tau = 1$ and no correction is required for that set.

Example 3. The data of Table 5–3 illustrate both kinds of ties. The data are for performance of a group of people, paired for intelligence and education, on a test with a maximum score of 20 points. A random one of each pair was trained before the test. Should we accept the null hypothesis of identical populations at the 5% level?

Table 5–3

Pair	Score Trained	Score Untrained	Difference	Absolute difference	Rank	Signed rank
1	11	11	0	0	—	—
2	10	10	0	0	—	—
3	9	8	1	1	1	1
4	12	5	7	7	9	9
5	3	10	-7	7	9	-9
6	11	4	7	7	9	9
7	15	10	5	5	$6\frac{1}{2}$	$6\frac{1}{2}$
8	6	11	-5	5	$6\frac{1}{2}$	$-6\frac{1}{2}$
9	8	4	4	4	$4\frac{1}{2}$	$4\frac{1}{2}$
10	2	6	-4	4	$4\frac{1}{2}$	$-4\frac{1}{2}$
11	12	9	3	3	3	3
12	18	16	2	2	2	2

$$W = 15$$

Solution. Pairs 1 and 2 give zero and are therefore not ranked; thus $n = 10$, not 12.

There are two sets of 2 ties and one set of 3, so $\tau_1 = 2$, $\tau_2 = 2$, and $\tau_3 = 3$. We get

$$W = 15, \qquad \sigma_W^2 = \frac{10(11)(21)}{6} - 2\left[\frac{3(2)1}{12}\right] - \frac{4(3)(2)}{12}$$

$$= 385 - 1 - 2 = 382,$$

and

$$z = 14.5/\sqrt{382} < 1.$$

The data provide little reason to reject the hypothesis of equality, since the observed result is within 1 standard deviation of the mean.

EXERCISES FOR SECTION 5–3

1. In a Wilcoxon signed rank test, $n = 6$. Under the null hypothesis, find $P(W \geq 15)$ using (a) Table A–11, (b) the normal approximation.

2. Repeat Exercise 1 with $n = 8$ and $W \geq 25$.

3. Repeat Exercise 1 with $n = 10$ and $W \geq 45$.

4. Use the normal approximation to check the results of Exercise 4, Section 5–2.

5. A manufacturer wishes to compare two methods of making light bulbs. He conducts four experiments, in each of which one kind of bulb is made by both method A and method B. If method A produces a higher percentage of bulbs meeting the standard in each of the four experiments, do the results indicate a significant difference in the methods (a) at the 5% level, (b) at the 10% level?

6. One hog in each of ten matched pairs is given diet X, and the other is given diet Y. At the end of the experiment, the following gains in weight were recorded.

Pair:	1	2	3	4	5	6	7	8	9	10
Diet X:	25	32	24	35	16	30	29	20	29	36
Diet Y:	21	30	19	21	27	29	24	27	20	21

Compute w, the sum of the signed ranks. Find $P(W \geq 27)$ (a) by using Table A–11, (b) by using the normal approximation.

7. In Exercise 6, should we conclude that diet X is superior to diet Y (a) at the 5% level of significance, (b) at the 10% level?

8. *Airline punctuality.* The punctuality of Airline X and Airline Y between North American cities at least 200 miles apart was measured by the percentage of flights arriving within 15 minutes of the scheduled time. The records for 13 consecutive months, beginning with July 1969, were as follows:

Month:	1	2	3	4	5	6	7	8	9	10	11	12	13
Airline X:	71	71	55	60	68	73	72	75	69	74	78	67	46
Airline Y:	78	79	63	72	72	72	73	72	71	71	80	72	56

Compute w and σ_W^2. Test for a difference between airline performance. Discuss the measure of punctuality. Might it favor some airlines? If you would prefer a different measure, suggest it and mention its advantages and disadvantages.

5–4 POWER OF THE WILCOXON SIGNED RANK TEST FOR SHIFT ALTERNATIVE WITH NORMALLY DISTRIBUTED DIFFERENCES

To illustrate the power function of the Wilcoxon signed rank test, we first need mathematical models for the sampling because we cannot compute power without a specific model for the alternative hypothesis. We adopt a two-stage sampling procedure to describe the measurements arising from matched pairs. Though this is but one of many possible models, it is instructive.

We proceed from two premises. The first is that the *pairs* differ, one from another. One pair of twins will grow faster than another, side-by-side pieces of cloth from one bolt are more alike than those from another bolt, and so on. Second, within a pair, the *individuals* are not quite alike even when they have the same treatment. We shall suppose that for each individual of a pair, the general level of measurement is Z_i, a number drawn from a distribution. In addition, we suppose that measurements for the individuals in the pair deviate from Z_i by amounts D_{i1} and D_{i2}, where the subscripts 1 and 2 refer to the first and second individuals of the pair. The total scores for the individuals will be denoted by

$$X_i = Z_i + D_{i1} \qquad \text{for individual 1 in pair } i,$$
$$Y_i = Z_i + D_{i2} \qquad \text{for individual 2 in pair } i. \tag{1}$$

Let us further suppose that among the n pairs

$$E(D_{i1}) = E(D_{i2}) = 0, \qquad i = 1, 2, \ldots, n,$$

and that both of the D's and the Z's are independent of one another.

We now have set up a complete sampling model under the null hypothesis of equal means. The model allows the possibility of correlation between X and Y, as we see below.

If the variance of Z is σ_Z^2, and if both D's have variance σ_D^2, then the correlation between X and Y is

$$\rho_{XY} = \frac{\sigma_Z^2}{\sigma_Z^2 + \sigma_D^2} \qquad (2)$$

(see Exercise 10). Thus, unless $\sigma_Z = 0$, the correlation $\rho_{XY} > 0$, and we have correlation created by the variation among pairs, stemming from variation in Z. Unless our model allowed for this, it would not be general enough to cover the cases where the Wilcoxon test might be used.

Since the difference between X_i and Y_i does not depend on Z_i, we have

$$E(X_i - Y_i) = 0, \qquad \sigma_{X_i-Y_i}^2 = 2\sigma_D^2. \qquad (3)$$

Alternative hypothesis, shift in mean. If one member of each pair of individuals, say the first, were given a special treatment that added a constant h to his score, then his score would be

$$X_i^* = h + Z_i + D_{i1}. \qquad (4)$$

If the second member of each pair were not treated, then the expected difference and variance would be

$$E(X_i^* - Y_i) = h; \qquad \sigma_{X_i^*-Y_i}^2 = \text{Var}(h + D_{i1} - D_{i2}) = 2\sigma_D^2. \qquad (5)$$

Finally, the average difference over the n pairs would have mean h and variance $2\sigma_D^2/n$. (See Appendix VI–1.)

Using the above assumptions, for purposes of illustration, we shall also assume that the D's are normally distributed, and under these conditions they illustrate the power of the Wilcoxon signed rank test to detect a treatment effect. Since Z_i does not appear in the differences $X_i^* - Y_i$, detecting a treatment effect is the same as detecting a shift in the mean. Is $h > 0$?

We show in Fig. 5–1 power curves for three sample sizes, $n = 4$, 10, and 25, for detecting shifts in the mean for normal distributions, where $\mu_{X_i} - \mu_{Y_i} = h$, $\sigma_{X_i-Y_i} = \sqrt{2}\sigma_D$, and we have arbitrarily chosen $\sigma_D = 1/\sqrt{2}$ to make $\sigma_{X_i-Y_i} = 1$.

The power functions are symmetrical about 0 shift, and the significance level is about 0.125, exactly that for $n = 4$. As the sample size increases from 4 to 10 to 25 pairs, the curves rise more steeply as the shift departs slightly from 0. For a shift of half a standard deviation, the sample of size 25 has about a 0.75 chance of detecting the departure from 0.

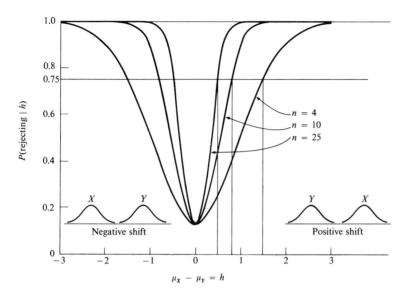

Fig. 5–1 Power of the Wilcoxon signed rank test for shifts of normal means, standard deviation 1, $n = 4, 10, 25$, based on simulation.

The standard deviation of $X_i - Y_i$ is the unit for the horizontal axis in Fig. 5–1. We simulated the results by drawing 1000 sets of n values of $X_i - Y_i$ for each difference h (0, ± 0.25, ± 0.5, ..., ± 3). Because we wanted to compare the results for a small sample, $n = 4$, with those for larger ones, we chose the rather large significance level 0.125.

Such curves are not easy to summarize, but we might note that the test is 75% certain to detect a difference of 1.5σ if $n = 4$, a difference of 0.8σ if $n = 10$, and a difference of 0.5σ if $n = 25$.

EXERCISES FOR SECTION 5–4

1. What value of h is associated with the null hypothesis in this section?

2. Use Fig. 5–1 to estimate the sample size required if we are to be 50% sure of detecting a difference of $\frac{1}{2}$ standard deviation.

3. Regarding 0.95 as "practically certain", what differences in means are practically certain to be detected by the three sample sizes used in Fig. 5–1?

Table 5–4 shows the results of a simulation like those described in the text for the shifting mean problem when $n = 4$ and when 1000 samples are drawn for each difference in means. The results of Table 5–4 are needed for Exercises 4–8.

Table 5–4

Simulation for power of Wilcoxon signed rank test (samples of size 4)

W_S: Sum of ranks with lower total	Difference of means $\mu_{X_i} - \mu_{Y_i}$												
	0.0	0.25	0.50	0.75	1.00	1.25	1.50	1.75	2.00	2.25	2.50	2.75	3.00
0	140	184	243	343	507	642	755	847	909	955	979	990	995
1	120	119	177	242	237	203	167	117	83	43	20	9	5
2	121	141	134	128	77	64	43	26	5	1	1	1	0
3	249	246	227	140	105	62	25	9	3	1	0	0	0
4	239	202	144	97	54	23	9	1	0	0	0	0	0
5	131	108	75	50	20	6	1	0	0	0	0	0	0
Total	1000	1000	1000	1000	1000	1000	1000	1000	1000	1000	1000	1000	1000

4. For $n = 4$, compute the exact probabilities of $W_S = 0, 1, 2, 3, 4, 5$ for the Wilcoxon signed rank test when the null hypothesis is true. (If you made the table for Exercise 3 in Section 5–2, it would help here.) Compare these probabilities with the counts produced by the simulation (Table 5–4) when the null hypothesis is true.

5. (Continuation) Use binomial theory to find how many standard deviations the result for $W_S = 0$ is from its expected count when $h = 0$.

6. If the null hypothesis is rejected when $W_S = 0$ or 1, what is the significance level as estimated by the simulation? What is the exact significance level?

7. Plot the points obtained from the simulation to show the power of the Wilcoxon test when $n = 4$ if rejection occurs when $W_S = 0$ or 1. Connect the plotted points with a smooth curve.

8. For $n = 4$, estimate the probability of rejection for the test rejecting when $W_S = 0$ or 1 when $h = 1.75$.

9. For the test that rejects when $W_S = 0$, use normal tables to compute the exact probability of rejection when the difference of means is 1 standard deviation. Compare this result with the simulated value.

10. Derive the correlation coefficient given in equation (2).

CHAPTER 6

Rank Correlation

6–1 CORRELATION COEFFICIENT; SPECIAL CASE

Frequently we want an index that expresses the degree of relation between two variables. Such an index is called a *correlation coefficient*. Those who have studied the ordinary product-moment correlation coefficient r (see Appendix IV–3) know that it is particularly useful for measuring the degree to which two variables are linearly related. It is therefore fair to say that the correlation measures the degree to which the points (x, y) cluster around a slanted straight line.

If we replace X and Y by transformed variables, like $X' = \sqrt{x}$ and $Y' = \log y$, then the correlation coefficient between X' and Y' will not ordinarily be the same as that between X and Y.

Sometimes the original observations are actually only orderings in X and orderings in Y, rather than measurements; and then we cannot compute r. Sometimes we are very uncertain about the choice of scale used for the original measurements, and we would be satisfied by any stretchings or contractions of this scale that do not change its order. Such a change of scale is called a *monotonic transformation*; for example, if x is positive, some monotonic transformed variables are $2x + 3$, x^2, e^x, $\log x$, and $\sqrt{x}$. Ratings often involve uncertainty about the scale. If one rates performance of people on a 100-point scale, he may well wonder if the change from 95 to 100 on the scale is equivalent to the change from 50 to 55, or if it is more nearly equivalent to the change from 25 to 75, say. These scale uncertainties make it desirable to have a measure that does not depend on the scales, provided that the measurements retain their order. One simple measure of correlation that has this invariance property is the *rank correlation coefficient*.

Example 1. Table 6–1 shows the quality ranks and average prices for 5 different makes of room air conditioners (adapted by permission of Consumers Union from *Consumer Reports*, June 1970).

Discussion. A quick inspection of the quality ranks and prices in the table suggests that top quality seems to go with high prices and low quality with low prices; but the agreement is not perfect.

Table 6–1

Quality and price in room air conditioners

Air conditioner	Quality rank	Average price
S	1	$196.20
T	2	166.75
U	3	151.91
V	4	160.77
W	5	163.74

If we rank the average prices, giving the highest price rank 1, the next highest price rank 2, and so on, we get Table 6–2.

Table 6–2

Ranks in quality and average price

Air conditioner	Quality rank	Price rank
S	1	1
T	2	2
U	3	5
V	4	4
W	5	3

When we use ranks to study the relation between quality and price, it is evident from Table 6–2 that, although there is some agreement, the agreement between the ranks, like that between quality and average price, is not perfect.

Perfect agreement in ranks would lead to Table 6–3. In this table, we let $d_i = x_i - y_i$, where x_i and y_i denote, respectively, the ranks in quality and average price of the ith air conditioner on the list.

Table 6–3
Perfect agreement in ranks

Air conditioner	Quality rank, x_i	Price rank, y_i	d_i^2
S	1	1	0
T	2	2	0
U	3	3	0
V	4	4	0
W	5	5	0
			$\sum d_i^2 = 0$

The complete opposite of perfect agreement is shown in Table 6–4.

Table 6–4
Maximum disagreement in ranks

Air conditioner	Quality rank, x_i	Price rank, y_i	d_i^2
S	1	5	16
T	2	4	4
U	3	3	0
V	4	2	4
W	5	1	16
			$\sum d_i^2 = 40$

The rankings in Table 6–4 would lead us to believe that high quality goes with low price, and vice versa. Such a result might be called complete disagreement in rank. Between the extremes of perfect agreement and complete disagreement in the ranks lie intermediate levels. To make the levels of agreement precise, we need to choose a way of measuring rank agreement. How shall we do this?

Several measures of rank agreement might be used. One such measure is the sum of the absolute values of the differences d_i in the ranks for quality and price. In this book, we shall use a measure based on $\sum d_i^2$, the sum of the squares of the differences d_i in the ranks for the two subjects.

If the number of ranked items is 5, the maximum sum of squares of differences in the ranks, as given in Table 6–4, is 40. Table 6–3 shows that the minimum sum of squares of differences in the ranks is 0. In addition to reflecting these facts, our measure of rank correlation, which we shall denote by r_S and call the *rank correlation coefficient*, should satisfy three criteria:

1. r_S should depend on, or be a decreasing function of, $\sum d_i^2$.
2. r_S should equal 1 for perfect agreement, that is, when $\sum d_i^2 = 0$.
3. r_S should equal -1 for complete disagreement, that is, when $\sum d_i^2 = 40$, its maximum value.

To satisfy these conditions, we design a special linear function of $\sum d_i^2$, as follows: Let

$$r_S = A + B \sum d_i^2,$$

where A and B are to be determined. If $r_S = 1$ when $\sum d_i^2 = 0$, then

$$1 = A + 0 \quad \text{and} \quad A = 1.$$

If $r_S = -1$ when $\sum d_i^2 = 40$, then

$$-1 = 1 + B(40) \quad \text{and} \quad B = -\tfrac{2}{40} = -\tfrac{1}{20}.$$

Therefore, if $n = 5$, we have

$$r_S = 1 - \tfrac{1}{20} \sum d_i^2. \tag{1}$$

To apply formula (1) to the ranks in Table 6–2, we arrange the data as in Table 6–5.

Table 6-5
Computation of sums of squares of deviations

Air conditioner	Quality rank, x_i	Price rank, y_i	$d_i = x_i - y_i$	d_i^2
S	1	1	0	0
T	2	2	0	0
U	3	5	-2	4
V	4	4	0	0
W	5	3	2	4
				$\sum d_i^2 = 8$

It follows that

$$r_S = 1 - \tfrac{1}{20} \sum d_i^2 = 1 - \tfrac{2}{5} = 0.6,$$

a positive value substantially above zero. This value of r_S tells us that the trend is generally positive. Roughly speaking, higher quality goes with higher price, but not invariably.

Deciding to base the correlation on $\sum d_i^2$ is a fundamental choice. But the chosen range, -1 to $+1$, is just a convention that many correlation coefficients satisfy. This is not fundamental because we could, if it were desirable, use the range -100 to 100 or 0 to 100 by linear transformations of r_S.

Another fundamental feature of the rank correlation, mentioned earlier, is that if we change the original data without changing their order, we do not change the ranks, and therefore we leave the correlation unchanged. For example, in Table 6–1, change S's price to $200, T's to $180, U's to $142, V's to $152, W's to $175, and the ranks are unchanged, and so is r_S. This general feature of the rank correlation coefficient is called invariance under monotonic transformations.

The capital S in the subscript of r_S is in honor of the psychologist Spearman, who seems to have invented the coefficient.

EXERCISES FOR SECTION 6–1

1. The following table gives class ranks in English and mathematics.

Student:	A	B	C	D	E
English rank:	1	2	3	4	5
Mathematics rank:	3	2	5	1	4

(a) Compute $\sum d_i^2$. (b) Compute the rank correlation coefficient.

2. Make up an exercise giving ranks of six students in two different subjects. Compute $\sum d_i^2$ for (a) perfect agreement, (b) complete disagreement, (c) a case other than (a) or (b).

3. In Exercise 2, how does your answer to part (a) compare with the sum in Table 6–3? How does your answer to part (b) compare with the sum in Table 6–4?

4. What general statement can you make about $\sum d_i^2$ when there is perfect agreement in the ranks?

5. Compute the maximum value of $\sum d_i^2$ for a sample of n students, given (a) $n = 3$, (b) $n = 4$, (c) $n = 5$, (d) $n = 6$.

6. Given $n = 6$, derive a formula for r_S, the rank correlation coefficient, that meets the same conditions used in deriving formula (1).

7. In an art contest, six paintings, denoted by A, B, C, D, E, and F, are given the following ranks by two different judges.

Painting:	A	B	C	D	E	F
Rank by judge X:	4	2	1	5	3	6
Rank by judge Y:	2	3	1	6	4	5

Find the rank correlation coefficient.

8. In Exercise 7, suppose that the ranks were as follows:

Painting:	A	B	C	D	E	F
Rank by judge X:	3	1	4	2	6	5
Rank by judge Y:	1	5	2	4	3	6

Without making the full calculation, try to decide whether r_S is higher or lower than in Exercise 7. Make a thoughtful guess at the value of r_S. Compute r_S.

9. To make a new correlation coefficient that depends on $\sum d_i^2$ but runs from -100 to 100, what transformation would you apply to r_S?

10. To make a new correlation coefficient that depends on $\sum d_i^2$ but runs from 0 to 100, what transformation would you apply to r_S?

6–2   RANK CORRELATION COEFFICIENT; THE GENERAL CASE

We shall get a formula for the rank correlation coefficient r_S when n items are ranked in two ways. Our method is essentially that used in dealing with the special case in Section 6–1. We shall define r_S so that it depends on $\sum d_i^2$, taking the value $+1$ when the ranks are in perfect agreement and the value -1 for complete disagreement. As before, we tailor a linear function of $\sum d_i^2$ that yields the desired results.

Let $r_S = A + B \sum d_i^2$, where A and B are to be determined. For perfect agreement in the ranks, $\sum d_i^2 = 0$ and $r_S = 1$. Hence, $A = 1$. For complete disagreement, $\sum d_i^2 = M$, its maximum possible value, and $r_S = -1$. Thus

$$-1 = 1 + BM \quad \text{and} \quad B = -2/M.$$

In order to complete our formula, we must find M, the maximum value of $\sum d_i^2$, which occurs when the ranks are in reverse order. We shall assume without proving it here that the reverse order does maximize $\sum d_i^2$. (The reader is asked to refer to Exercise 5 of Section 6–1 to recall the steps involved in finding $\sum d_i^2$.) The cases for odd n and even n require separate proofs. We deal with even n, leaving the odd n case as an exercise.

Let $n = 2m$, where m is a positive integer. For complete disagreement, the ranks appear thus:

x_i:	1	2	...	$m-1$	m	$m+1$	$m+2$	...	$2m-1$	$2m$
y_i:	$2m$	$2m-1$	...	$m+2$	$m+1$	m	$m-1$	...	2	1
d_i:	$-(2m-1)$	$-(2m-3)$	...	-3	-1	1	3	...	$2m-3$	$2m-1$

Hence

$$M = \sum d_i^2 = 2[1^2 + 3^2 + 5^2 + \cdots + (2m-1)^2].$$

The series in the square brackets can be represented as the sum of the squares of all the integers 1 through $2m - 1$ less the sum of the squares of the even numbers, 2 through $2m - 2$:

$$M/2 = 1^2 + 2^2 + \cdots + (2m-1)^2 - [2^2 + 4^2 + \cdots + (2m-2)^2]. \quad (1)$$

$$= 1^2 + 2^2 + \cdots + (2m-1)^2 - 4[1^2 + 2^2 + \cdots + (m-1)^2]$$

$$= \tfrac{1}{6}(2m-1)(2m)(4m-1) - 4(\tfrac{1}{6})(m-1)(m)(2m-1)$$

$$= \tfrac{1}{6}(2m-1)(2m)(2m+1).$$

Thus

$$M/2 = \tfrac{1}{6}(n-1)n(n+1) = \tfrac{1}{6}n(n^2-1).\tag{2}$$

It follows that

$$B = \frac{-2}{M} = \frac{-6}{n(n^2-1)}, \quad \text{and} \quad r_S = 1 - \frac{6\sum d_i^2}{n(n^2-1)}.$$

The foregoing discussion motivates the following definition.

Definition. *Rank correlation coefficient.* If a sample of n individuals is ranked twice so that the ith individual has an X-rank x_i and a Y-rank y_i, then the rank correlation coefficient is

$$r_S = 1 - \frac{6\sum d_i^2}{n(n^2-1)},\tag{3}$$

where $d_i = x_i - y_i$.

The form shown in equation (3) does not reveal that r_S is the ordinary product-moment correlation between the ranks. This fact is made explicit in the following theorem.

Theorem. The rank correlation r_S is the ordinary sample correlation coefficient (see Appendix IV–3)

$$r = \frac{\sum x_i y_i - n\bar{x}\bar{y}}{\sqrt{\sum (x_i - \bar{x})^2 \sum (y_i - \bar{y})^2}}\tag{4}$$

for the set of ranks (x_i, y_i).

Proof. We offer an outline of the proof, leaving the algebraic details as exercises.

Step 1. Since the ranks consist of the first n integers, we have

$$\bar{x} = \bar{y} = (n+1)/2$$

and

$$\sum (x_i - \bar{x})^2 = \sum (y_i - \bar{y})^2 = n(n^2-1)/12.$$

Step 2. Substitution in (4) gives

$$r = \frac{12\sum x_i y_i}{n(n^2-1)} - \frac{3(n+1)}{n-1}.\tag{5}$$

Step 3. To evaluate $\sum x_i y_i$, we note that

$$\sum d_i^2 = \sum (x_i - y_i)^2 = \sum x_i^2 - 2\sum x_i y_i + \sum y_i^2,$$

where

$$\sum x_i^2 = \sum y_i^2 = \sum_{i=1}^{n} i^2 = n(n+1)(2n+1)/6.$$

Hence

$$\sum x_i y_i = n(n+1)(2n+1)/6 - \tfrac{1}{2}\sum d_i^2. \tag{6}$$

Step 4. Substituting the result of (6) into (5) and simplifying, we get

$$r = 1 - \frac{6\sum d_i^2}{n(n^2-1)} = r_S. \tag{7}$$

EXERCISES FOR SECTION 6–2

1. Use formula (3) to check the results obtained in Exercises 6, 7, and 8 of Section 6–1.

2. The heights of 8 fathers and of their eldest sons were measured to the nearest inch. Results were as follows:

Height of father:	74	73	72	71	69	68	67	65
Height of eldest son:	73	71	76	70	72	69	67	68

Find the rank correlation coefficient.

3. At the end of the school year, a mathematics teacher assigns a percentage grade, based on the year's achievement, to each of his 10 students. Subsequently, the 10 students take a CEEB achievement test covering the material of the course. Here are the results of both gradings.

Student:	A	B	C	D	E	F	G	H	I	J
Year's grade:	94	92	87	83	79	76	71	65	61	52
CEEB grade:	800	724	784	792	712	612	584	672	562	512

Find the rank correlation coefficient.

4. Six contestants enter a beauty contest. They are ranked by two judges with the following results.

Contestant:	A	B	C	D	E	F
Judge X:	1	2	3	4	5	6
Judge Y:	4	6	2	1	3	5

Find r_S.

5. At the beginning of the 1969 baseball season, three enthusiasts, who prefer to remain anonymous, decided that each would predict the order of finishing for the 6 teams of the Eastern Division of the National League. It was agreed that, when the actual order of finishing was published, the prediction with the highest rank correlation coefficient would win a prize of $10, to be provided by the enthusiast whose prediction had the lowest rank correlation coefficient. Here are the facts.

Team	Actual order	A's order	B's order	C's order
Mets	1	5	4	5
Cubs	2	2	1	3
Pirates	3	1	2	1
Cards	4	3	5	4
Phillies	5	4	3	2
Expos	6	6	6	6

Who wins the prize and who provides it?

6. Six car salesmen take an aptitude test for salesmanship. At the end of a year their sales records are noted, and they are ranked by their test scores and by their sales volumes as follows:

Salesman:	A	B	C	D	E	F
Aptitude rank:	1	2	3	4	5	6
Sales rank:	3	5	1	6	2	4

Find r_S. Does this aptitude test appear to be useful for predicting performances of these salesmen?

7. The IQ's of 7 students are compared with their grades in physics as follows:

Student:	A	B	C	D	E	F	G
IQ:	129	127	118	116	111	109	102
Physics grade:	85	88	91	72	68	56	70

Find r_S.

8. In February 1968, *Consumer Reports* rated 9 electric fry pans in order of estimated overall quality. The order of rating and the price, to the nearest dollar, are tabulated as follows:†

Fry pan:	A	B	C	D	E	F	G	H	I
Quality rank:	1	2	3	4	5	6	7	8	9
Cost in dollars:	30	26	24	27	23	19	34	29	18

On the basis of this information, does it appear that quality is closely related to cost? Justify your answer.

9. Find the formula for the maximum $\sum d_i^2$ when n is odd. [*Hint:* Take $n = 2m + 1$.]

10. Give the algebraic details of step 1 in the theorem.

11. Give the algebraic details of step 2 in the theorem.

12. Give the algebraic details of step 3 in the theorem.

13. Give the algebraic details of step 4 in the theorem.

14. *How much does the midterm count?* Students often ask how much an examination

---

† Adapted by permission of Consumers Union from *Consumer Reports*, February 1968.

counts. The following data from a class of 17 students in statistics show the paired results for a midterm examination and for the final examination.

Midterm	Final	Midterm	Final
69	86	54	79
14	46	57	66
68	72	71	62
61	76	60	64
25	35	73	85
98	90	80	77
58	63	75	70
57	50	71	80
57	63		

a) Rank the midterm grades and the final grades, and compute the rank correlation coefficient, saving the squares.
b) How many students changed their rank by more than 2 between midterm examination and final examination?
c) How many students started out badly and improved a lot? How many started out well and fell a lot? You may wish to know that the man who went from 54 to 79 had a language problem because he had recently arrived from a foreign country.
d) If the performance of this class is typical of what happens between midterm examination and final examination (and it is), summarize in words what the data show about the ranks.
e) In view of all this, how would you answer the question "How much does the midterm count?", or since this is an unanswerable question, what does the midterm grade tend to imply about the grade on the final?

6–3 THE NULL DISTRIBUTION OF THE RANK CORRELATION COEFFICIENT, r_S

We have defined a statistic r_S to describe the agreement between two sets of ranks. Let us now see how r_S varies from one random sample to another drawn from the same population. Two cases need attention: (1) when only ranks in the samples are available; (2) when we have a joint distribution of the random variables X and Y, and the sample points (X_i, Y_i) are paired measurements that are in turn replaced by ranks.

1. *Ranks only.* When we have only ranks, independence means that for any ranking of the x-variable, all possible rankings of the y-variable are equally likely, and vice versa. We shall use (x_i, y_i) to denote the random variables that are the ranks of item i on the two variables.

2. *Joint distribution of measurements X, Y.* In the original population of measurements, we already have a definition of independence, namely

$$P(X \le a, Y \le b) = P(X \le a)P(Y \le b),$$

for every point (a, b) on the XY-plane. If X and Y are independent, it can be shown that for random samples of n pairs (X_i, Y_i) drawn from the distribution, the rankings have the "equally likely" property described in paragraph (1). In particular, if the (X_i, Y_i)-pairs are replaced by their ranks (x_i, y_i), then all possible rankings of the x-variable are equally likely, as are all possible rankings of the y-variable. Furthermore, for any given x-ranking, all rankings of y are equally likely, and vice versa. When we have independence, the correlation between the variables should be zero on the average. We want our definition of correlation to have this feature.

Because we have the possibility of original measurements and also of their rankings, in order not to proliferate notation, we use capital letters for the random variables corresponding to the measurements and lower-case letters for the random variables corresponding to the rankings. When we need to speak of specific values of these random variables, we shall let the context handle the matter so that we are not forced into additional notation.

If the two variables are independent in the population, then we can work out the distribution of r_S exactly for all possible samples of a given size, because *independence implies that all possible permutations of the ranks are equally likely.* The purpose of getting the probability distribution for r_S under the assumption of independence is this: The probability distribution enables us to judge the size of rank correlation coefficients that we may expect when there is no correlation between x and y in the population. If we study only a small sample, it is quite possible, even when the two variables are independent in the population, that a large value of r_S will occur by pure chance.

To gain some insight into the difference between the correlation observed in a sample and the correlation in the population, imagine that a sampling experiment is repeated hundreds of times. Suppose, for example, that we draw a sample of 5 students at random from the student population, that we rank their marks in English and history, and that we compute r_S for the sample. We perform this experiment on 1000 samples of 5, and we compute the average r_S for the 1000 samples. This average r_S would be very close to the true, but unknown, average rank correlation coefficient between the ranks in the population. Indeed, the rank correlation coefficient in the population is the long-run average rank correlation coefficient in such samples. Ordinarily, we observe only one sample.

To illustrate the ideas, we shall study the distribution of the rank correlation coefficient r_S for a few small samples taken from a large population having independence and therefore zero correlation between x and y.

First consider a sample with $n = 2$. If x and y are independent, then for a given order of the x's all orders of y are equally likely. The facts are conveniently assembled in Table 6–6.

Table 6–6
A sample of two

Rank of x	Possible ranks of y:	
	a	b
1	1	2
2	2	1
$\sum d_i^2$:	0	2
Probability:	$\frac{1}{2}$	$\frac{1}{2}$

We see that, for $n = 2$, $P(r_S = +1) = \frac{1}{2}$ and $P(r_S = -1) = \frac{1}{2}$. If we draw a random sample of 2 students and observe that the student who did better in English also did better in history, we have observed a perfect correlation ($r_S = +1$) between the ranks. However, on such slim evidence, we would be foolish to draw any serious conclusion about the general relation between English and history grades. Even if the grades were independent, we would get this high rank correlation coefficient half the time.

Next, consider a sample of 3 students randomly drawn from the student population, and compare their ranks. We now have $3! = 6$ possible arrangements of the ranking list of the history marks for each ranking of English marks. We assume independence, as before, inferring that, given any ranking of the English marks, each of the possible rankings of the history marks has probability $\frac{1}{6}$ of occurring. The possibilities are listed in Table 6–7.

Table 6–7
A sample of three

Rank of x		Possible ranks of y:					
		a	b	c	d	e	f
Ames	1	1	1	2	2	3	3
Carter	2	2	3	1	3	1	2
Boyd	3	3	2	3	1	2	1

The exercises for Section 6–3 provide a study of the distribution of r_S when $n = 3$, and when the variables x (English mark) and y (history mark) are independent.

In getting the variance of r_S, we shall lean heavily on the ordinary sample correlation coefficient, namely

$$r = \frac{\sum x_i y_i - n\bar{x}\bar{y}}{\sqrt{\sum(x_i - \bar{x})^2 \; \sum(y_i - \bar{y})^2}} \, ,$$

where $\bar{x} = \bar{y} = (n + 1)/2$ and

$$\sum(x_i - \bar{x})^2 = \sum(y_i - \bar{y})^2 = n(n^2 - 1)/12.$$

(See Theorem, Section 6–2.)

Theorem. *The mean and variance of r_S when x and y are independent.* If the variables x and y are independent, then

$$\mu_{r_S} = 0 \quad \text{and} \quad \text{Var}\,(r_S) = \frac{1}{n-1}, \quad n \geq 2. \qquad (1)$$

The following proof of this theorem is starred because the algebra, though not beyond the level of this book, is demanding. Several features of this method of proof recur in many problems that deal with rankings and with correlated variables, whether ranked or not. Consequently, the proof is well worth a serious effort.

★ *Proof.* From the theorem in Section 6–2, we have

$$r_S = \frac{12 \sum x_i y_i}{n(n^2 - 1)} - \frac{3(n + 1)}{n - 1}.$$

Step 1. Since the mean of r_S is

$$E(r_S) = \frac{12}{n(n^2 - 1)} E\left(\sum x_i y_i\right) - \frac{3(n + 1)}{n - 1}, \qquad (2)$$

we need the mean of $\sum x_i y_i$.

Step 2. We get $E(\sum x_i y_i)$ as follows: Since the x_i's and y_i's are ranks from 1 to n, and since x and y are independent,

$$E(x_i y_i) = E(x_i)E(y_i) = [(n + 1)/2]^2. \qquad (3)$$

Because the mean of the sum equals the sum of the n identical means, we have

$$E\left(\sum x_i y_i\right) = nE(x_i y_i) = n(n + 1)^2/4. \qquad (4)$$

Step 3. Using formulas (2) and (4), we can compute the mean of r_S,

$$E(r_S) = \frac{12}{n(n^2 - 1)} \cdot \frac{n(n + 1)^2}{4} - \frac{3(n + 1)}{n - 1} = 0, \qquad (5)$$

as required.

Step 4. Since the variance of r_S is

$$\text{Var}(r_S) = \frac{12^2}{n^2(n^2-1)^2}\,\text{Var}\left(\sum x_i y_i\right), \tag{6}$$

we need the mean of $\left(\sum x_i y_i\right)^2$.

Step 5. We get $E\left(\sum x_i y_i\right)^2$ as follows: The expansion of $\left(\sum x_i y_i\right)^2$ yields two kinds of terms:

1. n terms of the form $x_i^2 y_i^2$; and
2. $n(n-1)$ terms of the form $x_i x_j y_i y_j$, where $i \neq j$.

Terms of the first type are easily handled because

$$E(x_i^2 y_i^2) = E(x_i^2)E(y_i^2) \tag{7}$$

$$= \left[\frac{n(n+1)(2n+1)}{6n}\right]^2,$$

since $E(x_i^2) = E(y_i^2) =$ average of the squares from 1 to n. Therefore, the average of the *sum* of n terms $x_i^2 y_i^2$ is, once again, the sum of the n identical averages:

$$E\left(\sum x_i^2 y_i^2\right) = nE(x_i^2 y_i^2) = n(n+1)^2(2n+1)^2/36. \tag{8}$$

Now we must deal with the nasty terms in the expansion of $\left(\sum x_i y_i\right)^2$—those terms of the form $x_i x_j y_i y_j$, where $i \neq j$. Since the x's and y's are independent, we have

$$E(x_i x_j y_i y_j) = E(x_i x_j)\cdot E(y_i y_j). \tag{9}$$

The temptation to assume that $E(x_i x_j) = E(x_i)E(x_j)$ is almost irresistible. But there is a negative correlation between the ranks of the x's, so we must not be lured into making this assumption.

To find out how to evaluate $E(x_i x_j)$, let us make the $x_i x_j$ table for a special case, say $n = 5$, as shown in Table 6–8.

Table 6–8 shows all the products, in both orders, except the squares 1^2, 2^2, 3^2, 4^2, 5^2. If we expand $(1 + 2 + 3 + 4 + 5)^2$, we get all the products shown in the

Table 6–8

Table of products $x_i x_j$ for $n = 5$, $i \neq j$

$x_i \backslash x_j$	1	2	3	4	5
1	–	1×2	1×3	1×4	1×5
2	2×1	–	2×3	2×4	2×5
3	3×1	3×2	–	3×4	3×5
4	4×1	4×2	4×3	–	4×5
5	5×1	5×2	5×3	5×4	–

table *plus* the sum of the five squares. Consequently, the sum of the products $x_i x_j$, where $i \neq j$, is

$$(1 + 2 + 3 + 4 + 5)^2 - (1^2 + 2^2 + 3^2 + 4^2 + 5^2)$$

$$= \left[\frac{(5)(6)}{2} \right]^2 - \frac{(5)(6)(11)}{6} = 170.$$

And the average product $x_i x_j$ is $170/n(n - 1) = 170/20 = 8.5$.

Generalizing these ideas, we imagine an $n \times n$ product table; and the sum of the products $x_i x_j$, where $i \neq j$, is

$$\sum_{i \neq j} x_i x_j = (1 + 2 + 3 + \cdots + n)^2 - (1^2 + 2^2 + 3^3 + \cdots + n^2) \quad (10)$$

$$= \left[\frac{n(n + 1)}{2} \right]^2 - \frac{n(n + 1)(2n + 1)}{6}$$

$$= \frac{n(n + 1)}{12} [3n(n + 1) - 2(2n + 1)]$$

$$= \frac{n(n + 1)(n - 1)(3n + 2)}{12}.$$

Note that

$$\sum_{i \neq j} x_i x_j$$

means the sum of all terms in the $n \times n$ product table except those on the main diagonal. Since there are $n(n - 1)$ terms of the form $x_i x_j$, the average value of $x_i x_j$ is

$$E(x_i x_j) = \frac{n(n + 1)(n - 1)(3n + 2)}{12n(n - 1)} = \frac{(n + 1)(3n + 2)}{12}. \quad (11)$$

Because the terms $y_i y_j$ have the same average as the terms $x_i x_j$, we get

$$E(x_i x_j y_i y_j) = E(x_i x_j) E(y_i y_j) = \left[\frac{(n + 1)(3n + 2)}{12} \right]^2.$$

The average of the *sum* of $n(n - 1)$ of these terms is

$$E\left(\sum_{i \neq j} x_i x_j y_i y_j \right) = n(n - 1) \left[\frac{(n + 1)(3n + 2)}{12} \right]^2 \quad (12)$$

$$= \frac{n(n - 1)(n + 1)^2 (3n + 2)^2}{144}.$$

Don't weary in well doing! We are closing in! To get $E(\sum x_i y_i)^2$, recall that

$$(\sum x_i y_i)^2 = \sum x_i^2 y_i^2 + \sum x_i x_j y_i y_j, \quad \text{where} \quad i \neq j.$$

Using formulas (8) and (12), we have

$$E(\sum x_i y_i)^2 = E(\sum x_i^2 y_i^2) + E(\sum x_i x_j y_i y_j), \qquad i \neq j$$

$$= \frac{n(n + 1)^2(2n + 1)^2}{36} + \frac{n(n - 1)(n + 1)^2(3n + 2)^2}{144}.$$

We make an algebraic attack on the last two fractions, and after the smoke of battle clears away, we get

$$E(\sum x_i y_i)^2 = \frac{n^2(n + 1)^2(9n^2 + 19n + 8)}{144}. \qquad (13)$$

Step 6. With formulas (4) and (13), we are equipped to compute the variance of $\sum x_i y_i$:

$$\text{Var} (\sum x_i y_i) = E(\sum x_i y_i)^2 - [E(\sum x_i y_i)]^2 \qquad (14)$$

$$= \frac{n^2(n + 1)^2(9n^2 + 19n + 8)}{144} - \left[\frac{n(n + 1)^2}{4}\right]^2$$

$$= \frac{n^2(n + 1)^2}{144} [(9n^2 + 19n + 8) - 9(n + 1)^2]$$

$$= \frac{n^2(n + 1)^2(n - 1)}{144}.$$

Step 7. There remains only the job of using formula (14) in equation (6):

$$\text{Var} (r_S) = \frac{12^2}{n^2(n^2 - 1)^2} \cdot \frac{n^2(n + 1)^2(n - 1)}{144}$$

$$= \frac{1}{n - 1},$$

which is a surprisingly tidy result in view of the algebraic horrors that produced it.

EXERCISES FOR SECTION 6–3

1. In the foregoing example, compute the six rank correlation coefficients for the y-ranks a, b, c, d, e, and f, in Table 6–7.

2. Which of the six rank correlation coefficients in Exercise 1 are equal.

3. What is the probability of finding a perfect correlation in this example?

4. For a sample of four, find the exact probability distribution of r_S, assuming that x and y are independent in the population.

5. Use the results of Exercise 1 to compute the mean and variance of r_S and thus check the formulas of the theorem.

6–4 TABLES FOR THE DISTRIBUTION OF r_S: THE NORMAL APPROXIMATION

In Exercise 4 of Section 6–3, the probability distribution tells us that, if x and y are independent in the population, there is only 1 chance in 24 of getting a perfect correlation, or $r_S = +1$. Therefore, if we were to observe 4 pairs of ranks in perfect agreement, we might begin to believe that there is some relation between x and y in the population.

The task of constructing probability distributions for r_S when n is greater than 4 is soon impeded by the restrictions of the 24-hour day. Tables provide some relief. At the back of the book, we supply exact tables of the distribution of r_S when the population variables are independent, and when $n \leq 11$. These numbers, in Table A–13, have been derived by computing the sums of squares of differences $(\sum d_i^2)$ for all the possible permutations. Since a sample of 10 gives rise to $10! = 3,628,800$ permutations, it is easy to see why we do not offer a table for, say, $n \leq 30$. Probabilities are near 0.025, 0.05, and 0.10.

For samples of more than 10, we are by no means helpless. The normal distribution once more comes to our aid. The fact is this: For large samples, when x and y are independent in the population, observed values of r_S are approximately normally distributed about a mean $\mu = 0$, with standard deviation $\sigma = 1/\sqrt{n-1}$. The mean and standard deviation come from the theorem in the previous section. Therefore, if we observe $r_S = a$, independence gives (now regarding r_S as a random variable)

$$P(r_S \geq a) \approx P(Z \geq z),$$

where

$$z = \frac{a - 0}{1/\sqrt{n-1}} = a\sqrt{n-1},$$

and Z is a standard normal random variable.

Ties. Give ties the average of the tied ranks. This average will always be a whole number or a whole number plus a half. In squaring half-integers, it helps to use

$$(n + \tfrac{1}{2})^2 = n(n+1) + \tfrac{1}{4}; \qquad \text{thus} \qquad (3\tfrac{1}{2})^2 = 3 \times 4 + \tfrac{1}{4}.$$

Example 1. The ranks obtained by 10 students on a Scholastic Aptitude Test are compared with the ranks of their IQ's. The rank correlation coefficient is found to be 0.62. Under the assumption of independence in the population (the null hypothesis), find $P(r_S \geq 0.62)$ by using (a) Table A–13, (b) the normal approximation. On the basis of the results, would you reject the null hypothesis at the 5% level?

Solution. From Table A–13, we find that $P(r_S \geq 0.62)$ is between 0.048 and 0.024. The adjusted deviate for use in the normal table is

$$z = 0.62\sqrt{10-1} = 1.86.$$

Then $P(r_S \geq 0.62) \approx P(Z \geq 1.86) = 0.031$, from the normal table. More extensive tables than Table A–13 give the exact probability as 0.03.

Since the probability of getting a value of r_S at least as high as 0.62 is only 3% when there is no correlation, we reject the null hypothesis at the 5% level.

This example shows us how the normal approximation can be used in making a decision as to whether an observed value of r_S is real or fortuitous, a chance result obtained from a population in which there is no correlation.

Example 2. *Earthquakes* (C. Emiliani, C. G. A. Harrison, and M. Swanson). Following the detonation of an underground atomic device, counts of earthquakes were reported in successive eight-hour intervals as shown in Table 6–9. Use the rank correlation coefficient to test for a trend in numbers of earthquakes.

Table 6–9

x_i 8-hour interval	Number of earthquakes	y_i rank	$d_i = x_i - y_i$	d_i^2
1	33	2	-1	1
2	32	3	-1	1
3	30	4	-1	1
4	41	1	3	9
5	23	8	-3	9
6	17	$10\frac{1}{2}$	$-4\frac{1}{2}$	$20\frac{1}{4}$
7	27	$5\frac{1}{2}$	$1\frac{1}{2}$	$2\frac{1}{4}$
8	20	9	-1	1
9	15	12	-3	9
10	12	13	-3	9
11	27	$5\frac{1}{2}$	$5\frac{1}{2}$	$30\frac{1}{4}$
12	17	$10\frac{1}{2}$	$1\frac{1}{2}$	$2\frac{1}{4}$
13	25	7	6	36
			0	131

Solution. Using the data in the table, we find

$$r_S = 1 - \frac{6 \sum d_i^2}{n(n^2 - 1)} = 1 - \frac{6(131)}{13(168)} = 1 - 0.360 = 0.640.$$

The normal approximation gives

$$z = 0.640\sqrt{12} = 2.22.$$

The probability of a larger value is about 0.013 if there is no relation between the time and the number of earthquakes. It appears that the detonation period is followed by decreasing numbers of quakes; at least that was the result on the given occasion.

EXERCISES FOR SECTION 6–4

1. In Exercise 2 of Section 6–2, would you reject, at the 5% significance level, the null hypothesis of independence in the population?

2. In Exercise 5 of Section 6–2, would you reject, at the 20% significance level, the null hypothesis of independence in the population?

3. Work the problem of Example 1, given that $r_S = 0.52$.

4. Two judges each rank 10 essays in a competition. When the rankings are compared, a rank correlation coefficient of 0.43 is obtained. Use Table A–13 to assess the significance of r_S. Would you reject, at the 10% level, the null hypothesis of independence in the ranking?

5. Use the normal approximation to check your answer in Exercise 4.

6. The IQ ranks of 50 freshmen are compared with their class ranks in mathematics, and a rank correlation coefficient of 0.25 is obtained. Use the normal approximation to find $P(r_S \geq 0.25)$. Should we reject the null hypothesis of independence at the 5% level?

7. In June 1970, *Consumer Reports* gave quality rankings to 17 makes of room air conditioners. Table 6–10 lists the ranks and the average prices of the air conditioners (by permission of Consumers Union).

Table 6–10

Air conditioner	Quality rank	Average price
A	1	$196.28
B	2	169.20
C	3	181.51
D	4	166.75
E	5	172.54
F	6	160.38
G	7	151.91
H	8	160.70
I	9	170.89
J	10	160.77
K	11	168.43
L	12	171.28
M	13	163.74
N	14	173.48
O	15	171.06
P	16	162.21
Q	17	174.68

Find the rank correlation coefficient between quality and average cost. On the basis of the data, would you reject the null hypothesis of no correlation at the 25% level?

8. For people served by the same medical plan, Lewis (1969) reports the following data on the availability of hospital beds and the rate of performance of appendectomies in 10 regions of Kansas. [Let x denote the number of hospital beds per 1000 persons in the region, and let y denote the number of appendectomies per 10,000 persons covered by the plan.]

Region:	1	2	3	4	5	6	7	8	9	10
x:	4.71	4.90	5.53	7.44	7.43	5.62	5.63	5.67	5.75	3.96
y:	18.4	14.6	19.0	46.9	61.8	39.1	51.1	27.0	17.6	16.3

Compute the rank correlation between x and y. Would you reject the null hypothesis of independence at the 5% level? Suggest an explanation for your results.

6–5 POWER OF THE RANK CORRELATION COEFFICIENT; THE BIVARIATE NORMAL DISTRIBUTION

One special family of joint distributions for the random variables X and Y plays a major role in statistics; these distributions have the population product-moment correlation coefficient, ρ, as an important parameter. This family of distributions is called the *bivariate normal family*. We shall briefly describe it.

For one variable, we represent probabilities by areas above a line; for two variables, we represent probabilities by volumes above a plane. When X and Y are independent and $\sigma_X = \sigma_Y$, the joint distribution between X and Y has a density surface that looks like the mountain-shaped figure that one would generate if he

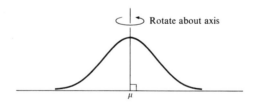

Rotate about axis

μ

were to rotate a normal density curve about the vertical line at the mean, μ. (The surface so generated is not actually the density surface of this bivariate normal, but it *looks* like it.) Such a surface has contours that are circles.

If $\sigma_X = \sigma_Y$ and the correlation $\rho = 0$, then the bivariate normal looks like the foregoing figure. *The bivariate normal has this special property: If $\rho = 0$, then X and Y are independent.* This is an extremely attractive feature of the distribution. If X and Y are correlated, then the contours of the mountain-shaped bivariate normal density surface are ellipses whose major axes, when projected on the XY-plane, are not parallel to either the x-axis or the y-axis (see figures). The more elongated the ellipses, the higher the correlation.

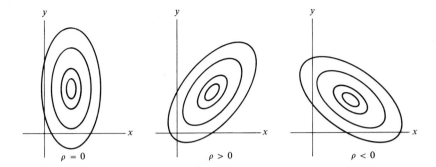

$\rho = 0$ $\rho > 0$ $\rho < 0$

To get a pair of random variables (X, Y) that are bivariately normally distributed, we start with three independent, standard normal random variables, U, V, W, and a constant, c. We then define X and Y thus:

$$X = U + cV, \tag{1}$$

$$Y = U + cW.$$

When $c = 0$, X and Y are perfectly correlated and $\rho = 1$. When c tends to infinity, X and Y become independent, because all the weight is then on the independent variables V and W.

It is left as an exercise to show that

$$\rho_{XY} = \frac{1}{1 + c^2}. \tag{2}$$

Thus any positive numerical value of ρ_{XY} can be obtained by suitable choice of c. The variance of X and of Y is $1 + c^2$. (See Appendix I–5.)

Example 1. *Random normal pairs.* Obtain four random normal pairs (x_i, y_i) from a bivariate normal with correlation $\rho = 0.2$.

Solution. To get $\rho = 0.2$, we solve $0.2 = 1/(1 + c^2)$ and find $c = \pm 2$. Let us use $c = +2$. Each (x, y)-pair requires 3 standard normal deviates. We get them from Table A–23. Let us use the first 3 deviates in each of the first 4 rows. The results may be tabulated as follows:

u_i	v_i	w_i	$x_i = u_i + 2v_i$	$y_i = u_i + 2w_i$
-0.56	$+1.51$	-0.35	2.46	-1.26
-0.55	$+0.91$	-0.55	1.27	-1.65
$+0.74$	-1.33	$+0.19$	-1.92	1.12
$+2.56$	-0.11	$+1.37$	2.34	5.30

We also have

$$\sigma_X^2 = \sigma_Y^2 = 1 + 2^2 = 5.$$

This method of generating bivariate normal pairs was used in our simulation, described below.

To see how well the rank correlation coefficient in small samples from a bivariate normal distribution detects positive correlation, we simulated 1000 samples of $n = 5, 10, 15$, and 20 for $\rho = 0.0, 0.20, 0.40, 0.60, 0.80, 0.90, 0.95$. The significance levels were between 0.04 and 0.05. This meant rejecting the null hypothesis $\rho = 0$ (a) if $r_S \geq 0.9$ for $n = 5$, (b) if $r_S \geq 0.564$ for $n = 10$, (c) if $r_S \geq 0.5$ for $n = 15$, and (d) if $r_S \geq 0.414$ for $n = 20$. The power curves displayed in Figure 6–1 show the results.

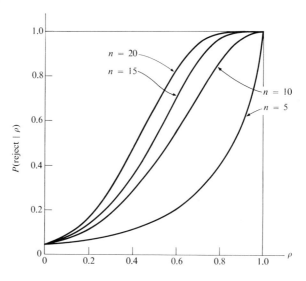

Fig. 6–1 Power of rank correlation coefficient for detecting positive correlation in a bivariate normal.

EXERCISES FOR SECTION 6–5

1. Use Figure 6–1 to find the value of ρ that makes it twice as likely to reject $\rho = 0$ when $n = 10$ as when $n = 5$.

2. Prove that $\sigma_X^2 = \sigma_Y^2 = 1 + c^2$.

3. Show that the correlation between X and Y of equation (1) is given by the result in equation (2).

4. Devise a method of constructing normally distributed random variables that are negatively correlated with a specified value of ρ.

CHAPTER 7

Rare Events: The Poisson Distribution

7–1 WHAT PROCESSES YIELD A POISSON DISTRIBUTION?

The distributions of the numbers of occurrences of events in fixed periods of time or space, or of counts of rare events, such as accidents, often conform approximately to a distribution called the Poisson.† Some examples of observations having Poisson or approximately Poisson distributions are:

a) the number of clicks on a Geiger counter in a fixed interval of time;
b) the number of wireworms in a random small circle of ground in a field;
c) the number of raindrops passing through a small hoop in 10 seconds during a rain of approximately fixed intensity;
d) the number of vacancies in the Supreme Court during a year;
e) the number of red corpuscles in the square grids in a blood sample prepared for counting;
f) the number of defects in a newly manufactured automobile (or airplane);
g) the number of children in a completed family during a particular time period;
h) the number of times an author uses the word *from* in a block of 200 words;
i) the number of cars passing a point in a road in a one-minute period;
j) the number of customers arriving at a check-out counter in a five-minute period;
k) the number of wrong numbers dialed in a telephone exchange in a one-minute period.

The versatility of the Poisson distribution recommends it. This distribution also approximates the binomial distribution in some special circumstances, which we describe in Section 7–3. In some circumstances, the Poisson can be approximated by the normal distribution.

We shall first study the Poisson distribution with the aid of a classical example about horse kicks. Besides bringing the Poisson to the public eye for the first time

---

† Named in honor of its discoverer, the French mathematician S. D. Poisson (1781–1840).

and thus becoming a fragment of history, this particular example exhibits a close tie with the binomial, brings out clearly a variety of features of the Poisson, and illustrates the important ideas of the distribution of accidents per unit time. Even though the Russians did use horse cavalry to good effect in World War II and General George Patton, the great American battle commander of that war, claimed that he could have profitably used some horse cavalry over certain terrain, we shall probably not have much horse cavalry to study further. However, accidental military vehicular deaths may follow the same pattern.

Example 1. *Fatal horse kicks: a tragic example.* Bortkiewicz's famous example gives the observed frequency distribution of the number of deaths from horse kicks in the Prussian army per corps-year for each of 200 corps-years. A corps-year represents one army corps for one year. The data are shown in Table 7–1, together with the numbers that would be "expected" on the assumption that the distribution is Poisson, with the same mean number of deaths per corps-year, 0.61, as in the 200 observed corps-years.

Table 7–1
Counts of deaths from horse kicks in 200 corps-years

Number of deaths, x	Observed frequency, f_x	Expected frequency	xf_x	x^2f_x
0	109	108.7	0	0
1	65	66.3	65	65
2	22	20.2	44	88
3	3	4.1	9	27
4	1	0.6	4	16
5	0	0.1	0	0
6	0	0.0	0	0
	200		$\Sigma x = 122$	$\Sigma x^2 = 196$

Discussion. One feature of a Poisson is that its mean, μ, and its variance, σ^2, are equal. Let us look at the sample values. By definition, we have the sample mean

$$\bar{x} = \Sigma x / 200 = 122/200 = 0.61$$

and the sample variance

$$s^2 = \frac{\Sigma(x - \bar{x})^2}{n - 1} = \frac{\Sigma x^2 - n(\bar{x})^2}{n - 1} = \frac{196 - 200(0.61)^2}{199} = 0.611.$$

The approximate equality of the sample mean and variance supports the theoretical fact mentioned; actually $\bar{x}$ and s^2 are much closer than we would expect from sampling variation.

The data of this example can be used to illustrate some ideas behind the Poisson distribution. First, the Prussian army corps were all of about the same size and composition; and one year's activities during the observation period were much like another's. Consequently, *the number of soldiers exposed to risk and the ways that they were exposed were much the same from one corps-year to another.*

Second—and this fact is more important mathematically than practically— *the event itself is rare, and yet it could in principle occur many times.* A corps contained thousands of soldiers, but as you can see from the table, only a few died from horse kicks; that is, the average probability of death from horse kicks is very small.

Third, we suppose that within the corps-year *the individual soldiers survive or die independently.* This independence idea arose earlier in connection with the binomial distribution. Indeed, we might regard a man-year as a single binomial trial, and the horse-kicks example might, at first glance, seem to come close to satisfying the criteria for a binomial distribution. If it did and if we knew the number of soldiers exposed, we could deal with the example in the usual binomial way. Since some soldiers might have had much less to do with horses than others, P(death from horse kicks) is not constant from soldier to soldier. Thus a binomial based on the total number of soldiers as n and a single proportion p might not fit very well. Even so, a binomial based on some n, not necessarily the number of soldiers, and some p might fit closely.

Table 7–1 shows that the Poisson distribution fits the facts of the observed frequency distribution, and this suggests that the Poisson may be a valuable distribution in such problems, whether the binomial applies or not.

7–2 THE POISSON PROBABILITY DISTRIBUTION: MEAN AND VARIANCE

A random variable X, having a Poisson distribution, can take only the values 0, 1, 2, 3, . . ., and the distribution is therefore *discrete*. The Poisson probability function has the form

$$f(x) = P(X = x) = \frac{e^{-m}m^x}{x!}, \qquad x = 0, 1, 2, \ldots, \tag{1}$$

where $x! = x(x - 1)(x - 2) \cdots 3 \cdot 2 \cdot 1$ if $x \geq 1$, $0! = 1$, e is the base of the natural logarithms, and m is a positive number that will turn out to be the mean of the distribution. (Some authors replace m by the Greek letter λ, others by the Greek letter μ. Although m is sometimes used as an index for integers, in this context m takes on any positive value.)

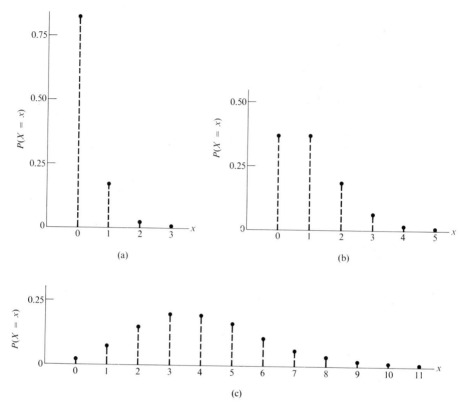

Fig. 7–1 (a) Poisson, $m = 0.2$; (b) Poisson, $m = 1$; (c) Poisson, $m = 4$.

To show that the sum of the probabilities is 1, we need to know an important fact about infinite series, namely that for any real number y,

$$e^y = \sum_{k=0}^{\infty} \frac{y^k}{k!} = \frac{y^0}{0!} + \frac{y^1}{1!} + \frac{y^2}{2!} + \frac{y^3}{3!} + \cdots. \tag{2}$$

Using this fact, we get, when y is replaced by m,

$$\sum_{x=0}^{\infty} P(X = x) = e^{-m} \left[\frac{m^0}{0!} + \frac{m^1}{1!} + \frac{m^2}{2!} + \cdots \right] = e^{-m} \cdot e^m = 1. \tag{3}$$

Theorem 1. *The mean of the Poisson distribution.* If X is a Poisson random variable, then

$$\mu_X = m. \tag{4}$$

Proof. By definition, we have

$$\mu_X = \sum_{x=0}^{\infty} xP(X = x) = e^{-m}\left[0 + 1\cdot\frac{m^1}{1!} + 2\cdot\frac{m^2}{2!} + 3\cdot\frac{m^3}{3!} + \cdots\right]$$

$$= e^{-m}\left[m^1 + \frac{m^2}{1!} + \frac{m^3}{2!} + \cdots\right]$$

$$= me^{-m}\left[\frac{m^0}{0!} + \frac{m^1}{1!} + \frac{m^2}{2!} + \frac{m^3}{3!} + \cdots\right]$$

$$= me^{-m}[e^m] = m.$$

Thus we have shown that *the mean of a Poisson distribution is its parameter, m.*

Theorem 2. *The variance of the Poisson distribution.* If X is a Poisson random variable, then

$$\text{Var}(X) = \sigma_X^2 = m. \tag{5}$$

Proof. By definition,

$$E(X^2) = \sum_{x=0}^{\infty} x^2 P(X = x) = e^{-m}\left[0 + 1\cdot\frac{m^1}{1!} + 4\cdot\frac{m^2}{2!} + 9\cdot\frac{m^3}{3!}\right.$$

$$\left. + 16\cdot\frac{m^4}{4!} + \cdots\right]$$

$$= me^{-m}\left[1 + 2\cdot\frac{m^1}{1!} + 3\cdot\frac{m^2}{2!} + 4\cdot\frac{m^3}{3!} + \cdots\right].$$

The expression in the square brackets equals

$$\left(1 + \frac{m^1}{1!} + \frac{m^2}{2!} + \frac{m^3}{3!} + \cdots\right) + \left(\frac{m^1}{1!} + 2\cdot\frac{m^2}{2!} + 3\cdot\frac{m^3}{3!} + \cdots\right) = e^m + me^m.$$

Hence

$$E(X^2) = me^{-m}[e^m + me^m] = m + m^2,$$

and

$$\sigma_X^2 = E(X^2) - \mu_X^2 = m + m^2 - m^2 = m.$$

The mean and the variance of any Poisson distribution are always identical! When you seen an empirical frequency distribution *based on counts* whose sample mean and variance are approximately equal, consider it possible that the distribution might be closely approximated by a Poisson distribution.

Let us return to the horse-kicks example. We fitted the distribution of death counts in the Prussian army corps to the Poisson distribution by using $\bar{x}$, the observed average, to estimate the Poisson mean m. To get $\bar{x}$, we used the xf_x column of Table 7–1. Note also that the sample variance

$$s^2 = \sum(x - \bar{x})^2/(n - 1),$$

as computed from Table 7–1, has for its first two decimal places the same value, 0.61, as $\bar{x}$. Thus the data exhibit an important property of the Poisson. Using tables of the Poisson distribution, or logarithms, we could compute

$$200\left[\frac{e^{-0.61}(0.61)^x}{x!}\right]$$

for each x and get the expected counts shown in Table 7–1.

Later, we shall introduce a statistical measure of the agreement (or "fit"). between the observed data and the Poisson distribution. In the horse-kicks example the agreement is extremely close. Indeed, according to the measure of goodness of fit, if observed data were drawn from an exact Poisson distribution, in 85% of such samples the fit would be worse than that found in Table 7–1. These facts suggest three things:

1. The Poisson distribution is a good approximation to the horse-kick data.
2. The man who collected the data just may have, for the sake of tidiness, selected them so as to make the similarity with the Poisson especially striking. The fit seems to be a little too good to be true.
3. Accident data may tend to be distributed like the Poisson.

Extensive tables are available for the Poisson distribution. In this book, two tables are provided. Table A–8 gives Poisson probabilities for individual terms and cumulatives at values of m ranging from 0.1 to 14, and for values of X ranging from 0 to 28. The cumulatives are the probabilities that the Poisson random variable X takes a value *equal to or greater than* the value of x given in the margin.

EXERCISES FOR SECTION 7–2

1. *Men on base.* In baseball, a home run may be hit with 0, 1, 2, or 3 men on base. The following data come from the National Baseball League.

x (number on base):	0	1	2	3
f_x (observed frequency of home runs):	421	227	96	21

Use these data to make a table like Table 7–1 in Section 7–1. Obtain $\bar{x}$, use the value of $\bar{x}$ as an estimate of a Poisson mean m, and then get the expected frequencies from the Poisson table. Associate $x = 3$ with the Poisson probability $P(X \geq 3)$. [We include the example to show that even a rather truncated distribution can be fairly closely approximated by a Poisson, and we put it first because it is small, rather than because it is representative.]

2. (Continuation) Use the fitted data of Exercise 1 to estimate the probability that, when a home run is hit, (a) there are no men on base, (b) there are fewer than two men on base.

3. To test whether the distribution of observed data in Exercise 1 is approximately Poisson, compute the mean and variance of the observed data and see if they are approximately equal.

4. *Cell counts.* William Gosset counted the number of yeast cells in each of the 400 unit volumes in a hemacytometer. His data were as follows, where X denotes the number of yeast cells in a unit volume, and f_x denotes the number of unit volumes in which x cells were observed.

x:	0	1	2	3	4	5	6	7	8	9	10	11	12	13	
f_x:	0	20	43	53	86	70	54	37	18	10	5	2	2	0	(Total 400)

For these data, compute $\bar{x}$, and enter the Poisson table with $m = \bar{x}$ to get the expected frequencies under the assumption that X is a Poisson random variable. (Tabulate as in Table 7–1, Section 7–1.)

5. Assuming a Poisson distribution for the data in Exercise 4, estimate from the fitted values (a) $P(X = 0)$, (b) $P(X \leq 3)$.

6. *Political writing.* In essays on political matters, Alexander Hamilton used the word *from*, in 247 blocks of length 200 words each, with the following frequencies.

x:	0	1	2	3	4	5	6	7 or more
f_x:	93	82	51	13	5	2	1	0

Here x is a value of the random variable X, the number of times the word *from* occurred in a block, and f_x denotes the number of blocks in which the word *from* occurred x times. Fit these data with the Poisson distribution by taking $m = \bar{x}$ and using the tables to get the expected frequencies.

7. Repeat Exercise 3 for the observed data in Exercise 6.

8. In a Poisson distribution, if $P(X = 0) = P(X = 1)$, find m.

9. Assume that X, the number of safeties per game in football, has approximately a Poisson distribution. If the number of safety touches in one season of professional football averages 0.2 per game, find (a) $P(X = 0)$, (b) $P(X \geq 3)$.

10. *Inventory management.* At the beginning of each month, a store manager wishes to order enough electric toasters to last the month. He cannot reorder, and he knows that on the average the store sells 4 toasters a month. If he has run out, what is the least number of toasters that he should order, if he wants to be 80% certain of not running out?

11. *Vision.* A light source emits photons according to a Poisson distribution with mean m. The mechanism that transmits the emitted photons to the eye is equally likely to transmit 0, 1, 2, ..., x photons, where x is the number of photons emitted by the source. Find a convenient expression for the probability that the eye receives exactly r photons from a given emission.

12. In James Madison's writings, uses of the word *can* in 200-word passages were counted, and the following frequencies were noted.

x:	0	1	2	3	4 or more
f_x:	211	44	6	1	0 (Total 262)

Find the mean and variance of X, the number of uses of the word *can* per passage, and use the information to decide if X is approximately a Poisson random variable.

13. *Likelihood ratios.* Suppose that the numbers of uses of the word *also* in 2000-word passages in the writings of Hamilton and Madison are known, and that these numbers are approximated by Poisson distributions with mean 0.6 for Hamilton and mean 1.3 for Madison. If a 2000-word passage, known to have been written by one of them and equally likely to have been written by either, contains 0 *also*'s, what are the odds in favor of Hamilton's having written the passage? What would the odds be if the passage contained 3 *also*'s?

14. Suppose that the army "rejects" tanks that have 3 or more major defects. How frequently will tanks be rejected if the mean number of major defects is 0.3 per tank?

15. If a typist averages 1.5 errors per page, what is the probability that 2 randomly chosen pages are free of errors?

16. The number of no-hit baseball games in a major league season approximately follows a Poisson distribution. If the probability that a randomly chosen year has 0 no-hit games is 0.05, find m and $P(2$ or more no-hit games$)$.

17. *A philatelic problem.* A certain issue of Bermuda postage stamps was printed in sheets of 60 stamps. During the postal life of the stamp, an ardent accumulator gathered thousands of the used stamps from many sources, soaked them off the paper, and randomly bundled them in 100-stamp lots. Subsequently, it was discovered that one stamp on every sheet exhibited a flaw much sought after by collectors. If a collector purchases one of the 100-stamp bundles, what is the probability that he gets (a) no flawed stamp, (b) three or more flawed stamps?

18. The World Health Organization reports that in 1965 the number of suicides in Scotland averaged 8.0 per week. Assuming that the number of suicides is a Poisson random variable, find how frequently, on the average, 12 or more suicides might occur in a given week.

19. *Telephone calls.* The telephone company estimates that, on the average, there will be 14 telephone lines in use between A-ville and B-burg at any time during the peak period. If the company wants to assure that a subscriber will find all lines busy in less than 5% of the peak periods, how many lines should be provided between the two towns?

20. *Arrivals.* Motorists arrive at a gas station at the rate of 30 per hour on the average. If the number of arrivals, X, is a Poisson random variable, what is the probability that during a five-minute interval no person arrives? Find $P(X \geq 7)$ during a ten-minute interval.

21. *Traffic accidents.* The number of accidents at Coroner's Curve averages 7 per week. What is the probability that 2 or more accidents will occur at this curve in any given day?

22. For the Poisson distribution, prove that $\sigma^2 = m$ by computing $E[X(X-1)]$ and using the fact that, for any distribution with a variance,

$$\sigma^2 = E[X(X-1)] + \mu - \mu^2.$$

23. Show that, if $P(X = x) = P(X = x + 1)$, then $m = x + 1$, and the converse. Use this fact to prove that, if $m = 4$, $P(X = 3) = P(X = 4)$. Check your results with the tables.

24. Prove that, if $P(X = x) \geq P(X = x + 1)$, then $x + 1 \geq m$, and the converse.

25. Prove that, if $P(X = x - 1) \leq P(X = x)$, then $x \leq m$, and the converse.

26. Use the results of Exercises 24 and 25 to show that $P(X = x)$ is maximized by choosing x to satisfy $m - 1 \leq x \leq m$.

27. Find $(x + 1)P(X = x + 1)/P(X = x)$. The constancy of this quotient is sometimes used to check the Poissonness of a set of data. To do this, plot $(x + 1)f_{x+1}/f_x$ against x when the f's are not too small. The resulting points should scatter randomly about a horizontal line. If they show a distinct trend, they cast doubt on the Poissonness.

28. Plot the data of Example 1, Section 7–1, in the manner suggested in Exercise 27. Comment on the result.

29. Plot the data of Exercise 4 (for $x > 0$) in the manner suggested in Exercise 27. Comment on the result.

7-3 THE POISSON APPROXIMATION FOR THE BINOMIAL

The mean and variance of the binomial distribution are, respectively,

$$\mu = np \qquad \text{and} \qquad \sigma^2 = np(1 - p).$$

If μ were equal to σ^2, then $1 - p$ would equal 1 or p would equal 0, and that fact would destroy the binomial for fixed n. If, instead, we think of a nonzero p growing smaller while n grows larger so that np is a constant, then p could tend toward 0 and $1 - p$ would *nearly* equal 1. And so we might expect that binomial distributions with large n's and small p's would closely approximate Poisson distributions with $m = np$.

In Table 7–2, we make the comparison using $n = 100$ and $p = 0.01$ for the

Table 7-2

Comparison of binomial
$n = 100$, $p = 0.01$ with
the Poisson $m = 1$

x	Binomial	Poisson
0	0.3660	0.3679
1	0.3697	0.3679
2	0.1849	0.1839
3	0.0610	0.0613
4	0.0149	0.0153
5	0.0029	0.0031
6	0.0005	0.0005
7	0.0001	0.0001

binomial and $m = np = 100(0.01) = 1$ for the Poisson. Such distributions might arise if lots of 100 items were manufactured by a process that randomly produced 1% defectives or, from the Poisson point of view, 1 defective per lot. Note the close agreement. If one had to make the distinction between the two distributions on the basis of a sample rather than four-decimal calculations, it would take tens of thousands of observations (tens of thousands of lots in the example) to detect the difference reliably. As a practical matter, it would be hopeless.

This example shows that the Poisson can be used to approximate the binomial when p is small and n is large. One might well ask: Who needs it? After all, we have binomial tables and the normal approximation, and so why would we want another approximation? Well, for one reason, the Poisson is a simpler distribution than the binomial, because the Poisson depends on only one constant, its mean *m*, rather than on two constants, *n* and *p*. Therefore, the Poisson can be tabled in much less space than that required for the binomial. Also, where appropriate, it is pleasant in theoretical work to use the simpler formulas of the Poisson. And finally, recall that in Section 2–6 we noted that the normal approximation may often, but not always, be used to give a good approximation for the binomial. As it fortunately turns out, the Poisson offers a close approximation for the binomial when *np* is small, just where the normal gives a poor fit.

Note that when we use the Poisson to approximate a binomial, we do not need to make a correction for continuity. The reason is that both the Poisson and the binomial are discrete distributions on the positive integers.

Example 1. A bettor buys a lottery ticket in each of 20 lotteries. If his probability of winning a prize in each of these lotteries is 0.01, find the approximate probability that he wins (a) no prize, (b) exactly 1 prize, (c) at least 1 prize.

Solution. Since we are dealing with a binomial experiment with $n = 20$ and $p = 0.01$, we may use the Poisson approximation with $m = np = \frac{1}{5}$. From the Poisson tables we get: (a) $P(X = 0) = 0.819$; (b) $P(X = 1) = 0.164$; (c) $P(X \geq 1) = 0.181$. These results check to within 1 in the third decimal with those given by the binomial tables. Note that *we do not need to know the separate values of n and p in order to use the Poisson; we need only the value of the binomial mean np.*

EXERCISES FOR SECTION 7–3

1. If B is a binomial random variable with $n = 15$ and $p = 0.1$, find $P(B = 3)$ and $P(B \geq 3)$ (a) by using the Poisson approximation, (b) by using binomial tables.

2. Repeat Exercise 1 for $n = 20$ and $p = 0.1$.

3. Repeat Example 1, given that the bettor buys one ticket in each of 100 lotteries and that the probability of winning a prize in each lottery is 1/200.

4. In a certain factory, the probability of an accident occurring on any given day is

1/500. Use the Poisson approximation to find the probability that 1000 days go by without an accident.

5. The chances are 1/100 that a certain type of fuse plug is defective. In a batch of 300 of these fuse plugs, what is the probability that none are defective? How large a randomly selected batch of fuse plugs would be required to ensure that the probability of getting at least 2 defectives is more than 1/4? [*Hint:* Use Poisson tables and interpolate.]

6. A baker plans to make 1000 almond cookies, and he would like the chance of producing a cookie without almonds to be less than 1 in 100; but within this limit, he doesn't want to buy more almonds. How many almonds should be put in the cookie mix? Use the Poisson tables with interpolation, and give your answer to the nearest 100 almonds.

7. The Tourist Bureau of the balmy isle of Trinuda reports that the island gets an average of 3 rainy days in April. Find the probability of no rainy days during a 20-day April vacation in two ways: (a) Assume a binomial experiment with all days considered as independent trials having the same chance of rain. (Use binomial tables.) (b) Use the Poisson approximation for the binomial experiment. What objections might be raised to the assumption of a binomial experiment?

7–4 DISTRIBUTION OF THE SUM OF TWO INDEPENDENT POISSON RANDOM VARIABLES

In some ways, the Poisson distribution has more generality than the binomial in spite of its simpler form. For example, the following theorem holds.

Theorem. *The sum of two independent Poisson random variables.* If X and Y are two independent Poisson random variables with means m and n, respectively, then the random variable $X + Y$ has a Poisson distribution with mean $m + n$.

The proof is embodied in Exercises 6 through 9.

Such a theorem is available for the binomial only when the p's are equal.

Example 1. *Defects in wire.* If 2 defects per 100 feet of wire are observed on the average, how would the number of defects be distributed for lengths of 1000 feet?

Solution. The number of defects in 100-foot lengths is a Poisson random variable with $m = 2$. We think of the 1000-foot lengths as made up of ten 100-foot lengths having $X_1, X_2, \ldots, X_{10}$ defects, respectively, with these random variables being independent. Then the number of defects in the 1000-foot length is

$$X = X_1 + X_2 + \cdots + X_{10}.$$

Since the X_i's are Poisson random variables, the foregoing theorem assures us that X is a Poisson random variable, and that the mean of X is $10(2) = 20$.

EXERCISES FOR SECTION 7-4

1. If 2 defects per 100 feet of wire are observed on the average, what is the chance of getting a random 10-foot length of this wire with no defects?

2. In Example 1, if X is the number of defects in a 500-foot length, find (a) $P(X = 10)$, (b) $P(X \geq 10)$.

3. A wire has exactly two kinds of imperfections, pinholes and nicks, produced by independent processes. If pinholes average 2 per 10-foot segment and nicks average 3 per 10-foot segment, find, for a 20-foot segment, the probability (a) of exactly 6 imperfections, (b) of 6 or more imperfections.

4. Flaws in steel plates occur on the average of 0.1 flaw per square foot of plate. What is the probability that a steel plate 3 feet by 10 feet will have no flaws? At least 1 flaw?

5. Generalize the theorem of the section to the sum of 3 independent Poisson random variables, and use the theorem of the section for the proof.

Given that X and Y are independent Poisson random variables with means m and n, respectively, and that $Z = X + Y$, then proofs of the next four exercises demonstrate that Z is a Poisson random variable with mean $m + n$.

6. Show that $P(Z = 0) = e^{-(m+n)}$.

7. Show that $P(Z = 1) = e^{-(m+n)} \cdot (m + n)/1!$.

8. Show that $P(Z = 2) = e^{-(m+n)} \cdot (m + n)^2/2!$.

9. Show that $P(Z = z) = e^{-(m+n)} \cdot (m + n)^z/z!$.

7-5 THE NORMAL APPROXIMATION FOR THE POISSON DISTRIBUTION

A binomial distribution is closely approximated by a normal distribution, provided that the binomial mean np is several standard deviations (say at least 3) away from both 0 and n. This condition guarantees enough distance available between the mean and the origin (and n) for the probability distribution to be able to adopt the shape of an approximately normal distribution. If there were only one standard deviation, say, then $\frac{1}{2}$ the density would have to be crammed into a space where only $\frac{1}{3}$ of it usually lies, and there wouldn't be room for normality to develop. One might then expect by analogy that the Poisson would approximate normality when m, its mean, is several standard deviations away from 0. Let us find out what such an approximation requires. If m is to be at least 3σ away from 0, then we must have

$$m \geq 3\sigma = 3\sqrt{m},$$

whence

$$m \geq 9.$$

Thus we might look for approximate normality in a Poisson distribution with $m \geq 9$ and try fitting a normal with mean $\mu = 9$ and $\sigma = \sqrt{9} = 3$.

Table 7–3

Comparison of exact Poisson probabilities for $m = 9$ with those obtained from the normal approximation

x	Normal approximation	Exact Poisson probabilities
0	0.0015	0.0001
1	0.0039	0.0011
2	0.0089	0.0050
3	0.0183	0.0150
4	0.0334	0.0337
5	0.0549	0.0607
6	0.0807	0.0911
7	0.1062	0.1171
8	0.1253	0.1318
9	0.1324	0.1318
10	0.1253	0.1186
11	0.1062	0.0970
12	0.0807	0.0728
13	0.0549	0.0504
14	0.0334	0.0324
15	0.0183	0.0194
16	0.0089	0.0109
17	0.0039	0.0058
18	0.0015	0.0029
19	0.0005	0.0014
20	0.0002	0.0006
21	0.0000	0.0003
22	0.0000	0.0001

Figure 7–2 shows, for these parameters, the normal curve and the rectangles whose heights represent the Poisson probabilities. Table 7–3 compares the normal approximation with the exact Poisson probabilities. In general, we approximate the Poisson probability,

$$P(X = x) = e^{-9}9^x/x!,$$

by the normal probability contained between $x - \frac{1}{2}$ and $x + \frac{1}{2}$. We enter the standard normal table with the standard normal deviates

$$\frac{x - 0.5 - \mu}{\sigma} \quad \text{and} \quad \frac{x + 0.5 - \mu}{\sigma},$$

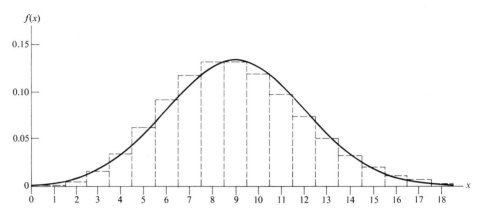

Fig. 7–2 Comparison of histogram for Poisson with $m = 9$ with normal probability density function, mean $\mu = 9$, standard deviation $\sigma = 3$.

where, in this example, $\mu = \sigma^2 = 9$. The 0.5's here are the usual continuity corrections that we use when approximating a discrete distribution by a continuous distribution. (See Section 2–2.) The following example illustrates the procedure.

Example 1. If X is a Poisson random variable with mean 9, use the normal approximation to find $P(X = 12)$.

Solution. Using $x = 12$, $\mu = 9$, and $\sigma = 3$, we compute the standard normal deviates as follows:

$$\frac{x - 0.5 - \mu}{\sigma} = \frac{12 - 0.5 - 9}{3} = 0.83333,$$

and

$$\frac{x + 0.5 - \mu}{\sigma} = \frac{12 + 0.5 - 9}{3} = 1.16667.$$

From the normal tables, we get

$$P(X = 12) = P(0 < Z < 1.16667) - P(0 < Z < 0.83333)$$

$$= 0.3783 - 0.2976 = 0.0807.$$

From Poisson tables, we find that the true probability is 0.0728. The larger the value of m, the closer the normal approximation fits.

EXERCISES FOR SECTION 7–5

1. If X is a Poisson random variable with $m = 16$, use the normal approximation to find (a) $P(X = 20)$, (b) $P(X \geq 18)$. Get a rough check on your results by extrapolating from the Poisson table.

2. If X is a Poisson random variable with mean 100, find (a) $P(X = 50)$, (b) $P(X \geq 50)$.

3. If newly manufactured automobiles have on the average 100 defects, approximately what percentage of the automobiles have fewer than 80 defects?

4. In Exercise 20 of Section 7-2, let X be the number of motorists arriving in a randomly chosen hour. Find $P(X < 30)$.

5. A switchboard in a large office building handles on the average 25 calls per minute, and its maximum capability is 35 calls per minute. Assuming that the number of calls per minute, X, is a Poisson random variable, find the chance that the switchboard, at a randomly chosen minute, will be overloaded.

6. If the number of clicks on a Geiger counter is, on the average, 6 per second, find the probability of getting at most 330 clicks in one minute. Assume a Poisson distribution, and use the normal approximation.

7-6 ESTIMATION

We already know, since $E(X) = m$, that the average of several Poisson observations drawn from the same distribution is an unbiased estimate of m. Let us suppose there are N Poisson measurements. Then the variance of the sample mean $\overline{X}$ is m/N, and $\sigma_{\overline{X}} = \sqrt{m/N}$.

From the approximate normality of $\overline{X}$, if we had to get approximate 95% confidence limits for m, we would like to use

$$\overline{X} - 1.96\sigma_{\overline{X}} \leq m \leq \overline{X} + 1.96\sigma_{\overline{X}}.$$

But $\sigma_{\overline{X}}$ is a function of the unknown m. We can get a further approximation by replacing the m in $\sigma_{\overline{X}}$ by $\overline{X}$ and get rough confidence limits

$$\overline{X} - 1.96\sqrt{\overline{X}/N} \leq m \leq \overline{X} + 1.96\sqrt{\overline{X}/N}.$$

For single values of X, tables and charts of confidence limits for m have been published.

By calculus methods, we can also find the maximum likelihood estimate of m.

Theorem. $\hat{m} = \overline{X}$. Maximum likelihood estimator. Let $X_1, X_2, \ldots, X_N$ be independent counts from a Poisson distribution with mean m. Then the maximum likelihood estimator of m is the sample mean:

$$\hat{m} = \overline{X}.$$

★ *Proof.* The likelihood function is for the specific outcomes $x_1, x_2, \ldots, x_N$:

$$L = \left(\frac{e^{-m}m^{x_1}}{x_1!}\right)\left(\frac{e^{-m}m^{x_2}}{x_2!}\right)\cdots\left(\frac{e^{-m}m^{x_N}}{x_N!}\right)$$

$$= \frac{e^{-mN}m^{x_1+x_2+\cdots+x_N}}{x_1! \, x_2! \cdots x_N!}.$$

We want the value of m that maximizes L. To find the maximum of a function we can also find the value of the variable that maximizes a monotonic function of it. For this purpose an attractive function of L is its natural logarithm:

$$\log L = -mN + (\Sigma x_i) \log m - \log x_1! \, x_2! \ldots x_N!.$$

Taking the derivative with respect to m gives

$$\frac{d \log L}{dm} = -N + \frac{\Sigma x_i}{m}.$$

Setting this equal to zero and solving for m gives the desired estimate,

$$\hat{m} = \Sigma x_i / N = \bar{x}.$$

Here we have expressed the result in terms of the specific outcomes $x_1, x_2, \ldots, x_N$. But we can also view $\hat{m}$ as a random variable,

$$\hat{m} = \bar{X}.$$

CHAPTER 8

The Exact Chi-Square Distribution

8–1 SUMS OF SQUARES OF STANDARD NORMAL DEVIATES

In our work thus far, we have dealt either with the exact discrete distribution of our statistics or with their normal approximations. Though frequently adequate, the normal distribution has not cornered the market for approximations. An extremely important distribution, called the chi-square distribution, or χ^2-distribution, will shortly be needed as an approximation in a variety of problems. (The lower-case Greek letter χ is pronounced "ky", rhyming with "sky".) Like the normal distribution, the χ^2-distribution is important in its own right, as well as for its uses as an approximation. We shall first develop the exact theory of the χ^2-distribution and then use this distribution as an approximation in the next few chapters.

Let us recall the following fact (see Appendix III–4):

If we have n independent measurements drawn from a normal distribution with mean μ and standard deviation σ, and if we adjust the *sample mean* $\overline{X}$ by subtracting the *population mean* μ and dividing by $\sigma_{\overline{X}}$ ($= \sigma/\sqrt{n}$), then we get a random variable

$$Z = \frac{\overline{X} - \mu}{\sigma_{\overline{X}}} = \frac{\overline{X} - \mu}{\sigma/\sqrt{n}},$$ (1)

which has exactly the *standard* normal distribution.

Thus the normal distribution describes exactly, among other things, *the behavior of sample means of independent normally distributed measurements.*

The χ^2-distribution, used not as an approximation but exactly, also deals with functions of normally distributed random variables. This distribution describes especially *the behavior of sums of squares of deviations from population means or from sample means.* To illustrate: From a normal distribution with mean μ

and standard deviation σ, let us draw a sample of n independent measurements $(X_1, X_2, \ldots, X_n)$. Then this sample gives rise to the random variable

$$\chi^2 = \left(\frac{X_1 - \mu}{\sigma}\right)^2 + \left(\frac{X_2 - \mu}{\sigma}\right)^2 + \cdots + \left(\frac{X_n - \mu}{\sigma}\right)^2, \qquad (2)$$

which has a chi-square distribution. We say "a" rather than "the" chi-square distribution because there are many of them. There is a different chi-square distribution for each value of n.

When the statistic χ^2 is formed as in (2), we say that it has "n degrees of freedom," where the n refers to the number of independent standard normal deviates that were squared to make it up. If we denote the standard normal deviates by

$$Z_i = \frac{X_i - \mu}{\sigma}, \qquad \text{where} \quad i = 1, 2, \ldots, n,$$

then we can write

$$\chi^2 = Z_1^2 + Z_2^2 + \cdots + Z_n^2.$$

The foregoing discussion leads to an important way of defining chi-square distributions.

Definition. *The χ^2-distribution with n degrees of freedom.* If $Z_1, Z_2, \ldots, Z_n$ are independent standard normal deviates, then the distribution of the random variable χ_n^2, where

$$\chi_n^2 = Z_1^2 + Z_2^2 + \cdots + Z_n^2, \qquad (3)$$

is called *the χ^2-distribution with n degrees of freedom.*

Hence the distribution of the chi-square random variable χ_n^2 is that of *the sum of the squares of n independent standard normal deviates.*

The definition of χ_n^2 implies a useful property of chi-square distributions:

Theorem 1. *Sums of independent chi-square variables—the reproductive law.* If χ_u^2 and χ_v^2 are two independent chi-square random variables, then their sum $\chi_u^2 + \chi_v^2$ has a chi-square distribution with $u + v$ degrees of freedom.

Proof. Consider $u + v$ independent standard normal random variables, Z_i, $i = 1, 2, \ldots, u + v$. The sum $Z_1^2 + Z_2^2 + \cdots + Z_u^2$ forms a χ_u^2 variable, $Z_{u+1}^2 + \cdots + Z_{u+v}^2$ forms a χ_v^2 variable, and their sum forms, by definition, a χ_{u+v}^2 variable. Note that χ_u^2 and χ_v^2 are independent because they are formed from distinct sets of independent random variables. (See *PWSA*, Appendix 2, pp.

465–467.) The distribution of the sum of two random variables is unique; since we have constructed the distribution of the sum in one way, as above, that distribution must be the unique correct one.

Example 1. *Horizontal dispersion.* After carefully adjusting the sights of his rifle, an experienced marksman shoots at a standard target consisting of a bull's-eye and concentric rings. His horizontal errors are independently normally distributed about the center, and his standard deviation at the given distance is 2 ring-widths. The departures of his first five shots, in ring-widths, are -3.5, -1.4, -0.3, 1.1, 2.4, where minus indicates left of center and plus indicates right of center. Is this performance especially good or bad?

Solution. One method of appraising the marksman's performance is to examine the behavior of the sum of the squares of the deviations from the mean. Since $\mu = 0$ and $\sigma = 2$, we compute the statistic

$$\chi_5^2 = \Sigma \left(\frac{x_i - \mu}{\sigma} \right)^2 = \frac{1}{\sigma^2} \Sigma (x_i - \mu)^2$$

$$= \tfrac{1}{4}[(-3.5)^2 + (-1.4)^2 + \cdots + (2.4)^2] = 5.32.$$

Later we shall see that a chi-square based on 5 deviations has expected value 5 and standard deviation $\sqrt{2(5)} \approx 3.2$. Consequently, the observed value has fallen quite close to the mean value, and we find nothing unusual about the set of departures.

Example 2. The marksman of Example 1 makes independent, normal errors both horizontally and vertically. His standard deviation in the vertical dimension is 3 ring-widths. The coordinates of his five shots are $(-3.5, +4.2)$, $(-1.4, -0.9)$, $(-0.3, -3.6)$, $(1.1, 2.8)$, $(2.4, -1.8)$. If you consider errors in both dimensions, are his shots unusually close or unusually spread out?

Solution. In measuring horizontal and vertical distances of a shot from the center of the target, our unit is the ring-width as used for the standard deviation. This unit makes it easy to combine information from both coordinates.

Once again we use the sum of the squares of deviations from the mean to get a chi-square random variable. Applying the foregoing theorem, we get

$$\chi_{10}^2 = \frac{(-3.5)^2}{4} + \frac{(4.2)^2}{9} + \cdots + \frac{(2.4)^2}{4} + \frac{(-1.8)^2}{9}$$

$$= \frac{21.27}{4} + \frac{42.49}{9} = 5.32 + 4.72 = 10.04.$$

We shall soon show that the expected value of χ_{10}^2 is 10; and so the results are surprisingly close to the theoretical mean.

EXERCISES FOR SECTION 8–1

1. In Example 1, if the marksman had had a standard deviation of 1 instead of 2, but the observations had been the same, how large would χ_3^2 have been? Would his results have seemed far from those expected?

2. In Example 1, what was μ? In Example 2, there must have been two values of μ. What were they?

The following sets of independent measurements are drawn from a normal distribution with $\mu = 1$ and $\sigma = 2$. For each set, find χ_n^2.

3. $-3, 4, 7.$

4. $-6, -2, 1, 9.$

5. $-1.2, -0.3, 1.5, 3.0, 4.1.$

6. Rework Example 1, given that $\mu = 0$ and $\sigma = 3$.

7. Rework Example 2, given that $\mu = 2$ and $\sigma = 4$ for vertical errors.

8. A manufacturer of light bulbs maintains that his bulbs have an average life of 200 hours with standard deviation 25 hours. A sample of 3 bulbs is selected at random for testing. One bulb lasts 180 hours, 1 lasts 210 hours, and 1 lasts 230 hours. Is this result unusual? [Use the facts that the mean of χ_3^2 is 3 and the standard deviation $\sqrt{6}$. Assume that the bulb-lives are approximately normally distributed.]

9. In Exercise 8, if the bulb-lives had been 150 hours, 250 hours, and 300 hours, would you say the result was reasonable?

10. How would the square of the Z in equation (1) be distributed?

8–2 CHI-SQUARE DISTRIBUTION WITH ONE DEGREE OF FREEDOM

The chi-square distribution with one degree of freedom, according to equation (3) of the preceding section, is simply the distribution of the square of one standard normal random variable Z:

$$\chi^2 = Z^2. \tag{1}$$

Since values of Z run from $-\infty$ to $+\infty$, values of χ^2 run from 0 to $+\infty$.

To visualize the probability density graph of χ^2, imagine that the probability density graph of Z is folded about the line $z = 0$ so that the left half falls on the right half (Fig. 8–1). Then the probability originally near $-z$ $(z > 0)$ is combined with that near $+z$, because the sign is lost in the squaring as we go from Z to Z^2. Furthermore, since numbers less than 1 produce smaller numbers when squared, after the foldover the probability in the normal distribution that lay originally between -1 and 1

1. now lies between 0 and 1, and
2. becomes much denser on the intervals near 0.

For example, the probability between -0.1 and $+0.1$ in the normal distribution fits in the narrow interval between 0 and 0.01 in the χ^2-distribution. Consequently, the new ordinates must average $2(0.1)/0.01 = 20$ times as high as those in the normal distribution.

Numbers greater than 1 produce larger numbers when squared. Thus, in the χ^2-distribution, the probability density decreases beyond 1, as compared with that of the normal. For example, consider two numbers z_1 and z_2, where $z_1 > z_2 > 1$. Then

$$z_1 + z_2 > 2,$$

and since $z_1 - z_2 > 0$, we can multiply both sides by $z_1 - z_2$ to get

$$(z_1 - z_2)(z_1 + z_2) > 2(z_1 - z_2).$$

Thus

$$z_1^2 - z_2^2 > 2(z_1 - z_2).$$

This means that the probability between z_2 and z_1 and between $-z_2$ and $-z_1$, after the foldover, becomes less dense because it is spread over the new interval from z_1^2 to z_2^2, which is longer than the total length of the two old intervals, $2(z_1 - z_2)$.

The details of these changes in density are pictured in Fig. 8–1. The diagram on the left shows the distribution of Z, a standard normal random variable, and the diagram on the right shows the distribution of Z^2, or χ_1^2.

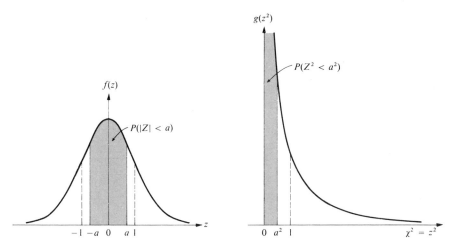

Fig. 8–1 The standard normal probability density function, f, and the density of Z^2, g.

The probabilities represented by the shaded areas are equal because when $a > 0$,

$$|Z| < a \qquad \text{if and only if} \qquad Z^2 < a^2.$$

The probability distribution of χ_1^2 is usually tabulated in one of two ways:

1. For a given value $a^2 > 0$, the cumulative probability is shown: We tabulate a^2 against $P(\chi_1^2 \leq a^2)$.

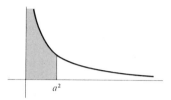

2. Alternatively, the probability in the tail of the distribution to the right of a^2 is given: We tabulate a^2 against $P(\chi_1^2 > a^2)$.

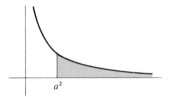

Either tabulation can readily be converted into the other, since

$$P(\chi_1^2 > a^2) = 1 - P(\chi_1^2 \leq a^2).$$

For the purposes of this book, we shall use the second method of tabulation, which is illustrated for χ_1^2 in Table 8–1.

Table 8–1
Probability distribution for $\chi_1^2 = Z^2$

a^2:	0.00016	0.0039	0.016	0.064	0.45	1.07	1.64	2.71	3.84	6.63
$P(Z^2 > a^2)$:	0.99	0.95	0.90	0.80	0.50	0.30	0.20	0.10	0.05	0.01

Since

$$P(Z^2 > a^2) = P(Z < -a \quad \text{or} \quad Z > a) = P(|Z| > a), \qquad a > 0,$$

and since Z is a standard normal deviate, the probabilities in Table 8–1 correspond to those in the two tails of a standard normal distribution. The following table, obtained from the normal table A–1 at the back of the book, checks this fact for several values of a.

Two-tailed normal table

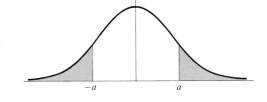

a^2:	0.0039	0.016	1.64	2.71	3.84
a:	0.062	0.126	1.28	1.65	1.96
$P(|Z| > a)$:	0.95	0.90	0.20	0.10	0.05

Example 1. A machine in good adjustment produces axles with a mean diameter of 1.340 inches (the specification) and standard deviation 0.003 inches. What is the probability that this machine produces a sample of 4 pieces whose average departs from the specification by more than 0.002 inches?

Solution. We assume 4 independent measurements drawn from an approximately normal distribution having $\mu = 1.340$ and $\sigma = 0.003$. Since we are dealing with the deviation of a sample mean from the population mean, we may use the χ^2-distribution with one degree of freedom. We have

$$\chi_1^2 = Z^2 = \left(\frac{\overline{X} - \mu}{\sigma_{\overline{X}}}\right)^2,$$

and evaluating this quantity gives

$$z^2 = \left(\frac{\overline{x} - \mu}{\sigma/\sqrt{n}}\right)^2 = \left(\frac{0.002}{0.003/2}\right)^2 = \left(\frac{4}{3}\right)^2 = 1.78.$$

Table 8–1 shows that

$$0.10 < P(Z^2 > 1.78) < 0.20.$$

Interpolation gives

$$P(Z^2 > 1.78) = 0.187,$$

which means that we may expect a deviation more extreme than 0.002 about 19% of the time.

Alternative solution. Note that Z is approximately normally distributed. (We assume that the average of 4 measurements randomly drawn from an approximately normal distribution is itself approximately normally distributed.) If we enter the two-tailed normal table at the back of the book (Table A–1) with $z = \frac{4}{3} \approx 1.33$, we find, with considerable accuracy, that

$$P(|Z| > 1.33) = 0.1836.$$

The difference between the results in the two solutions is entirely a matter of interpolation. If we had interpolated in the χ^2 table with more care or if we had a finer table, the answers would have been identical.

EXERCISES FOR SECTION 8–2

1. Use Table 8–1 to evaluate the following.
   a) $P(\chi_1^2 > 0.00016)$
   b) $P(Z^2 > 1.07)$
   c) $P(Z^2 \le 1.07)$
   d) $P(Z^2 > 2)$
   e) $P(Z^2 \le 2)$

2. Use the normal table to evaluate $P(Z^2 > 3)$ and $P(Z^2 > 0.25)$.

3. Use the normal table to check Table 8–1 for probability level 0.01.

4. Use the normal table to evaluate $P(\chi_1^2 \le 0.064)$ and $(P(\chi_1^2 \le 1.07)$.

5. In Example 1, use Table 8–1 to find the probability that the average diameter of a sample of 4 pieces departs from the specifications by more than 0.003 inches.

6. In Example 1, find the probability that the average diameter of a sample of 25 departs from the specification by more than 0.002 inches.

7. When a single die is tossed, the mean number of dots on the top face is 3.5 and the variance is 35/12. If 100 dice are tossed, find the probability that the sample average departs from the population mean by more than 0.2.

8. Scores on a CEEB test have mean 500 and standard deviation 100. What is the probability that a random sample of 25 test scores has an average that differs from 500 by more than 50.

8–3   TABLES FOR χ^2-DISTRIBUTIONS

In Section 8–2, we discussed two methods of tabulating chi-square variables and exhibited a table for the χ^2-distribution with one degree of freedom. Table A–4 at the back of the book gives the probability distributions for $\chi_1^2, \chi_2^2, \ldots, \chi_{30}^2$, that is, for χ^2-distributions having from 1 to 30 degrees of freedom. In addition, the distributions for selected numbers of degrees of freedom up to 1000 are shown. That table shows $P(\chi_n^2 > a)$.

The body of the table gives values of a, the side lists values of n, and the top gives corresponding probabilities. Example 1 illustrates the use of Table A–4.

Example 1. Use Table A–4 to evaluate the following.

a) $P(\chi^2_{10} > 4.87)$ b) $P(\chi^2_{10} \leq 4.87)$
c) $P(\chi^2_{16} > 25)$ d) $P(10.85 < \chi^2_{20} < 31.41)$

Solution
a) Table A–4 gives $P(\chi^2_{10} > 4.87) = 0.90$.
b) $P(\chi^2_{10} \leq 4.87) = 1 - P(\chi^2_{10} > 4.87) = 0.10$.
c) The table shows that

$$0.05 < P(\chi^2_{16} > 25) < 0.10.$$

By interpolation, we get

$$P(\chi^2_{16} > 25) = 0.074.$$

d) The table gives

$$P(\chi^2_{20} > 10.85) = 0.95 \quad \text{and} \quad P(\chi^2_{20} > 31.41) = 0.05.$$

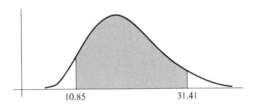

Hence

$$P(10.85 < \chi^2_{20} < 31.41) = 0.90.$$

This means that the value of χ^2_{20} will lie between 10.9 and 31.4 about 90% of the time.

EXERCISES FOR SECTION 8–3

Evaluate the following.
1. $P(\chi^2_6 > 3.07)$
2. $P(\chi^2_{12} > 10)$
3. $P(\chi^2_7 > 2)$
4. What percent of the values of χ^2_{10} lie between 3.06 and 21.16?
5. What percent of the values of χ^2_{25} lie between 15 and 35?

8–4 SHAPES OF χ^2-DISTRIBUTIONS; THE NORMAL APPROXIMATION

As the number of degrees of freedom increases, we get a sequence of χ^2-distributions. Imagine the graphs of the χ^2-distributions for $n = 1, 2, 3, \ldots$ as successive frames of a moving picture film. When we run the film, the graphs of the distributions will appear to move to the right. Figure 8–2 shows the curves representing

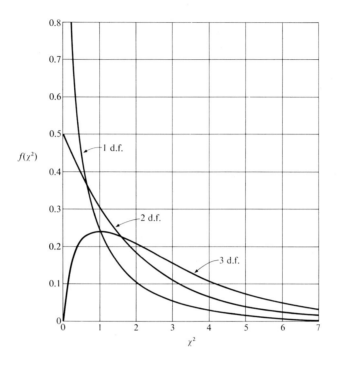

Fig. 8–2 Shapes of the densities of the chi-square distributions with $n = 1, 2$, and 3 degrees of freedom. For $n = 1$, the density is unbounded at the origin; for $n = 2$ the function is monotonic, but with finite value at $\chi^2 = 0$.

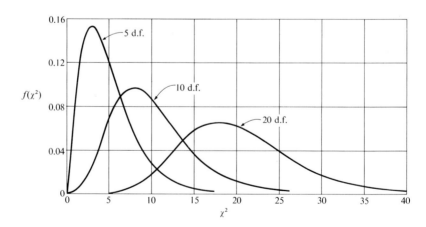

Fig. 8–3 Densities of the χ^2-distributions with 5, 10, and 20 degrees of freedom. Note that as the number of degrees of freedom increases, the shape of the normal distribution begins to emerge.

the densities for $n = 1, 2$, and 3 degrees of freedom, and Fig. 8–3 shows those for $n = 5, 10$, and 20.

From an inspection of these curves, it appears that, as n increases, the density graph of the chi-square distribution becomes more bell-shaped, like that of the normal. Indeed, this result is to be expected because, for a large value of n, a χ^2 random variable is the sum of many identically distributed independent random variables. Such a sum approaches normality by virtue of the Central Limit Theorem. (See Appendix III–4.)

In order to make use of the normal approximation for a χ^2-distribution with large n, we need to know the mean and the variance of the χ^2-distribution. We shall therefore proceed to find them.

Case $n = 1$. Since $\chi_1^2 = Z^2$, the mean value is

$$E(\chi_1^2) = E(Z^2) = \sigma_Z^2 + \mu_Z^2 = 1,$$

because $\mu_Z = 0$ and $\sigma_Z = 1$, since Z is a standard normal variable.

To get the variance of Z^2, we need to know the value of $E(Z^4)$. We shall not attempt to deduce this value, but we shall accept from advanced calculus the fact that for the standard normal $E(Z^4) = 3$. [We do not now need expected values of higher powers of Z, but it may be of interest to know that $E(Z^6) = 5 \times 3 = 15$, $E(Z^8) = 7 \times 5 \times 3 = 105$, etc., and $E(Z^{2k+1}) = 0$, where k is a positive integer.]

For one degree of freedom, we have

$$\text{Var } \chi_1^2 = \text{Var } Z^2 = E(Z^4) - [E(Z^2)]^2 = 3 - 1 = 2.$$

Case $n = 2$. Since $\chi_2^2 = Z_1^2 + Z_2^2$, the mean value is

$$E(\chi_2^2) = E(Z_1^2 + Z_2^2) = 1 + 1 = 2.$$

The variance is

$$\text{Var } \chi_2^2 = \text{Var } (Z_1^2 + Z_2^2).$$

Since the variance of a sum of independent random variables is the sum of their variances, we have

$$\text{Var } \chi_2^2 = \text{Var } Z_1^2 + \text{Var } Z_2^2 = 4.$$

Theorem. *The mean and the variance of χ_n^2.* If χ_n^2 denotes a chi-square random variable with n degrees of freedom, then

$$E(\chi_n^2) = n, \tag{1}$$

and

$$\text{Var } (\chi_n^2) = 2n. \tag{2}$$

Proof. By definition, we can represent X_n^2 as

$$X_n^2 = Z_1^2 + Z_2^2 + \cdots + Z_n^2,$$

where the Z_i's are independent standard normal deviates. Hence the mean is

$$E(X_n^2) = E(Z_1^2 + Z_2^2 + \cdots + Z_n^2)$$

$$= 1 + 1 + \cdots + 1 = n.$$

The variance of X_n^2 is

$$\text{Var}(X_n^2) = \text{Var}(Z_1^2 + Z_2^2 + \cdots + Z_n^2)$$

$$= \text{Var } Z_1^2 + \text{Var } Z_2^2 + \cdots + \text{Var } Z_n^2$$

$$= 2n.$$

We are now able to use the normal approximation for X_n^2 when n is large.

Example 1. Evaluate $P(X_{25}^2 > 32)$, using (a) Table A–4, (b) the normal approximation.

Solution
a) Table A–4 shows that

$$0.10 < P(X_{25}^2 > 32) < 0.20,$$

and interpolation gives $P(X_{25}^2 > 32) = 0.164$.
b) Since $n = 25$, the chi-square distribution has mean 25 and variance 50. The normal deviate corresponding to 32 is

$$z = \frac{32 - 25}{\sqrt{50}} = 0.990.$$

Since we are dealing with continuous and not discrete measurements, we need no continuity correction. Therefore, from Table A–1 we get

$$P(X_{25}^2 > 32) \approx P(Z > 0.990) = 0.161,$$

very close to the result given by the chi-square table.

Table A–4, discussed in Section 8–3, may be useful for small numbers of degrees of freedom. Recall that the X^2-distribution was defined as the sum of the squares of n independent standard normal random variables. (See Section 8–1.) Since in practical work—as opposed to theory—this normality assumption is realized only approximately, the probability levels in Table A–4 are then also approximate. When n is small, the normality assumption is an important consideration; but as n gets larger, the Central Limit Theorem dominates the situation, and the normality assumption is no longer critical. For many purposes, it is sufficient to know the mean and the standard deviation of X_n^2.

Example 2. (Refer to Example 1 of Section 8–1.) Before adjusting his sights, a marksman scores 3 shots with horizontal deviations 2.4, 8.0, and 6.2 ring-widths. If the marksman's standard deviation for horizontal deviations is 2, do you think that he should reset his sights?

Solution. To judge the importance of these deviations, we find

$$\chi_3^2 = \left(\frac{2.4 - 0}{2}\right)^2 + \left(\frac{8.0 - 0}{2}\right)^2 + \left(\frac{6.2 - 0}{2}\right)^2 = 27.05.$$

Since χ_3^2 has mean 3 and standard deviation $\sqrt{6} = 2.4$, the value 27.05 is about 10 standard deviations above the mean. The probability of a larger χ^2 value with a properly adjusted rifle in the hands of a marksman in good shape is very small indeed. It appears that the sights need adjustment (or perhaps that the marksman has different glasses) and that the bullets are hitting to the right of the bull's-eye.

Example 3. *Replacement parts.* The useful life of a component like a battery in a space vehicle may be short compared with the length of the mission, and so replacements are required. Let L be the useful life in hours of a battery, where $L/5$ has the chi-square distribution with 2 degrees of freedom. Is it likely that the original battery plus three replacements will be adequate for a 50-hour mission?

Solution. For a single component, the mean of $L/5$ is 2, and so $\mu_L = 10$; $\sigma_{L/5}^2 = 2(2) = 4$, and so $(1/5^2)\sigma_L^2 = 4$, and $\sigma_L^2 = 100$. Let L_1, L_2, L_3, L_4 be the lengths of life of four batteries, and let the total life $T = L_1 + L_2 + L_3 + L_4$. Then $\mu_T = 4\mu_L = 40$, and $\sigma_T^2 = 4(100)$, $\sigma_T = 20$. We note that the mean is 10 hours short of the mission requirement—very bad news. Computing the standard deviate, we get

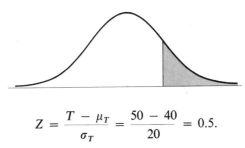

$$Z = \frac{T - \mu_T}{\sigma_T} = \frac{50 - 40}{20} = 0.5.$$

The probability of completing the mission with a live battery is

$$P(T \geq 50) = P(Z \geq 0.5) \approx 0.3.$$

Finally, it is a rare day when anyone except possibly a physicist has as many as 100 degrees of freedom, genuinely independent. With very large numbers of degrees of freedom, some method of analysis other than χ^2 should usually be adopted. Nevertheless, in theoretical work, it is sometimes useful to have tabulations for high values of n. For this reason, a few high values are included in Table A–4.

EXERCISES FOR SECTION 8–4

Using Table A–4, evaluate each of the following.

1. $P(\chi_8^2 > 3.49)$

2. $P(\chi_{30}^2 > 38)$

3 and 4. Evaluate the probabilities in Exercises 1 and 2 using the normal approximation.

5. Scores on a CEEB test have mean 500 and standard deviation 100. A class of 6 students get the following scores: 490, 550, 580, 640, 680, 712. Is this group above average at the 5% level of significance. [*Note:* You must assess direction *and* magnitude.]

6. A random sample of 20 independent measurements $(x_1, x_2, \ldots, x_{20})$ is drawn from a standard normal population. In what interval must $\sum_{i=1}^{20} x_i^2$ lie in order to be neither significantly large nor significantly small at the 5% level.

7. If components have lives L, where $L/10$ measured in hours is distributed like χ_5^2, find the mean and variance of the total life of three components.

8–5 LOSING DEGREES OF FREEDOM

Up to now we have constructed χ^2 variables using squares of standard normal random variables. But in studying variation *we ordinarily use sample means $\overline{X}$ rather than population means μ* for measuring deviations because we often do not know μ. For example, the sample variance is defined as

$$S^2 = \frac{(X_1 - \overline{X})^2 + (X_2 - \overline{X})^2 + \cdots + (X_n - \overline{X})^2}{\text{denominator}},$$

where the denominator may be n or $n - 1$ or something else, depending on our usage. It may seem natural to divide by n, and some authors follow this plan. However, the divisor $n - 1$ makes our tables more convenient to use and gives us a sample variance S^2 that is an unbiased estimate of σ^2. In this book, we shall use the divisor $n - 1$ and define the sample variance as

$$S^2 = \sum(X_i - \overline{X})^2/(n - 1). \tag{1}$$

Let us focus attention on the numerator in the special case where $n = 2$. Then $(X_1 + X_2)/2 = \overline{X}$, and the numerator can be written

$$\left[X_1 - \frac{X_1 + X_2}{2}\right]^2 + \left[X_2 - \frac{X_1 + X_2}{2}\right]^2 = \tfrac{1}{4}(X_1 - X_2)^2 + \tfrac{1}{4}(X_2 - X_1)^2$$

$$= \tfrac{1}{2}(X_1 - X_2)^2. \tag{2}$$

If the X's come independently from a normal distribution with mean μ and variance σ^2, then

$$\mu_{X_1 - X_2} = \mu_{X_1} - \mu_{X_2} = \mu - \mu = 0$$

and

$$\text{Var}\,(X_1 - X_2) = \sigma_{X_1}^2 + \sigma_{X_2}^2 = \sigma^2 + \sigma^2 = 2\sigma^2.$$

Recall that the difference of two independent normal random variables is normally distributed. (See Appendix V–4.)

Indeed, if we divide the left side of (2) by σ^2, we have the square of a standard normal random variable, thus:

$$\frac{(X_1 - \bar{X})^2 + (X_2 - \bar{X})^2}{\sigma^2} = \frac{(X_1 - X_2)^2}{2\sigma^2} = \frac{(X_1 - X_2)^2}{\sigma_{X_1 - X_2}^2} = \chi_1^2, \qquad (3)$$

which is a chi-square variable with one degree of freedom.

Recall that

$$\frac{(X_1 - \mu)^2 + (X_2 - \mu)^2}{\sigma^2} = \chi_2^2. \qquad (4)$$

Thus *replacing μ by $\bar{X}$ leaves us with a chi-square distribution, but this distribution has 1 instead of 2 degrees of freedom.*

In expression (4), X_1 and X_2 are free to take values at will except for the probabilistic constraints of the normal distribution. And so they vary in 2 dimensions, the X_1- and the X_2-dimension, or in the $X_1 X_2$-plane. When a point can take any value in a plane, we sometimes say it has two degrees of freedom. However, in expression (3) we have imposed a linear constraint on X_1 and X_2. They add up to $2\bar{X}$: $X_1 + X_2 = 2\bar{X}$. Once $\bar{X}$ is fixed, the point (X_1, X_2) is constrained to a *line* in the $X_1 X_2$-plane. A variable point that may move only on a line is sometimes said to have one degree of freedom, even if the line is in a space of 2 or more dimensions.

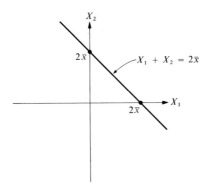

In other words, the number of degrees of freedom can be thought of as the dimension of the space in which the sample point is free to move. In expression (4), the dimension is 2; in expression (3), the dimension is only 1. When we have 3 measurements, the sample point (X_1, X_2, X_3) takes values in a 3-dimensional space if unrestricted; but if we impose a linear constraint, such as

$$X_1 + X_2 + X_3 = 3\bar{x},$$

we reduce the number of degrees of freedom by 1, for then the X_i's are constrained to move in a plane.

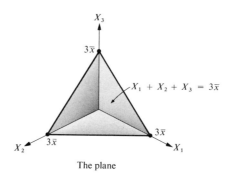

The plane

Each additional distinct linear constraint can potentially reduce the number of degrees of freedom by an additional 1. We shall speak more of this later.

Although the discussion above has dealt with only linear constraints, non-linear constraints frequently arise; for example, when we fit a variance, we lose one degree of freedom.

We now generalize, without proof, the important fact that was illustrated in the foregoing discussion: If, in defining a χ^2 random variable, we use the sample mean $\bar{X}$ instead of the population mean μ, we reduce the number of degrees of freedom by 1. Thus

$$\frac{(X_1 - \bar{X})^2 + (X_2 - \bar{X})^2 + \cdots + (X_n - \bar{X})^2}{\sigma^2} = \frac{(n-1)S^2}{\sigma^2}$$

is a χ^2 random variable with $n - 1$ degrees of freedom.

To sum up:

If S^2 is the variance of a random sample of n measurements taken from a normal population having variance σ^2, then

$$\chi^2_{n-1} = \frac{(n-1)S^2}{\sigma^2} \tag{5}$$

is a χ^2 random variable with $n - 1$ degrees of freedom.

If we are sure of the normality, we can use formula (5) along with a χ^2 table to set confidence limits on σ^2.

Example 1. A manufacturer of light bulbs claims that the bulb-lives are normally distributed with a mean of 200 hours and a variance of 625 hours^2. A random sample of 10 bulbs turns out to have a mean bulb-life of $\bar{x} = 190$ hours with variance $s^2 = 400$ hours^2. Should this result for the sample variance cause the manufacturer to alter his claim?

Solution. Since the sample has 10 measurements, we are dealing with a χ_9^2-distribution. Let us look at the 90% confidence interval for σ^2. Table A–4 tells us that

$$P(\chi_9^2 > 3.33) = 0.95 \quad \text{and} \quad P(\chi_9^2 > 16.92) = 0.05.$$

Consequently,

$$P(3.33 < \chi_9^2 < 16.92) = 0.90,$$

or more specifically,

$$P\left(3.33 < \frac{(n-1)S^2}{\sigma^2} < 16.92\right) = 0.90.$$

After the experiment is performed, we replace S^2 by the actual numerical result s^2, and we say that we have 90% confidence that

$$3.33 < \frac{(n-1)s^2}{\sigma^2} < 16.92.$$

(See Appendix VII–1 or Section 2–2 for a discussion of "confidence" ideas.) Since $(n-1)s^2 = 9(400) = 3600$, we can rearrange the last inequality and get

$$1081 > \sigma^2 > 213.$$

For this problem, the 1081 and 213 are called the upper and lower 90% confidence limits, and the interval 213 to 1081 is called the 90% confidence interval for σ^2. Since the variance claimed, namely 625, is within the 90% confidence interval, the claim is not upset by these results. The sample mean, incidentally, is only about one or two standard deviations of the mean away from the announced value, 200.

Although Example 1 exhibits an application of the χ^2-distribution, this method of setting confidence limits on σ^2 is not much used because of the strong dependence of the result on normality. In this chapter, our main purpose is to introduce the χ^2-distribution. Its more practical uses are discussed in later chapters.

Summary. *Interpretations of n, the number of "degrees of freedom" in χ^2.* The number of degrees of freedom in a χ^2-distribution is variously interpreted to be the following.

1. The number of independent standard normal deviates the sum of whose squares equals the χ^2 random variable.

2. The number ordinarily used to describe the χ^2-distribution and to pick it from a table; it is essentially *the* parameter of the χ^2-family of distributions.
3. The number of dimensions of the space in which the sample point

$$(X_1, X_2, \ldots, X_k)$$

can move. Note that k may exceed n.
4. The mean of the χ^2-distribution; and twice this number, $2n$, is the variance of the χ^2-distribution.
5. The divisor required to make a sum of squares of deviations from the mean (μ or $\overline{X}$, as appropriate) become an unbiased estimate of σ^2. [But unlike interpretations 1, 2, and 4, this interpretation does not depend on the underlying normality of the measurements.]

Examples: $\sum_{i=1}^{n} (X_i - \mu)^2/n$ is an unbiased estimate of σ^2.

$\sum_{i=1}^{n} (X_i - \overline{X})^2/(n - 1)$ is an unbiased estimate of σ^2.

Example 2. Refer to Example 2 of Section 8-4. Solve that example using deviations from $\overline{X}$ rather than from the center of the target.

Solution. $\overline{X} = (2.4 + 8.0 + 6.2)/3 = 5.53$. Then

$$\chi_2^2 = \frac{(2.4 - 5.53)^2 + (8.0 - 5.53)^2 + (6.2 - 5.53)^2}{4} = 4.09.$$

Table A-4 with interpolation gives $P(\chi_2^2 > 4.09) = 0.137$. This suggests that the pattern of shots is displaced to the right, and that the clustering about the sample mean, although looser than expected, is not extreme enough to make us believe that σ^2 has increased.

EXERCISES FOR SECTION 8-5

1. A sample of 21 measurements drawn from a normal population has $s^2 = 10$. Set 90% confidence limits on the variance of the population, σ^2.

2. Repeat Exercise 1 for a sample of 25 with variance 20.

3. Repeat Example 1 for a random sample of 20 bulbs having a mean bulb-life of 203 hours and a variance of 25 hours^2.

4. The variance of the IQ's of the students in X-ville is 400. A teacher wishes to determine whether the IQ's in his class of 26 are more variable than those of the population as a whole. The variance of the IQ's in his class is 450; is this result significant at the 10% level?

5. A manufacturer of automobile tires claims that his tires have a mean mileage of 30,000 miles with a standard deviation of 1200 miles. A sample of 8 tires gives a standard deviation of 1500 miles. Set 90% confidence on the population standard deviation. Is the difference large enough to reject the manufacturer's claim at the 10% level?

CHAPTER 9

Applying Chi-Square to Counted Data: Basic Ideas

Target. This chapter prepares the reader to deal with the practical applications of chi-square presented in the next two chapters. It offers a low-voltage introduction to:

1. the idea of a contingency table;
2. a scheme of notation;
3. a statistic that measures goodness-of-fit;
4. the approximation of the discrete chi-square statistic by the continuous χ^2-distribution;
5. the relation of the sign test to the chi-square statistic;
6. the continuity correction for the chi-square statistic with one degree of freedom.

In short, this chapter is designed to clear the way for the more general chi-square applications in Chapters 10 and 11.

9–1 THE IDEA OF GOODNESS-OF-FIT: PARTICULAR CASE

The chi-square distribution can be used to assess agreement, or goodness-of-fit, between a body of data and a theoretical model. In discussing this application of the χ^2-distribution, we shall frequently refer to contingency tables, so we begin by describing them.

Contingency tables. A contingency table is a set of counts tabulated in some systematic way. If counts come to us in unordered categories, we make a table by assigning an order, either arbitrarily or by using some device. This procedure may lead to a one-way table, $1 \times c$ (1 row with c categories). Frequently, contingency tables are counts ordered into a rectangular array by using two sets of characteristics and requiring more elaborate tabular arrangements. If a table consists of r rows and c columns, we call it an $r \times c$ *contingency table*. We read $r \times c$ as "r by c" in this context.

We shall now illustrate the use of the chi-square distribution to assess good-ness-of-fit by discussing an example.

Example 1. *Mendelian genetics: a 1 × 3 contingency table.* In one of Mendel's genetic experiments, a simple genetic theory suggests that the outcomes of a certain cross would be three kinds of offspring, symbolized by *AA*, *Aa*, and *aa*, having probabilities $\frac{1}{4}$, $\frac{1}{2}$, and $\frac{1}{4}$, respectively. Mendel's reported counts of offspring (pea plants) were as follows:

$$35\ AA,\quad 67\ Aa,\quad 30\ aa;\qquad \text{total } 132.$$

How can we devise a test to see if the theoretical model fits the results?

Discussion. We need to find a statistic that measures the discrepancy between observed results and those expected under the proposed model. For example, the expected number of *AA* offspring is

$$132P(AA \text{ offspring}) = 132(\tfrac{1}{4}) = 33.$$

The data may be tabulated as follows:

	AA	*Aa*	*aa*	Total
Observed counts, O_i:	35	67	30	132
Expected counts, E_i:	33	66	33	132

The table and past experience suggest a number of possible measures of departures between the observed counts O_i, $i = 1, 2, 3$, and the expected counts E_i, $i = 1, 2, 3$. For example, we might use the sum of the absolute values $|O_i - E_i|$, or the sum of the squared differences $(O_i - E_i)^2$. The measure most commonly used in practice is the chi-square statistic,

$$\chi^2 = \sum_{i=1}^{3} \frac{(O_i - E_i)^2}{E_i}. \tag{1}$$

There are two reasons for this choice:

1. General theorems promise that *statistic (1) will have approximately the chi-square distribution with 2 degrees of freedom*, provided that the O_i's come from a multinomial distribution with the E_i's as expected values. We do not prove these general theorems in this book, but we shall shortly offer evidence to make this fact plausible. Since the most important property of any statistic is its distribution, it is convenient that the chi-square approximation enables us to work with statistic (1) without having to undertake much new theoretical development.

2. From a variety of theoretical points of view, the chi-square test is a "good" test, and this is the more important reason.

Let us compute statistic (1) for Mendel's data. We get

$$\chi^2 = \sum_{i=1}^{3} \frac{(O_i - E_i)^2}{E_i} = \frac{(35 - 33)^2}{33} + \frac{(67 - 66)^2}{66} + \frac{(30 - 33)^2}{33}$$

$$= \frac{27}{66} \approx 0.41.$$

Accepting the fact that, when the model is correct, our statistic has a chi-square distribution, we face two basic questions:

1. *How many degrees of freedom are there?* Given that the total must be 132, we can freely assign the numbers of two categories of offspring, say O_1 and O_2; but then the number O_3 is determined because

$$O_1 + O_2 + O_3 = 132.$$

We have a linear constraint on the number of counts: As soon as two O_i's are chosen, the third is determined. Hence, although we have three categories of counts, our statistic has approximately the chi-square distribution with only $3 - 1 = 2$ degrees of freedom. *The linear constraint costs us one degree of freedom.*

In Chapter 8 we used χ^2 to refer to a random variable distributed exactly according to a chi-square distribution; but in Chapters 9, 10, and 11 we use it for the statistic based on $\sum(O_i - E_i)^2/E_i$, which is called "chi-square". The justification for this inconsistency is that we use the exact continuous chi-square distributions as approximations for the complex and varying discrete distributions of the chi-square statistic.

The expected value of the chi-square statistic. Let us compute the expected value of the chi-square statistic for this problem. Since each O_i is binomially distributed, it has mean $E_i = np_i$ and variance $np_i(1 - p_i)$, where p_i is the probability associated with cell i. Since, by definition, the variance of O_i is $E(O_i - np_i)^2$, we have

$$E(O_i - np_i)^2 = np_i(1 - p_i),$$

and the expected value of χ^2 for our problem is

$$E(\chi^2) = \frac{np_1(1 - p_1)}{np_1} + \frac{np_2(1 - p_2)}{np_2} + \frac{np_3(1 - p_3)}{np_3}$$

$$= 3 - (p_1 + p_2 + p_3) = 2.$$

Note that the value of n and the specific values of the p_i's were not involved in the calculation. The method used in this special case can be readily extended to yield the following theorem.

Theorem. Given a multinomial distribution with c categories, where O_i and E_i are, respectively, the observed and the expected numbers of counts for the ith category, $i = 1, 2, \ldots, c$. Then the expected value of the chi-square statistic

$$\chi^2 = \sum_{i=1}^{c} \frac{(O_i - E_i)^2}{E_i}$$

is $c - 1$.

The proof is left as an exercise. The result, $c - 1$, agrees with the idea of one constraint.

2. *How do we use the information from the chi-square approximation to make decisions about the goodness-of-fit?* The general idea is that large values of χ^2 correspond to large departures from the model. Hence large values of our statistic indicate a poor fit. But how large and how poor? Suppose, for example, that we compute $\chi^2_2 = k$, and the chi-square table of probability levels (Table A–4) shows that $P(\chi^2_2 \geq k) = 0.01$. This means that there is only a small chance (1 in a 100) of getting a result equal to k or more if the assumed model is the true one. Under such circumstances, we feel that our model very likely does not fit the facts. We usually have some level of significance, say 5%, in mind. Then, if $P(\chi^2 \geq k) \leq 0.05$, we reject, at the 5% level, the hypothesis that the model fits the facts.

Our reasoning is that the proposed model has a substantial chance of being incorrect. If the model is incorrect, then we are using the wrong mean values, say E_i^*'s, instead of the correct E_i's. Hence the average value of O_i is E_i instead of E_i^*, and the squares of the departures $(O_i - E_i^*)^2$ will, on the average, be larger than $(O_i - E_i)^2$; and χ^2 will therefore be larger than it would have been for the correct mean value, E_i. (See Exercise 5.) Therefore, too large a chi-square value is a strong symptom of an incorrect model. We discuss this further in Section 10–5.

We know that the mean value of a chi-square random variable equals its number of degrees of freedom. So moderate values of a chi-square statistic are those near its number of degrees of freedom, which in this example is 2. *This mean gives us a clear standard against which to measure departures from the model.* Values of the statistic near its mean suggest a fit whose departure from the E_i's is appropriate, in view of the sampling variation that the multinomial forces on the data, namely the O_i's.

Can the fit be suspiciously close? Although very small values of our statistic mean that the O_i's are close to the E_i's, these small values may also mean that the data do not exhibit the proper variability for the model. If the value of our χ^2 is too small, the indication is that the fit is too good, and we may look for some bias or even some tampering with the data. For example, suppose that we get $\chi^2_5 = 0.55$; the tables give $P(\chi^2_5 \geq 0.55) = 0.99$, and this means that even if the model were exactly correct, we would get a smaller value only 1% of the time. Such a result seems too good to be true, though it can happen.

In Example 1, we have $\chi_2^2 = 0.41$. If this χ_2^2 has a chi-square distribution, we can look up 0.41 in Table A–4 at the back of the book. We find that *chi-square values larger than 0.41 can be expected about 80% of the time*. We regard the fit as satisfactory, and we would therefore regard the probabilities $\frac{1}{4}, \frac{1}{2}, \frac{1}{4}$ as reasonable.

One complaint about Mendel's experiments was that the results fitted the theory better than random variation would allow! The point is that the Mendelian model implies *not only averages but variability* that stems from the multinomial distribution of the O_i's. If the variability in the experimental data is consistently too small for the theory, then something is wrong. The variability should be present if the rest of the theory is correct. Some have suggested that Mendel may have had a too helpful research assistant. With only one small set of data, such as we have in Example 1, we could not reach any such conclusion. We found the fit close, but not excessively close.

When we assess goodness-of-fit and find the resulting discrepancies very small, we begin looking for the reason, just as we would if the fit were very poor. Frequently computational errors are caught in this manner. For example, in computing the chi-square, someone may have divided each squared difference by E^2 instead of by E, or divided all by n. But sometimes no reason is to be found, though looking was wise.

EXERCISES FOR SECTION 9–1

1. *A 1 × 2 contingency table with p = $\frac{1}{2}$.* William Longcor tested plastic dice whose dots had been made by drilling out and filling in small holes. For each 20,000 tosses, he recorded the number of dots showing on the top face as even or odd. For a typical die of this kind, he observed 10,200 evens and 9800 odds. Use the chi-square method of this section to see if the 50-50 model [$P(\text{even}) = P(\text{odd}) = \frac{1}{2}$] for one perfectly balanced die fits Longcor's results.

2. Test Longcor's data in Exercise 1 by using the sign test and the binomial distribution, as discussed in Chapter 2.

3. To test a die, an investigator made 600 tosses and recorded the number of dots showing on top, as follows:

Number of dots:	1	2	3	4	5	6	Total
Observed count:	110	112	88	113	92	85	600

Test to see if these data fit the multinomial model for perfectly balanced dice. [All faces equally likely: $p = \frac{1}{6}$.]

4. *Mendel's data.* In another of Mendel's experiments, similar to that of Example 1, he got the following counts of offspring of pea plants:

AA	Aa	aa	Total
78	175	79	332

Use chi-square to assess the $\frac{1}{4}, \frac{1}{2}, \frac{1}{4}$ theory for these data.

5. Given a random variable X with mean μ. If X is an estimate of m, then $b = \mu - m$ is called the *bias* of the estimate, and $X - m$ is the *error* of the estimate. Prove that $E(X - m)^2 = \sigma_X^2 + b^2$, thus showing that the mean square error is the variance of the estimate plus the square of its bias. [*Hint:* Use the relation $X - m = (X - \mu) + (\mu - m)$.]

6. (Continuation) Use the results of Exercise 5 to find the expected value of χ^2 when E_i are replaced by E_i^*, $i = 1, 2, 3$, as discussed in the text.

7. Prove the theorem stated in this section.

9–2 APPROXIMATION OF THE EXACT DISCRETE DISTRIBUTION BY THE CONTINUOUS CHI-SQUARE

The χ^2 statistic (1) in Section 9–1 stems from *counts*, and its distribution is therefore discrete; but the χ^2-distribution studied in Chapter 8 is continuous. To show how the discrete chi-square distribution is fitted by its continuous approximation, we exhibit two examples—one involving a small sample and the other a larger sample.

Small sample. Consider the chi-square distribution arising from the multinomial distribution with $n = 10$ and $p_1 = 0.2$, $p_2 = 0.3$, and $p_3 = 0.5$. From our multinomial table, Table A–7, we can get the probability associated with each possible triplet of outcomes (x_1, x_2, x_3). For example, the triplet $(2, 3, 5)$ has probability 0.0850. When the chi-square is computed for this triplet, we get $\chi^2 = 0$ since the expected values are also $(2, 3, 5)$.

Different triplets sometimes have identical values for chi-square; for example, the triplets with the next largest chi-squares are $(2, 4, 4)$ and $(2, 2, 6)$. Thus, for triplet $(2, 4, 4)$,

$$\chi^2 = \frac{(2-2)^2}{2} + \frac{(4-3)^2}{3} + \frac{(4-5)^2}{5} = \frac{8}{15} \approx 0.53,$$

and for triplet $(2, 2, 6)$,

$$\chi^2 = \frac{(2-2)^2}{2} + \frac{(2-3)^2}{3} + \frac{(6-5)^2}{5} = \frac{8}{15}.$$

From tables of the continuous chi-square distribution, $P(\chi_2^2 \leq 0.53) = 0.233$. On the other hand, Table A–7 gives $P(\text{triplet } 2, 4, 4) + P(\text{triplet } 2, 2, 6) = 0.0638 + 0.0709 = 0.1347$, which yields a cumulative of $0.1347 + 0.0850 = 0.2197$.

The results of the calculations for the set of possible triplets are tabulated in Table 9–1. We used a more detailed chi-square table than Table A–4 to get the approximating cumulative given in the rightmost column because we were interested in the third decimal. The exact and approximating cumulatives can be compared both in Table 9–1 and in Fig. 9–1.

Table 9–1

Comparison of the exact cumulative distribution of the χ^2 statistic for $n = 10$, $p_1 = 0.2$, $p_2 = 0.3$, $p_3 = 0.5$, with the approximation based on the chi-square distribution having 2 degrees of freedom

Triplet		χ^2	P	Cumulative Exact	Approximate
2 3 5		0	0.0850	0.0850	0
2 4 4	2 2 6	0.53	0.1347	0.2197	0.233
1 3 6	3 3 4	0.70	0.1276	0.3473	0.295
1 4 5	3 2 5	0.83	0.1205	0.4678	0.340
1 2 7	3 4 3	1.63	0.0846	0.5524	0.557
1 5 4	3 1 6	2.03	0.0698	0.6222	0.638
2 5 3	2 1 7	2.13	0.0644	0.6866	0.655
0 4 6	4 2 4	2.53	0.0550	0.7416	0.718
0 3 7	4 3 3†	2.80	0.0480	0.7896	0.753
0 5 5	4 1 5	3.33	0.0380	0.8276	0.811
1 1 8	3 5 2	3.63	0.0333	0.8609	0.837
0 2 8	4 4 2	4.13	0.0260	0.8869	0.873
1 6 3	3 0 7	4.30	0.0228	0.9097	0.883
2 0 8	2 6 2‡	4.80	0.0162	0.9259	0.909
0 6 4	4 0 6	5.20	0.0148	0.9407	0.926
5 2 3		5 63	0.0091	0.9498	0.940
5 1 4	§	6.03	0.0076	0.9574	0.951
5 3 2		6.30	0.0054	0.9628	0.957
0 1 9	4 5 1	6.53	0.0083	0.9711	0.962
1 0 9	3 6 1	6.70	0.0063	0.9774	0.965
5 0 5		7.50	0.0025	0.9799	0.976
1 7 2		7.63	0.0039	0.9838	0.978
5 4 1		8.03	0.0016	0.9854	0.982
0 7 3		8.13	0.0033	0.9887	0.983
2 7 1		8.53	0.0016	0.9903	0.986
0 0 10	4 6 0	10.00	0.0012	0.9915	0.993
23 triplets		$\chi^2 > 10$ 0.0085		1.	1.

† Triplets at or below this line form the 0.2584 region.
‡ Triplets at or below this line form the 0.0903 region.
§ Triplets at or below this line form the 0.0502 region.

The agreement is especially poor for small values of chi-square, notably at 0 and at 0.83. For larger values of chi-square, the agreement is *absolutely* fairly close. If we had cumulated from right to left, the *relative error* would sometimes be large.

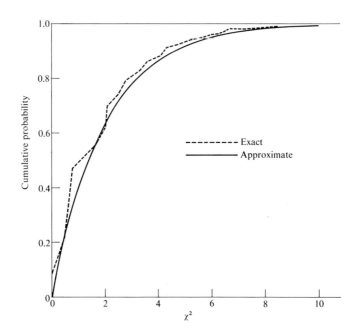

Fig. 9–1 Comparison of the exact cumulative distribution of the χ^2 statistic for $n = 10$, $p_1 = 0.2$, $p_2 = 0.3$, $p_3 = 0.5$, with the approximation based on the chi-square distribution having 2 degrees of freedom.

The power of chi-square. Let us look at the power of the chi-square test based on $n = 10$, taking probabilities (0.2, 0.3, 0.5) as the null hypothesis. From Table A–7, we can compute the power of the chi-square test for various levels of significance and for various alternatives. We do this by going to the column appropriate for the given alternative and adding up the probabilities in the critical region which is formed of all the triplets at or below the given symbols in Table 9–1. For example, given the alternative probabilities (0.1, 0.1, 0.8), we find from Table A–7 that the sum of the probabilities on or below the 5% (0.0502) line is 0.3775; and for the alternative probabilities (0.2, 0.2, 0.6), the sum is 0.0736.

Table 9–2 shows the results of the various computations. As we would expect, probabilities close to the null hypothesis, like (0.2, 0.2, 0.6). give power only slightly different from the significance level; but probabilities further away from the null hypothesis, like (0.1, 0.1, 0.8), give much higher power, even for $n = 10$.

One point of special interest arises. Note that the probabilities $(\frac{1}{6}, \frac{2}{6}, \frac{3}{6})$ have power 0.0465. Since $0.0465 < 0.0502$, we have found an alternative with this special property: The null hypothesis is not so likely to be accepted when it is true as is this alternative. When this condition holds, the *test* is said to be *biased*. It would be interesting to know for what alternatives this occurs and which alternative would give the smallest power. We have studied only the p's in our table.

Table 9–2

Some values of the power function for chi-square tests when $n = 10$, null hypothesis (0.2, 0.3, 0.5), and significance levels 0.2584, 0.0903, and 0.0502

			Power of χ^2 test with significance level		
p_1	p_2	p_3	0.2584	0.0903	0.0502
0.1	0.1	0.8	0.7407	0.4587	0.3775
0.1	0.2	0.7	0.5017	0.1837	0.1514
0.1	0.3	0.6	0.3676	0.0910	0.0588
0.1	0.4	0.5	0.3786	0.1384	0.0691
0.2	0.2	0.6	0.3307	0.1294	0.0736
0.2	0.3	0.5†	0.2584‡	0.0903‡	0.0502‡
0.2	0.4	0.4	0.3509	0.1675	0.1016
0.3	0.3	0.4	0.3932	0.2067	0.1446
$\frac{1}{3}$	$\frac{1}{3}$	$\frac{1}{3}$	0.5150	0.3018	0.2298
$\frac{1}{4}$	$\frac{1}{4}$	$\frac{2}{4}$	0.2969	0.1285	0.0775
$\frac{1}{6}$	$\frac{2}{6}$	$\frac{3}{6}$	0.2674	0.0884	0.0465

† Null hypothesis.
‡ Significance level.

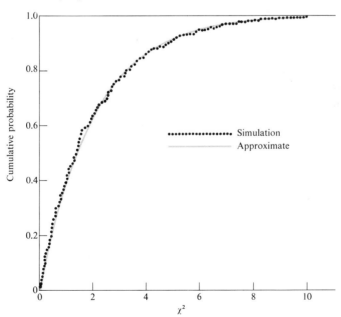

Fig. 9–2 Comparison of the cumulative distribution of the χ^2 statistic obtained by simulation for $n = 132$, $p_1 = 0.25$, $p_2 = 0.50$, $p_3 = 0.25$ with the approximation based on the chi-square distribution having 2 degrees of freedom.

Larger samples. To illustrate the agreement between the discrete distribution and the continuous chi-square for larger samples, we have obtained a simulation for Example 1 of Section 9–1, which has a sample size $n = 132$ and two degrees of freedom. Figure 9–2 compares the discrete and continuous cumulative. For this example, the agreement is much closer than that for $n = 10$.

9–3 IDEA OF GOODNESS-OF-FIT: GENERAL DISCUSSION

We now generalize the ideas discussed in Section 9–1. In all problems involving those ideas, we are given a set of categories, finite or infinite. These categories may have no special arrangement, or they may be arranged in a line or in a rectangular array or even in some more complicated pattern. Associated with each category C_i, there is an observed count O_i of items falling into that category, where $i = 1, 2, 3, \ldots, k$. The category also has associated with it a probability, formulated under the assumption of a binomial distribution or a multinomial distribution or some other hypothesis appropriate to the problem. The procedure is as follows:

a) Using the data and our theoretical model, *we construct a theoretical count E_i for each category i.* These E_i's are the expected values of the counts O_i under the model. In the example of Section 9–1, we knew from the outset the parameters $\frac{1}{4}, \frac{1}{2}, \frac{1}{4}$, which were provided by the Mendelian theory. Hence, for the three categories, the E_i's were $n/4$, $n/2$, $n/4$.

b) Using the results of the discussion in Section 9–1, *we compute for each category C_i a measure of departure* of the observed counts from the hypothetical results provided by the model under test, as follows:

$$\frac{(\text{Observed} - \text{Expected})^2}{\text{Expected}} = \frac{(O_i - E_i)^2}{E_i}, \qquad i = 1, 2, 3, \ldots, k.$$

c) *We sum this quantity over the k categories* and obtain the statistic

$$\chi^2 = \sum_{i=1}^{k} \frac{(O_i - E_i)^2}{E_i}. \tag{1}$$

In Section 9–1, we gave two reasons for choosing (1) as our statistic for measuring goodness-of-fit. One reason is that the statistic (1) has a distribution that is approximately chi-square with $k - 1$ degrees of freedom.

What is being tested? In every instance, we may think of the underlying mathematical model as a multinomial distribution. For a physical representation of this model, imagine a set of k boxes, each with its own assigned probability. These probabilities must add to 1, but they are not necessarily equal and not necessarily known. We have a set of n balls. Each ball is independently tossed into a box, and its chance of falling into a given box is equal to the probability assigned to that box. At the end of the experiment, we have k counts—the number of balls in each of the k boxes. *These counts are what we analyze.*

The foregoing is not the only possible model for these contingency tables, but it offers a good, standard way of thinking about them. When we test to see if our counts agree with such a model, *we test not only the probabilities but also the independence of tossing.* If we throw 10 balls and all go into the same box, and if we have independence, we are confident from the outcome that that box has a high probability. But consider this alternative model for tossing: We choose the box for the first ball according to the given probabilities, but then we deliberately put the next 9 balls into the same box. In this model, the independence fails. If we reject a given set of probabilities, we must understand that we may be rejecting for reasons other than the probabilities themselves, such as for lack of independence.

A plausibility argument; the case of two categories. To give plausibility to the assertion that statistic (1) has an approximate chi-square distribution, let us study the binomial case where we have only two categories. Suppose that we want to test the 50-50 model for successes and failures in n binomial trials. Our data may be tabulated thus:

	Successes	Failures	Total
Observed, O_i:	O_1	O_2	n
Expected, E_i:	$E_1 = (\tfrac{1}{2})n$	$E_2 = (\tfrac{1}{2})n$	n

We shall show that our chi-square statistic is the square of the standard, approximately normal, deviate

$$\frac{O_1 - (n/2)}{\sqrt{n(\tfrac{1}{2})(\tfrac{1}{2})}}.$$

Since $E_i = n/2$ and $O_2 = n - O_1$, we have

$$\chi^2 = \sum_{i=1}^{2} \frac{(O_i - E_i)^2}{E_i} = \frac{(O_1 - (n/2))^2}{n/2} + \frac{(n - O_1 - (n/2))^2}{n/2}$$

$$= \frac{2(O_1 - (n/2))^2}{n/2} = \frac{(O_1 - (n/2))^2}{(\tfrac{1}{2})(n/2)}$$

$$= \left[\frac{O_1 - (n/2)}{\sqrt{n(\tfrac{1}{2})(\tfrac{1}{2})}} \right]^2.$$

Since O_1 has a binomial distribution with $p = \tfrac{1}{2}$ and sample size n, we have $\mu = E(O_1) = np = n/2$ and $\sigma = \sqrt{npq} = \sqrt{n(\tfrac{1}{2})(\tfrac{1}{2})}$. Also, by the Central Limit

Theorem (Appendix III–3), the random variable

$$Z = \frac{O_1 - \mu}{\sigma} = \frac{O_1 - (n/2)}{\sqrt{n(\tfrac{1}{2})(\tfrac{1}{2})}}$$

is, for large n, approximately normally distributed with mean 0 and standard deviation 1. Thus we have shown, for this special case, that

$$\sum_{i=1}^{2} \frac{(O_i - E_i)^2}{E_i} = Z^2$$

is approximately a chi-square random variable with one degree of freedom, since it is the square of a standard deviate that is approximately normal. (See Section 8–1.)

9-4 THE 1 × 2 CONTINGENCY TABLE; GENERAL p

Research problems readily arise where p is not equal to $\tfrac{1}{2}$, its value in the previous section; and p may still be known or hypothesized. What then? Our procedure is much as before. If p is the probability of success, and $q = 1 - p$ that of failure, then, using the notation of Section 9–1, we have $E_1 = np$, $E_2 = nq$, and the following table:

	Successes	Failures	Total
Observed, O_i:	O_1	O_2	n
Expected, E_i:	np	nq	n

Our statistic is

$$\chi^2 = \frac{(O_1 - np)^2}{np} + \frac{(O_2 - nq)^2}{nq}. \tag{1}$$

In the exercises, you are asked to show by algebra that we can rewrite equation (1) as

$$\chi^2 = \frac{(O_1 - np)^2}{npq} = \left[\frac{O_1 - np}{\sqrt{npq}}\right]^2 = Z^2, \tag{2}$$

where

$$Z = \frac{O_1 - np}{\sqrt{npq}}. \tag{3}$$

For a fixed p, not 0 or 1, we know that, as n gets large, Z tends to the standard normal random variable. Hence we have again shown that, for this case with large values of n, χ^2 has approximately a chi-square distribution with one degree of freedom.

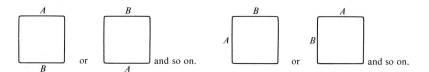

Example 1. *Table seating* (Joel Cohen). When two people eat lunch together at a small square table, there are 4 ways to sit opposite and 8 ways to sit at right angles at a corner. Thus, under random seating, $\frac{2}{3}$ of the couples would be at corners and $\frac{1}{3}$ would be opposite. At a cafeteria, 134 pairs were observed seated at corners and 63 opposite in 197 luncheon seatings. Test these data for departure from the hypothesis of random seating. (Joel Cohen suggested the model for data gathered by a psychologist.)

Solution. We tabulate the data thus:

	Corners	Opposites	Total
Observed, O_i:	134	63	197
Expected, E_i:	$131\frac{1}{3}$	$65\frac{2}{3}$	197

Then

$$\chi^2 = \frac{(2\frac{2}{3})^2}{131\frac{1}{3}} + \frac{(2\frac{2}{3})^2}{65\frac{2}{3}} = 0.162.$$

From the table of chi-square probabilities, we find that

$$P(\chi_1^2 \geq 0.162) = 0.69.$$

The result shows close agreement indeed. Agreement this close or closer would occur only 31% of the time if the hypothesis of random seating were exactly correct.

Possible lack of independence. Could there be anything about the gathering of these data that might violate our conditions? Yes, it is possible that the observations were taken in the same lunchroom so that many of the same people were observed every day. If so, we might not have 197 independent observations, for the same pairs might always sit the same way. Such observations would depend a lot on the ratio of preferences of the basic cadre of lunchers, whose number might be rather small. We should do well to inquire more deeply into the method of gathering the data. It would be reasonable to ask people if they have preferences and to record results for the same people for several days.

Note that, if we had not done the initial counting of possible seating positions, we might have been led to suppose that pairs of people had a strong preference for sitting at corners rather than opposite. Note also that, though random seating can account for the results, it is also possible that some strongly like to sit opposite and others to sit adjacent, and that the close fit with randomness is a mere happenstance. It may also be that the seating is essentially random for a given pair of people on the first day, but that they tend to persist in the same seating thereafter.

EXERCISES FOR SECTION 9–4

1. *An ESP test.* In an ESP experiment, an investigator uses a test in which the probability of giving a correct response by pure chance is $\frac{1}{5}$. In 25 trials, subject X gives 11 correct responses. Test at the 1% level, the hypothesis that the subject got his results by pure chance.

2. Repeat Exercise 1, given 9 correct responses.

3. For the data in Exercise 1, use the normal approximation of the binomial to find the probability of getting 11 or more correct responses by pure chance.

4. A player played 12 bridge hands and got no aces in 8 of them. He complained of poor shuffling. With good shuffling, the probability of getting at least one ace is approximately 0.7. At the 5% level, would you reject the null hypothesis of fair shuffling?

5. Check the result in Exercise 4 by using binomial tables.

6. When tossed 150 times, a die shows 5 or 6 on its top face 43 times. Would you reject, at the 5% level, the null hypothesis $P(5 \text{ or } 6) = \frac{1}{3}$?

7. A radio manufacturer claims that, on the average, 95% of his radios are nondefective. A retailer finds 2 defective radios in a shipment of 20. Would you reject the manufacturer's claim at the 5% significance level?

8. *Changers.* In a large panel study running over a three-month period, 160 families changed their coffee brand—68 from cheap to expensive and 92 from expensive to cheap. Test the hypothesis that the two changes are equally likely.

9. *Chi-square for binomial counts.* If the number of successes in a binomial experiment of n trials is x, and $P(\text{success}) = p$, we have the following information.

	Success	Failure	Total
Probability	p	$1 - p$	1
Count	x	$n - x$	n

Write out the formula for the chi-square goodness-of-fit test.

10. (Continuation) Show that the formula obtained in Exercise 9 reduces to

$$\left(\frac{\bar{p} - p}{\sqrt{pq/n}}\right)^2 ,$$

where $\bar{p} = x/n$, and $q = 1 - p$.

11. (Continuation) Interpret the formula in Exercise 10 in terms of a normal deviate.

9–5 CONTINUITY CORRECTION FOR ONE DEGREE OF FREEDOM

When we use a continuous normal distribution to approximate the discrete binomial distribution, we make a continuity correction before entering the normal table. For example, to find $P(X \geq x)$, where X is the count, we enter the normal table with the deviate

$$z_1 = \frac{x - \frac{1}{2} - np}{\sqrt{npq}}. \qquad (1)$$

When we want a two-tailed result rather than a one-tailed one, we get the distance between x and the mean np, namely $|x - np|$, and reduce it by $\frac{1}{2}$. This gives us a $\frac{1}{2}$ correction toward the mean. Then we compute the deviate

$$z_2 = \frac{|x - np| - \frac{1}{2}}{\sqrt{npq}}, \qquad (2)$$

get the one-tailed probability from the normal table, and multiply it by 2.

When we use the chi-square approximation to get two-tailed probabilities, the table itself takes care of the multiplication by 2. We enter the chi-square table with the adjusted value

$$z^2 = \frac{(|x - np| - \frac{1}{2})^2}{npq}. \qquad (3)$$

In the notation of the present chapter, x would be replaced by O_1. This correction improves the chi-square approximation for the probability, exactly as it does when we use the normal approximation to the binomial.

Example 1. If $n = 20$, $p = \frac{1}{2}$, and $O_1 = 15$, compute the probabilities for the corrected and uncorrected χ_1^2.

Solution. We find that

$$\text{uncorrected } \chi_1^2 = \frac{(O_1 - E_1)^2}{npq} = \frac{(15 - 10)^2}{5} = 5.$$

Using $\sqrt{5} = 2.24$, we can enter the normal tables and get good accuracy. We find

$$P(Z \geq 2.24) = 0.013,$$

so

$$P(\chi_1^2 \geq 5) = 0.026.$$

On the other hand,

$$\text{corrected } \chi_1^2 = \frac{(|15 - 10| - 0.5)^2}{5} = 4.05,$$

and we find, similarly, that

$$P(\chi_1^2 \geq 4.05) = 2P(Z \geq 2.01) = 0.044.$$

Thus the continuity correction makes quite a difference: The probability of a more extreme deviation goes from 0.026 to 0.044, an increase of about 69%. Binomial tables give the correct probability as 0.042.

Although many think that the correction doesn't matter, it is a good idea to make it. Unless n is enormous, as in the Longcor dice example, the correction will make a substantial improvement in approximating the correct probability.

An exception. If $|x - np| \leq \frac{1}{2}$, then the foregoing correction would make $|x - np| - \frac{1}{2}$ negative or zero. When we are dealing with the normal, this would make the computed probability greater than or equal to $\frac{1}{2}$ before the doubling; and we would get a two-tailed probability greater than or equal to 1. *If x is within $\frac{1}{2}$ unit of the mean np, the two-tailed probability of a result as extreme as, or more extreme than, the observed x is 1,* because all other observed values must then be at least as far from np as x.

Note. More than one degree of freedom. Empirical and mathematical investigations show that it is better not to use this correction with more than one degree of freedom. It overcorrects.

EXERCISES FOR SECTION 9–5

Rework the following exercises from Section 9–4, using the corrected value of chi-square. In each exercise, report the percentage change in the value of chi-square as a result of the correction.

1. Exercise 4.

2. Exercise 7.

3. Show that the percentage reduction in χ^2 induced by the $\frac{1}{2}$ correction is

$$100 \left(\frac{1}{|O_1 - np|} - \frac{1}{4|O_1 - np|^2} \right), \qquad O_1 \neq np.$$

4. Use the formula in Exercise 3 to find the percentage reduction occurring in χ^2 in Example 1.

CHAPTER 10

Applying Chi-Square: 1 x *k* Tables

10–1 THE 1 × *k* CONTINGENCY TABLE; KNOWN *p*'s

Sometimes an experiment produces counts in a number of categories for which theory may predict probabilities:

Category:	1	2	3	$\cdots$	*k*	Total
Probability:	p_1	p_2	p_3	$\cdots$	p_k	1

When $k = 2$, we have a binomial distribution with its 2 categories; for larger values of k, we have the multinomial distribution with k categories.

In Example 1 of Section 9–1, we discussed Mendel's experiment that led to a multinomial distribution. Situations where we believe that several categories are equally likely commonly give rise to multinomial models and $1 \times k$ contingency tables.

Primes. Consider an interval consisting of N successive positive integers beginning with integer m. Then as m takes on larger and larger values, the primes in the intervals become rarer on the average. We plan to study the *local* distribution of primes rather than their long-run behavior. To see how orderly they are locally, we use a multinomial distribution for a baseline model, as described in Example 1. We shall regard the primes as a set of data.

Example 1. *Distribution of primes.* Sometimes it is remarked that the primes exhibit very chaotic behavior. Let us see just how chaotic their behavior is.

We divide an interval with N successive integers into k subintervals, each containing N/k integers (N an integral multiple of k). Let the interval contain n primes. We ask: Are the counts of primes in the subintervals distributed like the counts in a sample of n from an equally likely multinomial with k categories?

The 44 primes in the interval from 1369 to 1680 distribute themselves as follows when the interval is divided into 8 equal parts.

Eighths:	1st	2nd	3rd	4th	5th	6th	7th	8th	Total
Observed:	3	6	7	6	5	6	7	4	44
Expected:	5.5	5.5	5.5	5.5	5.5	5.5	5.5	5.5	44

How well does the multinomial model fit these data?

Discussion. We compute

$$\chi_7^2 = \frac{1}{5.5}\left[(3 - 5.5)^2 + (6 - 5.5)^2 + \cdots + (4 - 5.5)^2\right] = 2.55.$$

The 8 categories imply 7 degrees of freedom because we have fixed the total count at 44. The observed chi-square value 2.55 is well below the mean of χ^2 for the equally likely multinomial, which is 7. Since

$$P(\chi_7^2 \geq 2.55) = 0.92,$$

more than 90 samples out of 100 drawn from 8 equally likely categories would produce chi-square values larger than 2.55, even if the multinomial model were exactly correct. The agreement is somewhat too close for the model: In this interval the primes are distributed more evenly into eighths than they should be under an equally likely model. When we look at additional primes in other intervals, we find again and again that they divide too evenly for a multinomial model. Table 10–1 shows this fact clearly. Not one of the chi-square values in Table 10–1 is as high as 7, which is the mean of χ_7^2 and a value that we expect to have exceeded about 43% of the time.

Table 10–1
Distribution of primes into 8 intervals of equal length between squares of successive primes

Primes		Eighths								Total	
a	a^2	1st	2nd	3rd	4th	5th	6th	7th	8th	N	χ^2
47	2209	10	11	10	8	8	8	15	10	80	3.80
53	2809	9	11	8	9	10	9	13	9	78	1.79
59	3481	2	6	4	4	4	4	3	5	32	2.50
61	3721	10	12	12	12	9	15	9	11	90	2.44
67	4489	8	7	10	7	9	4	11	10	66	4.30
71	5041	2	6	3	4	3	4	4	4	30	2.53

We conclude that the multinomial distribution does not provide a good fit for the local distribution of primes, at least in the range under study. On intervals of modest length, the primes are more evenly distributed than we expect from multinomial theory. This was not obvious before we started. Example 1 is unusual because ordinarily we find chi-square values that are much too high for the sampling model to have produced.

The theorem in Section 9–1 assures us that the chi-square mean for a 1 × *k* contingency table is $k - 1$, which is the desired mean value of chi-square when there are $k - 1$ degrees of freedom.

No continuity correction. Reminder: For more than one degree of freedom, we do not recommend a continuity correction.

EXERCISES FOR SECTION 10–1

1. A die is tossed 60 times with the following results.

Number of dots on top:	1	2	3	4	5	6	Total
Observed frequency:	14	12	9	12	6	7	60

   At the 5% level, do these data fit a multinomial model with equally likely probabilities for the 6 faces?

2. *Genetic theory.* For a genetic experiment in the crossbreeding of plants, theory prescribes 4 types of offspring, *A*, *B*, *C*, and *D*, with probabilities $\frac{9}{16}, \frac{3}{16}, \frac{3}{16}, \frac{1}{16}$, respectively. The observed counts are as follows;

Type of offspring:	*A*	*B*	*C*	*D*	Total
Observed frequency:	150	62	51	25	288

   Do these data support the theory?

3. In order to test a device for randomly generating the digits 0, 1, 2, 3, 4, 5, 6, 7, 8, 9, one thousand trials are made. Frequencies obtained for the 10 digits are, respectively, $n_0, n_1, n_2, \ldots, n_9$. In order for these frequencies not to reach the 1% level when the chi-square test is used, what number must the sum

$$\sum_{i=0}^{9} n_i^2$$

   be less than?

4. *Independent experiments.* Chi-squares from independent experiments can be added, and the resulting random variable has a chi-square distribution whose number of degrees of freedom equals the sum of the numbers of degrees of freedom of the component chi-squares. Apply this fact to the six chi-squares of Table 10–1 to find how frequently a total fit at least this good would occur, if the multinomial model were true and the results for the intervals independent.

10–2 THE 1 × *k* TABLE; FITTING A PARAMETER

In the previous section of this chapter and in Chapter 9, the probabilities were given to us by a hypothesis. Unfortunately, science is not always so convenient; in some instances, we have to *estimate* the parameters of a distribution from the data. Thus, in the Mendelian example of Section 9–1, if we had not been given that $P(AA$ offspring$) = \frac{1}{4}$, we would have had to estimate this probability from the reported counts.

Example 1. *Home runs.* The number of men on base when a home run is hit may be approximately distributed like a binomial with $n = 3$ and unknown p. Fit a binomial distribution to the following National Baseball League data, and use the chi-square approach to see how well the binomial model fits the facts.

Number on base:	0	1	2	3	Total
Observed frequency:	421	227	96	21	765

Solution. To obtain the fitted counts, we must have an estimate of p, say $\bar{p}$. We then get the binomial probabilities and compute

$$765[\binom{3}{i}(\bar{p})^i(1 - \bar{p})^{3-i}] \quad \text{where} \quad i = 0, 1, 2, 3.$$

The observed data give, as average number of men on base,

$$n\bar{p} = 3\bar{p} = 0\left(\frac{421}{765}\right) + 1\left(\frac{227}{765}\right) + 2\left(\frac{96}{765}\right) + 3\left(\frac{21}{765}\right),$$

whence

$$3\bar{p} = 0.63 \quad \text{and} \quad \bar{p} = 0.21.$$

What is this $\bar{p}$? It could be considered as an estimate of the average probability that a base is filled. The binomial model implies that each base is independently filled (success) or not filled (failure) by a base runner, and that P(base filled) ≈ 0.21 for each base. The model might fit well without being correct in all details.

Interpolating in the binomial tables for $\bar{p} = 0.21$, we compute the "expected values" or fitted counts as follows:

Number on base:	0	1	2	3	Total
Observed frequency:	421	227	96	21	765
Fitted counts:	378.7	298.4	80.3	7.7	765

We could use these results to compute chi-square in the usual way. However, the term for 3 men on base gives the huge contribution of $(21 - 7.7)^2/7.7 = 23.0$ to our chi-square. We need go no further to see that the binomial model gives a poor fit, because the other deviations can only make a large chi-square larger. (In Exercise 1, Section 7–2, we saw that the Poisson distribution fits much better.)

How many degrees of freedom are there in the χ^2 of this example? We imposed one constraint when we forced the counts to add to 765. In addition, we forced the fitted counts to have the same weighted average as the observed counts. These two constraints remove 2 degrees of freedom, leaving us with X_2^2. So our chi-square value, 23, is very far from the mean value, 2.

Number of degrees of freedom. In general, the number of degrees of freedom is the number of categories that can receive counts minus the number of functionally independent parameters fitted from the data.

Usually we fit the grand total as 1 parameter. In addition, in Example 1, we fitted $\bar{p}$ to p, thus making 2 fitted parameters for 4 categories and leaving $4 - 2 = 2$ degrees of freedom.

Example 2. Members of a class of 189 students closed their eyes and "randomly" wrote down, every 5 seconds, numbers in sequence. These numbers were chosen from among the integers 1, 2, 3. Let us consider adjacent pairs of digits. If the class were indeed able to make random calls, all 9 possible pairs of digits (1, 1), (1, 2), (1, 3), ... would be equally likely. The following table gives the frequencies recorded for the sixth and seventh numbers in the sequence.

		Seventh number			
		1	2	3	
	1	21	19	23	63
Sixth number	2	29	18	21	68
	3	21	25	12	58
		71	62	56	189

Analyze these data.

Solution. Let us first see if the number pairs fit an equally likely model. Since the total count is fixed at 189, our chi-square has $9 - 1 = 8$ degrees of freedom. Under the equally likely hypothesis, the expected value for each cell is $\frac{189}{9} = 21$. Then

$$X_8^2 = \tfrac{1}{21}[0^2 + 2^2 + 2^2 + 8^2 + 3^2 + 0^2 + 0^2 + 4^2 + 9^2]$$

$$= \tfrac{178}{21} = 8.5.$$

Since we have 8 degrees of freedom, our chi-square value 8.5 is just above the mean. And so for the sixth and seventh numbers the pairs are nearly equally likely, except possibly for the avoidance of (3, 3) and the liking of (2, 1).

Example 3. (Refer to Example 2.) Early in the sequence, people avoid writing down the duplicates (1, 1), (2, 2), (3, 3). We might ask: Are the three cells (1, 1),

(2, 2), (3, 3) equally likely and the other six cells also equally likely? The data for calls 3 and 4 follow.

Fourth number called

		1	2	3	
	1	12	32	20	64
Third number 2		21	10	23	54
called	3	34	26	11	71
		67	68	54	189

Does the suggested model fit the facts?

Solution. The suggested model is that the 3 main-diagonal cells are equally likely and the 6 off-diagonal cells are also equally likely, but with different probabilities. If we estimate the counts on the main diagonal with $(12 + 10 + 11)/3 = 11$ and the rest of the counts with $\frac{156}{6} = 26$, then we get

$$\chi_7^2 = \tfrac{1}{11}[1 + 1 + 0] + \tfrac{1}{26}[25 + 64 + 0 + 36 + 36 + 9] = 6.72.$$

We have $9 - 2 = 7$ degrees of freedom because we fitted 2 parameters. Our value χ_7^2 is 6.72, which is very near the average 7. The proposed model may be a satisfactory model for early calls since the chi-square value is near its mean. We would, of course, like to see the calculations repeated for other early numbers in the sequence.

Note. Observe that one could readily miscount the number of parameters fitted in Example 1. We seem to have fitted 3 parameters—the grand total, n, the average count for the main diagonal, $\bar{d}$, and the average count for the rest of the cells, $\bar{r}$.
But $\bar{d}$, $\bar{r}$, and n are related by the linear constraint

$$3\bar{d} + 6\bar{r} = n.$$

Therefore, we actually fitted only 2 functionally independent parameters.

Example 4. *Three tries to a success.* An experiment consists of trials with probability p of success, and it continues until either a success occurs or three trials have occurred. Of 100 such experiments, 44 ended on trial 1, 19 on trial 2, and 37 on trial 3. Estimate p, and use the chi-square test to measure the fit of the data to the theoretical model.

Solution. The probabilities that the experiment will end on the first trial, the second trial, or the third trial are, respectively, p, qp, or q^2. One way to estimate p is to equate the observed and the theoretical mean number of trials. The theoretical mean is

$$\mu = 1 \times p + 2 \times qp + 3 \times q^2 = p + 2p(1 - p) + 3(1 - p)^2 = 3 - 3p + p^2.$$

Since the observed mean is $\bar{x} = (1 \times 44 + 2 \times 19 + 3 \times 37)/100 = 1.93$, we get

$$3 - 3p + p^2 = 1.93. \tag{1}$$

If we solve the quadratic equation (1), we find the estimate $\hat{p} = 0.414$.

Table 10–2 shows the data along with the theoretical distribution of the number of trials. The expected numbers of counts have been fitted, using the estimate $\hat{p} = 0.414$.

Table 10–2
100 experiments of three tries to success

Experiment ends on trial	1	2	3	Total
Probability	p	qp	q^2	1
Observed counts	44	19	37	100
Expected counts (fitted with $\hat{p} = 0.414$)	41.4	24.3	34.3	100
Observed − Expected	2.6	− 5.3	2.7	

Since we have 3 cells with a fitted total, n, and a fitted value of p, our chi-square has $3 - 2 = 1$ degree of freedom. We get

$$\chi_1^2 = \sum (\text{Observed} - \text{Expected})^2/\text{Expected}$$
$$= 2.6^2/41.4 + (-5.3)^2/24.3 + 2.7^2/34.3 = 1.53,$$

and $P(\chi_1^2 \geq 1.53) = 0.22$. Our interpretation is that the data fit well; over one-fifth of the time we would get a poorer fit. (See also Exercise 8 of this section and the discussion of Section 11–6.)

EXERCISES FOR SECTION 10–2

1. *Home runs.* Refer to Exercise 1, Section 7–2. Show that the Poisson distribution gives a better fit than the obvious binomial with $n = 3$, for the data of Example 1.

2. *World Cup soccer.* In the 1966 series of World Cup football (soccer) matches, 32 games (each involving 2 teams) were played. The numbers of goals per game scored by the competing teams, with observed frequencies, were as follows:

Number of goals scored by a team:	0	1	2	3	4	5	> 5
Observed frequency:	18	20	15	7	2	2	0

Fit these data with a Poisson distribution, and use chi-square to measure the goodness-of-fit. [Group 4 or more goals into one category.]

3. *Binomial distribution, $n = 3$.* A supposedly binomial experiment with $n = 3$ and unknown p was carried out 80 times. The following distribution of x was observed.

x:	0	1	2	3	Total
Frequency:	22	39	16	3	80

Estimate *p*, use binomial tables to get fitted counts, and use chi-square to measure goodness-of-fit. Interpret the results.

4. *Three tries to a success.* See Example 4. An experiment consists of trials with $P(\text{success}) = p$. The trials continue until either a success occurs or 3 trials have occurred. A set of 100 experiments produced the following data.

Experiment ends on trial:	1	2	3	Total
Observed counts:	34	21	45	100

Estimate *p*, and use chi-square to test the fit of the data to the model used in Example 4.

5. (Continuation) Test the compatibility of the two sets of data, in Exercise 4 and Example 4, with one another *and* with the model used in Example 4 by fitting a common *p* and then computing chi-square for goodness-of-fit to each set and summing. How many degrees of freedom will there be?

6. See Example 2. In a classroom experiment, students are asked to choose at random one of the numbers 1, 2, 3 and then choose again at random. If they do choose randomly, then we expect $\frac{1}{9}$ of the response pairs to appear in each of the 9 cells of a 3 × 3 table. The following table gives the results obtained on the *i*th and $(i + 1)$st trials.

		\multicolumn{4}{c}{Number chosen on trial *i* + 1}
		1	2	3	Total
Number chosen	1	78	178	190	446
on trial *i*	2	148	76	179	403
	3	209	155	97	461
		435	409	466	1310

Use chi-square to measure the departure of the table from the $\frac{1}{9}$ expectation.

7. (Continuation) Test to see if the model used in Example 3 fits the data of Exercise 6.

8. In the model of Example 4, let the counts in the 3 categories be O_1, O_2, and O_3. Consider the set of experiments as a sequence of 100 trinomial trials with

$$p_1 = p, \qquad p_2 = qp \qquad \text{and} \qquad p_3 = q^2.$$

Show that the maximum likelihood estimate of *p* is

$$\hat{p} = \frac{O_1 + O_2}{O_1 + 2O_2 + 2O_3}.$$

[*Hint:* Recall Theorem 1 of Section 2–9.]

10–3 POOLING CELLS

When one is fitting expected values to counts in many cells, some clumps of cells may have small expected values and only occasional counts. For example, in fitting counts to a Poisson distribution, we have potentially infinitely many cells, because this distribution has positive probability for every integer. If $p_i = P(X = i)$ for the Poisson, then the expected count is np_i. For large values of i, the probability p_i tends to zero. Instead of summing the values $(O_i - np_i)^2/np_i$ over all the cells for which O_i equals zero, we group the cells beyond a certain value so as to get a pooled count of at least 2 or 3 into the pooled cell. Otherwise we would have an infinite number of degrees of freedom, and the long tail of the Poisson distribution with all its O_i's equal to zero would contribute to our χ^2 only

$$\sum_{i=k}^{\infty} (np_i)^2/np_i = \sum_{i=k}^{\infty} np_i = nP_k,$$

where

$$P_k = \sum_{i=k}^{\infty} p_i.$$

It seems better to face squarely the fact that we can compare the theoretical and empirical distributions only over the region where counts tend to occur, that is, where the probabilities should produce substantial counts. If we get no counts and can expect none, we cannot do much more than note little disagreement between the expected and the observed values of the random variable. Further, the pooling makes the agreement between the χ^2 approximation and the exact distribution closer.

Example 1. *Poisson distribution of radiation counts* (Joseph Berkson). Physical theory suggests that the presence of radioactive material gives rise to numbers of counts on a Geiger counter that should have a Poisson distribution. From an elaborate experiment designed to test this theory, Berkson gives the following data. The counts are the numbers of particles counted in short time intervals of equal length, 12,169 intervals in all.

Count X	Frequency Observed	Expected	Contribution to chi-square
0	5,267	5,266.4	0
1	4,436	4,410.8	0.14
2	1,800	1,847.1	1.20
3	534	515.7	0.65
4	111	108.0	0.08
≥ 5	21	20.8	0
	12,169	12,168.9	2.07

Discussion. We can roughly approximate Berkson's calculations by using $\bar{x}$, the average observed count, as our Poisson mean:

$$(0 \times 5267 + 1 \times 4436 + \cdots)/12{,}169 = 0.837\ldots.$$

Then, using Poisson probability tables, we compute the expected values; for example, if $X = 0$, the expected number of counts is $12{,}169 P(X = 0)$. Contributions to chi-square are computed using $(O_i - E_i)^2/E_i$, as usual.

Every time we fit an independent constant, we lose one degree of freedom. Since the mean and the total count were fitted, there are $6 - 2 = 4$ degrees of freedom. We have pooled all counts starting with $X = 5$.

From the tables of chi-square probabilities, we get

$$P(\chi_4^2 \geq 2.07) = 0.72.$$

This means that we may expect a χ^2 value of 2.07 or larger about 72% of the time. It is most unusual for such a large count—over 12,000—to be distributed so closely to a delicate theoretical result that no difference is detectable. This was a most careful physical experiment precisely designed to detect departures from the Poisson. We observe that the fit is excellent: The data support the theory.

EXERCISES FOR SECTION 10–3

1. *Normal approximation* (G. Iversen *et al.*). To see whether certain statistics were approximately normally distributed, values of 267 supposedly independent, standard normal deviates z were tabulated. The table gives the observed counts and the fitted counts in various intervals. Since theory gave the means and standard deviations leading to the z's, the only loss in degrees of freedom is due to the expected counts being forced to add to 267.

Class interval

	$-\infty$ to -3	-3 to -2	-2 to -1	-1 to 0	0 to 1	1 to 2	2 to 3	3 to ∞	Total
Observed:	0	7	31	94	86	40	6	3	267
Expected:	0.4	5.7	36.3	91.1	91.1	36.3	5.7	0.4	267.0

   a) Use tables to check the expected count for the interval 0 to 1.
   b) Decide whether to pool the 3 to ∞ interval with 2 to 3 and the $-\infty$ to -3 interval with -3 to -2. Whatever you decide, state the resulting number of degrees of freedom.
   c) Compute χ^2, report the probability level, and interpret.

2. In Example 1, find the value Berkson used for the Poisson mean by dividing the expected value for $X = 0$ into the expected value for $X = 1$. First, show theoretically that this technique works.

3. Use the approximate value found for the mean in Exercise 2 to compute the expected
number of counts when $X = 2$.

10-4 GOODNESS-OF-FIT FOR CONTINUOUS DISTRIBUTIONS

Occasionally, one wishes to assess an observed distribution to see if its shape
approximates that of a specific theoretical continuous distribution.

Example 1. *Logarithms of separation factors.* R. Keisch, R. L. Feller, A. S.
Levine, and R. R. Edwards studied the authenticity of oil paintings. To distinguish
paintings of the eighteenth century from those of the nineteenth and twentieth
centuries, they examined the distribution of the logarithm of the separation coefficient of lead. This coefficient measures the ratio of the initial lead-210 concentration to radium-226 concentration in the sample. The investigators asked:
Are the original measurements of log separation factors on authentic pictures in
the nineteenth and twentieth centuries approximately normally distributed?
(*Science*, March 10, 1967, Vol. 155, No. 3767, pp. 1238–1242.) How shall we test
for normality?

Solution. In adapting the data to this example, we use 49 measurements with
mean 1.86 and standard deviation 0.60.
   One method of analyzing the data is to divide the horizontal axis of the density
graph of the normal distribution with mean 1.86 and standard deviation 0.60 into
7 segments with equal probabilities. Then the expected count in each segment is
$\frac{49}{7} = 7$. The observed and expected counts are shown in Table 10–3.

Table 10–3

Frequency distribution of the
logarithms of the separation factors
for a normal distribution divided
into sevenths ($\bar{x} = 1.86$, $s = 0.60$)

Intervals	Observed number	Expected number
$-\infty$ to 1.21	9	7
1.22 to 1.51	4	7
1.52 to 1.74	5	7
1.75 to 1.96	10	7
1.97 to 2.19	8	7
2.20 to 2.49	6	7
2.50 to $+\infty$	7	7
	49	49

For k equally likely categories and a sample of size n, the formula for chi-square is

$$\chi^2 = \sum_{i=1}^{k} \frac{(O_i - (n/k))^2}{n/k} \tag{1}$$

$$= [4 + 9 + 4 + 9 + 1 + 1 + 0]/7 = 4.$$

There are 4 degrees of freedom, because we have 7 categories and we have fitted the total count $n = 49$, the mean $\bar{x} = 1.86$, and the standard deviation $s = 0.60$. This leaves us with $7 - 3 = 4$ degrees of freedom. Since the mean value of χ_4^2 is 4, the result in this example represents a close fit. Indeed, the tables of chi-square probabilities give $P(\chi_4^2 \geq 4) = 0.41$, which means that we may expect chi-square values of 4 or more about 41% of the time.

EXERCISE FOR SECTION 10–4

1. *Chi-square for equally likely categories.* If n balls are randomly tossed into k boxes and n_i balls are observed in box i, show that the chi-square formula for testing that the boxes are equally likely [formula (1), Section 10–4] can be simplified to

$$\chi_{k-1}^2 = \frac{k \sum n_i^2}{n} - n.$$

[*Note:* This formula is easier to use than formula (1) if n/k is not an integer.]

10–5 CHI-SQUARE WHEN THE NULL HYPOTHESIS IS FALSE

The Big Ideas

1. When the assumed probabilities are false, larger sample sizes increase the mean value of χ^2, thus increasing the probability of detecting the lack of agreement. We derive a formula that makes this explicit.

2. When one uses the χ^2 statistic to compare the fit, or lack of it, for several sets of counts with the same assumed cell probabilities, the comparison is interfered with by sample sizes that vary from set to set. The issue is not the significance of the departure but the amount. Dividing each χ^2 by the sample size gives a statistic that makes the comparisons more meaningful, as we shall see.

For a multinomial model with c categories, suppose that the probabilities being tested are p_i^*, $i = 1, 2, 3, \ldots, c$, but the true probabilities are p_i, $i = 1, 2, 3, \ldots, c$. Thus the null hypothesis is a multinomial distribution with probabilities p_i^*. Unless $p_i = p_i^*$ for all i, the null hypothesis is false. Then the chi-square statistic is

$$\chi^2 = \sum_{i=1}^{c} \frac{(O_i - np_i^*)^2}{np_i^*}. \tag{1}$$

What is the expected value of this statistic? Note that $E(O_i) = np_i$, and so the bias in O_i for estimating np_i^* is $n(p_i - p_i^*)$. Recall from Exercise 5 of Section 9–1 that the mean square error of a biased estimate is the variance of the estimate plus the square of its bias. Hence, if $q_i = 1 - p_i$,

$$E(O_i - np_i^*)^2 = np_i q_i + n^2(p_i - p_i^*)^2. \tag{2}$$

Consequently, the expected value of the statistic in equation (1) is

$$E(\chi^2 \mid p_i, p_i^*, i = 1, 2, \ldots, c) = \sum_{i=1}^{c} \frac{np_i q_i}{np_i^*} + \sum_{i=1}^{c} \frac{n^2(p_i - p_i^*)^2}{np_i^*}$$

$$= \sum_{i=1}^{c} \frac{p_i q_i}{p_i^*} + n \sum_{i=1}^{c} \frac{(p_i - p_i^*)^2}{p_i^*}. \tag{3}$$

We see that the mean value increases with n, but in order to see more clearly how it does so, we look at a special case.

In the special case where the p_i^*'s are equal, then $p_i^* = 1/c$,

$$E(\chi^2 \mid p_i, p_i^* = 1/c, i = 1, 2, \ldots, c) = nc \sum p_i^2 - n + c \sum p_i q_i$$

$$= nc \sum p_i^2 - n + c - c \sum p_i^2$$

$$= (n - 1)c \sum p_i^2 - (n - c). \tag{4}$$

We see that dividing χ^2 by n, when n is large compared with c, would in this case give a statistic whose mean value depended little on n.

Often we are not especially eager to get the probability level for a value of chi-square. Instead, *we want to compare several chi-squares*, as is to be done in Example 1.

Releasing ourselves from the condition $p_i^* = 1/c$, let us consider the case where n is large compared with c. We state without proof that if some of the differences, $|p_i - p_i^*|$, are large compared with $1/n$, then $\sum p_i q_i / p_i^*$ is small compared with $n \sum (p_i - p_i^*)^2 / p_i^*$. Consequently, dividing the statistic χ^2 by n will make its mean approximately independent of n. The value χ^2/n approximately estimates $\sum (p_i - p_i^*)^2 / p_i^*$, and it is this measure that is essentially being compared when we have several sets of counts with the same p_i^* but possibly different p_i's. And so, again, if we wish to compare several values of χ^2, as in Example 1, we may profitably divide by n.

Example 1. *Contextuality and common words* (Kučera and Francis, *Computational Analysis of Present-Day American English*, Brown University Press, 1967, pp. 275–293). From 15 kinds of writing, or genres, Kučera and Francis analyzed 500 samples of about 2000 words each. The samples were distributed as follows (*ibid.*, p. xix):

Genre	No. of samples
A. Press reportage	44
B. Press editorial	27
C. Press review	17
D. Religion	17
E. Skills and hobbies	36
F. Popular lore	48
G. Belles lettres	75
H. Miscellaneous (governmentese)	30
J. Learned and scientific	80
K. Fiction: General	29
L. Fiction: Mystery and detective	24
M. Fiction: Science	6
N. Fiction: Adventure and Western	29
P. Fiction: Romance and love	29
R. Humor	9
Total number of samples	500

Kučera and Francis wanted to see whether the rates of use of common words like *the*, *of*, *and*, and *to* varied from one genre to another. The natural null hypothesis is that the expected frequencies E_i^* are proportional to the number of words in the samples in genre i. How shall we compare, from one genre to another, the sensitivity of the frequency of words to genre?

Discussion. Kučera and Francis computed chi-square across the 15 genres for the counts of each of the 100 most frequently used words. Table 10–4 gives the data and summarizes the computations for the first 2 words. For each word, the first row in the table gives the counts in the genre, the second row gives the expected count under the assumption that the count is proportional to the number of words in the genre samples, and the third expresses the count as a percentage of the total number of words in the genre. Chi-squares with 14 degrees of freedom are computed in the usual way, and the results are tabulated in the rightmost column. For each of the 100 words studied, the chi-square value is highly significant. (Recall that the mean of χ_{14}^2 is 14 and the standard deviation is $\sqrt{28} \approx 5.29$.)

In the rightmost column of Table 10–4, we have given values of $1000\chi^2/n$, where n is the number of times the word was used. We have seen in our theoretical analysis that if it is not true that the occurrences are distributed among the genres proportionally to the total usage, then the more a word is used (the larger the value of n), the bigger the chi-square, on the average. This is of course desirable for testing significance. It pleases us that larger samples make it easier to detect a departure from random distribution into genres. *Once we reject the hypothesis of*

Table 10-4

Distribution of occurrence of the most frequent word-types

Word		A	B	C	D	E	F	G	H	J	K	L	M	N	P	R	Total n	x^2 and $\dfrac{1000x^2}{n}$
	Frequency	6385	3961	2370	2480	4757	6976	10758	4621	12536	3792	2817	723	3780	2988	1027	69971	688.7
the	Expected frequency proportional to genre size	6122.5	3764.8	2441.6	2386.1	5009.1	6711.0	10500.6	4311.5	11183.3	4030.3	3331.3	831.9	4033.7	4050.9	1261.5		9.8†
	Ratio, as % of frequency to total number of words in genre	7.19	7.26	6.70	7.17	6.55	7.17	7.07	7.39	7.73	6.49	5.83	5.99	6.46	5.09	5.62		
	Frequency	2858	1994	1340	1505	2411	3696	6382	3059	7454	1423	913	329	1327	1202	518	36411	2362.6
of	Expected	3185.9	1959.1	1270.5	1241.6	2606.5	3492.2	5464.2	2243.6	5819.4	2097.3	1733.5	432.9	2098.9	2107.9	656.4		64.9†
	Ratio	3.22	3.65	3.79	4.35	3.32	3.80	4.19	4.89	4.60	2.44	1.89	2.73	2.27	2.05	2.83		

Source: H. Kučera and W. N. Francis, *Computational Analysis of Present-Day American English*, Brown University Press, 1967, p. 277.

† $1000x^2/n$

Table 10–5

Distribution of $1000\chi^2/n$ of 100 most frequent word-types in the corpus

0–20		60–80		150–200		500–1400	
6.2	to	60.6	if	150.6	what	537.4	he
6.9	and	62.8	than	151.6	them	565.8	had
9.8	the	63.4	any	162.9	would	671.0	my
10.3	with	63.8	first	165.1	now	721.6	down
11.5	from	64.9	of	170.5	we	747.6	I
12.6	a	67.2	it	185.1	also	771.5	him
		68.7	were	196.5	can	858.3	back
		71.2	well			871.9	you
		72.2	after			894.5	me
20–40		72.4	where	**200–300**		999.2	said
23.4	been	72.7	be	214.2	then	1141.4	her
23.8	as	74.4	they	216.5	is	1148.3	your
24.7	on	77.0	even	219.0	new	1397.7	she
25.0	one			224.3	these		
26.1	in			226.0	are		
26.9	that			231.2	was		
30.0	an	**80–100**		290.7	could		
37.0	for	81.7	there	296.9	such		
38.8	some	86.5	but				
		87.0	about				
		90.3	other				
40–60		91.8	its	**300–500**			
40.2	more	95.1	before	302.6	his		
45.7	time	97.4	so	307.3	has		
45.7	two			309.8	did		
45.9	made			317.0	will		
46.2	at	**100–150**		329.5	may		
46.6	only	101.1	no	331.4	out		
46.7	have	103.3	must	340.6	our		
49.2	much	106.0	by	364.1	up		
49.8	all	106.8	into	372.2	man		
50.8	their	111.2	way	471.3	like		
53.7	not	117.3	over				
54.6	through	117.3	years				
57.5	when	122.0	do				
60.0	this	125.3	or				
		127.0	who				
		128.7	which				
		134.7	most				
		147.8	many				

Computed from Table A1, H. Kučera and W. N. Francis, *Computational Analyses of Present-Day American English*, pp. 277–293.

random distribution, or distribution proportional to genre sizes, we may want to know which words seem to depart more from the random distribution and which less. We speak of some words as more "contextual" than others, meaning that their rates vary more from genre to genre than others.

For words of comparable departure, the larger the number of occurrences, the larger the word's chi-square. Dividing by n fixes this imbalance. We multiply by 1000 to get nice numbers and to permit ourselves to talk about departure as measured for samples of 1000. Values of $1000\chi^2/n$ are keyed with dagger symbols.

The keyed numbers in the rightmost column of Table 10–4 tell us that the word *of* is more contextual than *the*. It is not especially contextual compared with other words among the 100 studied. Table 10–5 gives contextuality scores for the 100 words studied. For example, *is* produced a value of 217 for $1000\chi^2/n$, *was* produced 231; *he* produced over 500, *she* nearly 1400, *him* about 770, and *me* about 900.

This sort of comparison is fairly satisfactory when each item (word, in this example) to be compared has the same number of categories and ought, without contextuality, to be similarly distributed.

Remember that we use

χ^2 *for significance testing*, and
χ^2/n *for interpretation of departure from the random distribution* when n is large compared with c.

EXERCISES FOR SECTION 10–5

1. Show that when $p_i = 1/c$, formula (4) gives the expected value $c - 1$.

2. Table 10–5 shows ten groups of words. Select from these words all examples of each of the following parts of speech: (1) verbs; (2) nouns; (3) possessive pronouns (their, its, his, our, my, her, your); (4) prepositions; (5) conjunctions. (If a listed word lends itself to more than one of the 5 given parts of speech, use the classification that comes first on the foregoing list.) Then make a table to show the number of each of the 5 parts of speech in each of the ten groups in Table 10–5.

3. (Continuation) Divide the scores in Table 10–5 into three categories: low (less than or equal to 60), medium (between 60 and 200), and high (over 200). Then make a table showing the number of each of the 5 parts of speech in each of the 3 categories. Keep this table for later use.

4. Use the Mann–Whitney test to compare the scores in Table 10–5 for possessive pronouns with those for prepositions.

5. Is there evidence in Table 10–5 that conjunctions are more contextual than prepositions?

6. After making such detailed studies as you believe appropriate, write a report discussing the relation between degree of contextuality and part of speech.

CHAPTER 11

Applying Chi-Square: Two-Way Tables

11–1 2 × 2 CONTINGENCY TABLES; SOME ORIGINS

We illustrate through an example the three most common ways in which 2 × 2 contingency tables arise, even though these three ways do not exhaust the possibilities.

1. *Two binomials with row totals (or column totals) fixed.* A random sample of *n* males aged 16–19 is classified as employed or unemployed, and a random sample of males aged 20–24 is similarly classified. This information leads to estimates of the proportions, or percentages, in the two groups that are employed. For example, the results might be tabulated thus:

Type 1 table

		Counts			Percentages		
		Un-employed	Employed	Totals	Un-employed	Employed	Totals
Age	16–19	80	320	400	20	80	100
	20–24	36	564	600	6	94	100

In this example, note that the sample sizes 400 and 600 may be deliberately chosen, and that *they may not reflect the relative population sizes* of the two age groups. Nevertheless, the percentages are *unbiased estimates* of the proportions unemployed in the separate age groups.

2. *Four-category multinomial table with grand total fixed.* Imagine drawing *one* random sample of size *n* from the population of males aged 16 through 24.

Divide them into the same four categories exhibited in the tables of type 1 above. The result might look like this:

Type 2 table

		Counts			Row percentages			Column percentages	
		Unem-ployed	Em-ployed	Total	Unem-ployed	Em-ployed	Total	Unem-ployed	Em-ployed
Age	16–19	90	360	450	20	80	100	73	41
	20–24	33	517	550	6	94	100	27	59
		123	877	1000			Total	100	100

In this example, it makes sense to calculate percentages either way, because we drew the sample randomly from all the ages 16 through 24. We chose only the number in the sample, 1000.

3. *Tables with fixed row and column totals.* Again, we randomly sample males from 16 through 24 years old. But in this example, we divide them by age into the younger half versus the older half; and we divide them by employment into the quarter who have been least employed in the last six months versus the three-quarters who have been more fully employed. The tables might then look like this:

Type 3 table

	Counts			Percentages			Percentages	
	Least em-ployed	Most em-ployed	Total	Least em-ployed	Most em-ployed	Total	Least em-ployed	Most em-ployed
Younger half	200	300	500	40	60	100	80	40
Older half	50	450	500	10	90	100	20	60
	250	750	1000			Total	100	100

This approach differs from the others in that we have deliberately chosen the splits (500–500 and 250–750) ourselves. No sampling error is involved in these totals. In the type 2 table, either the total 123 for unemployed or the 450 for age 16–19 could on another sample produce a different number, but here the margins have been fixed. Instead of obeying binomial distributions as in type 1 or multi-nomials as in type 2, these type 3 tables follow the hypergeometric distribution (see *PWSA*, p. 317). We shall not discuss the exact small-sample distribution.

11–2 2 × 2 CONTINGENCY TABLES; THE CHI-SQUARE TEST FOR INDEPENDENCE

For large samples, some possible ways of analyzing 2 × 2 contingency tables are:
a) to set confidence limits on the differences between the observed and the expected proportions; or
b) to use the exact hypergeometric distribution (Fisher's exact test, see *PWSA*, p. 317, Example 2); or
c) to make a chi-square test for independence.
The confidence-limits approach has advantages, but many people like the chi-square test. And since we are studying chi-square in this chapter, we shall use the χ^2 statistic to test for independence.

The notation for counts in a 2 × 2 table is well established:

Table of counts
$$(n = a + b + c + d)$$

		Second attribute V	Not-V	Total
First attribute	U	a	b	$a + b$
	Not-U	c	d	$c + d$
	Total	$a + c$	$b + d$	n

If we divide each entry in the table of counts by n, we get a table of proportions.

Table of proportions

		Second attribute V	Not-V	Total
First attribute	U	a/n	b/n	$(a + b)/n$
	Not-U	c/n	d/n	$(c + d)/n$
		$(a + c)/n$	$(b + d)/n$	1

These proportions may be treated similarly to probabilities. Recall that U and V are independent events if and only if

$$P(U \cap V) = P(U) \cdot P(V).$$

Hence, *if we assume independence*, we have for cell (U, V) the estimated probability

$$\bar{P}(U \cap V) = \frac{a + b}{n} \cdot \frac{a + c}{n} = \frac{(a + b)(a + c)}{n^2};$$

and the *estimated expected count* for this cell under this assumption is

$$n\bar{P}(U \cap V) = (a + b)(a + c)/n.$$

Similarly for the other cells in the table: the estimated expected count for a given cell is

$$(\text{Row total})(\text{Column total})/n.$$

Thus, a, b, c, and d are the observed counts O_i; and the expressions such as $(a + b)(a + c)/n$ are used as the expected counts E_i, *if one assumes that independence holds.*

Let us apply the χ^2 method to test for independence in our type 3 table in Section 11–1. In the following table, the upper number in each cell is the observed count, and the lower number is the expected count under the assumption of independence, when we regard the row and column totals as fixed:

	25% least employed	75% most employed	Total
Younger half	200 125	300 375	500
Older half	50 125	450 375	500
	250	750	1000

We get the expected counts by using the foregoing rule; for example, in the upper left cell, we have $(500)(250)/1000 = 125$. Our statistic is

$$\chi^2 = \frac{(75)^2}{125} + \frac{(-75)^2}{375} + \frac{(-75)^2}{125} + \frac{(75)^2}{375}$$

$$= 45 + 15 + 45 + 15 = 120.$$

Note that, except for sign, all cell differences, O_i E_i, are alike: It is a fact that, *in a 2 × 2 table, cell differences are always equal in absolute value.*

How many degrees of freedom do we have? We began with four counts, $a, b, c,$ and d; but we held the grand total fixed, and then, for purposes of estimation, we held the row totals and the column totals fixed. The situation may be pictured thus:

		Total
		fixed
		fixed
Total fixed fixed		fixed

If we insert a number into any one of the cells, it will automatically determine the other three numbers in the cells that have been left blank. Consequently, we are "free", in the sense of degrees of freedom, to assign one number. There is therefore just one degree of freedom. (One may ask: Are we "free" when in principle all numbers must be positive or zero in a contingency table? For the determination of degrees of freedom, this restriction is not an issue. What matters is the dimensionality of the space that the random variables, O_i, can move in, not the size of the region.)

Since, in a 2 × 2 contingency table, all absolute differences $|O_i - E_i|$ are equal, using the foregoing notation, we may write

$$\chi^2 = (a - E_1)^2 \left[\frac{1}{E_1} + \frac{1}{E_2} + \frac{1}{E_3} + \frac{1}{E_4} \right],$$

or

$$\chi^2 = \left(a - \frac{(a + b)(a + c)}{n} \right)^2 \left[\frac{n}{(a + b)(a + c)} + \frac{n}{(a + b)(b + d)} \right.$$

$$\left. + \frac{n}{(c + d)(a + c)} + \frac{n}{(c + d)(b + d)} \right]. \tag{1}$$

This χ^2 is functionally related to usual tests of significance for the difference of two proportions. (See *PWSA*, p. 320.) If we let

$$\bar{p}_1 = \frac{a}{a + b}, \qquad \bar{p}_2 = \frac{c}{c + d}, \qquad \bar{p} = \frac{a + c}{n},$$

then to assess the difference $\bar{p}_1 - \bar{p}_2$ and thus test whether $p_1 - p_2$ differs from zero, we look at the critical ratio

$$z = \frac{\bar{p}_1 - \bar{p}_2}{\sqrt{\bar{p}(1 - \bar{p}) \left(\dfrac{1}{a + b} + \dfrac{1}{c + d} \right)}}. \qquad (2)$$

Or we might have taken proportions by columns rather than rows, and then we would get the middle expression of equation (3) below.

How does the chi-square value (1) relate to these usual test statistics for differences in proportions? It takes some gumption plus algebraic skill to show that, if we square the value of z given in (3) below, we get the chi-square value (1). And it takes more of the same to show that the two forms of z in (3) are equal: This equality tells us that we get the same result whether we test for a difference in proportions for rows or for columns.

$$z = \frac{\dfrac{a}{a + c} - \dfrac{b}{b + d}}{\sqrt{\dfrac{a + b}{n} \cdot \dfrac{c + d}{n} \left(\dfrac{1}{a + c} + \dfrac{1}{b + d} \right)}}$$

$$= \frac{\dfrac{a}{a + b} - \dfrac{c}{c + d}}{\sqrt{\dfrac{a + c}{n} \cdot \dfrac{b + d}{n} \left(\dfrac{1}{a + b} + \dfrac{1}{c + d} \right)}}. \qquad (3)$$

These results tell us that the χ^2 test gives us essentially the same results as we would get from the normal test based on the differences of two proportions.

Although there may be good sense in setting confidence limits on a difference in proportion, for example,

$$\bar{p}_1 - \bar{p}_2 \pm 2 \sqrt{\bar{p}(1 - \bar{p}) \left(\frac{1}{a + b} + \frac{1}{c + d} \right)},$$

testing for independence in a 2 × 2 contingency table may be unsound. So we give the following warning.

Warning. Rushing to compute a chi-square for a contingency table may be a bad habit. First, *we should have some grounds for thinking that the null hypothesis of independence is of interest before we test for it.* Second, the calculation is often just a time-filler and a ritual, and thus it may prevent us from thinking of the sort of analysis most needed. (This need may be a measure of degree of association, like a correlation coefficient!) Another possibility is that a larger table should be taken apart and studied in more sensible groupings. Perhaps in two-way tables larger than 2 × 2 there is an order for the categories of the random variables, and the simple chi-square approach does not take account of trends. In any case, one should not let the chi-square calculation for a null hypothesis like independence prevent careful thought about the appropriate analysis.

For example, suppose that you are studying the connection between time of year and temperatures in northern climates. If you have a table of mean temperatures for the four seasons and the counts represent years, you would not expect temperatures and seasons to be independent. You might be interested in just how the counts are related, but you wouldn't test for independence (not if in Boston you annually stifle in August and get your ears frozen in February). Similarly, if a pollster asks opinions from a set of people and shortly thereafter asks the same people the same questions again, he expects them to hold much the same opinions on both occasions, unless some important event has intervened. In the absence of any such important event, it would be strange for the pollster to waste his time testing to see if there is a relation between the two sets of opinions. The question for him is this: How marked is the relation? Or how does the departure from 100% agreement with the original opinions develop?

It may be worth noting that in some problems there are varying degrees of nullness that might be considered. Some examples for the 2 × 2 table are:

a) all four cells are equally likely;

b) the row totals and column totals each split 50–50, but the variables may not be independent;

c) the margins are not fixed and there may be correlation (as in the type 2 table mentioned above).

EXERCISES FOR SECTION 11–2

1. Interpret the chi-square value, 120, obtained for the 2 × 2 table tested in this section.

2. Give the algebraic details that lead to formula (1) from the equation that immediately precedes it.

3. *Correlation of cardiac symptoms* (L. B. Ellis and D. E. Harken). For 322 patients who are essentially cardiac invalids, the following table shows the distribution of the relation between the degree of mitral insufficiency and the degree of valvular calcification. Test for association between the two symptoms.

		Valvular calcification		
		Low	High	
Mitral	Low	142	75	217
insufficiency	High	45	60	105
		187	135	322

4. *Sterile bandages.* After surgeons began using clean instead of dirty bandages for wounds, the surgeon Lister went further and tried bandaging with antiseptic methods, which we shall call sterile bandaging. After surgery, data like those shown in the following table were found.

Condition of wound	Bandaging	
	Sterile	Ordinary
Well	14	9
Blood poisoning	1	6

How strong is the evidence in favor of sterile bandaging in these data?

5. *Hail suppression* (Paul Schmid). In the Swiss hail suppression experiment, seeding was intended to reduce the amount and frequency of hail. Seeding was randomly carried out on storm days. The following data were obtained on days with maximum wind velocity between 40 and 80 km/hr.

	Days without seeding	Days with seeding
Hail	5	15
No hail	55	57
Total	60	72

Test whether hail days and no-hail days are independent of seeding. What do you conclude?

★6. Show that χ^2 from equation (1) is identical with z^2 from the rightmost expression in equation (3).

★7. Show that the two forms of z in equation (3) are identical.

11–3 CORRECTION FOR CONTINUITY

Just as in Section 9–5 we recommended a continuity correction for chi-square with one degree of freedom, we now recommend a correction for continuity in the test for independence in a 2 × 2 table. As before, this correction sometimes makes a substantial difference and improvement.

If we use E_i to denote the expected values, then our corrected statistic is

$$\text{corrected } \chi^2 = (|a - E_1| - \tfrac{1}{2})^2 \left[\frac{1}{E_1} + \frac{1}{E_2} + \frac{1}{E_3} + \frac{1}{E_4} \right].$$

As in Section 9–5, if $|a - E_1| - \tfrac{1}{2}$ is negative, we do not make the correction. Let us compute the uncorrected and the corrected chi-squares for a simple 2 × 2 table.

Example 1. *Chi-square with and without the continuity correction.* Compute χ^2 and corrected χ^2 for the following table.

6	4	10
2	8	10
8	12	20

Solution. We have $E_1 = (10)(8)/20 = 4$. Similarly, $E_2 = 6$, $E_3 = 4$, and $E_4 = 6$. Then

$$\chi^2 = (6 - 4)^2(\tfrac{1}{4} + \tfrac{1}{6} + \tfrac{1}{4} + \tfrac{1}{6}) = 4(\tfrac{5}{6}) = 3.33,$$

and

$$\text{corrected } \chi^2 = (2 - \tfrac{1}{2})^2(\tfrac{5}{6}) = (\tfrac{9}{4})(\tfrac{5}{6}) = 1.875.$$

Note that the difference in these chi-square values is substantial. Similarly, the probability levels will differ considerably. To appraise these differences accurately, one would take the square roots and use a two-tailed normal probability. Recall that $\chi_1^2 = Z^2$, where Z is the standard normal random variable. Hence, $\chi_1^2 \geq 3.33$ when $Z^2 \geq 3.33$ or when $Z \geq \sqrt{3.33}$ or $Z \leq -\sqrt{3.33}$.

EXERCISES FOR SECTION 11–3

Rework the following exercises in Section 11–2 using the continuity correction.

1. Exercise 3.

2. Exercise 4.

3. Exercise 5.

4. Make an accurate appraisal of the difference in the two chi-square values of Example 1, as suggested at the end of the example.

11–4 $r \times c$ CONTINGENCY TABLES

Occasionally it makes sense to test for independence in a contingency table larger than 2×2. Usually it doesn't, however, because independence is a very special relation between two attributes.

Example 1. *Mendel's data.* In the following table, the letters at the top and left side are symbols for categories of garden peas. Test the data for independence.

	AA	*Aa*	*aa*	Total
BB	38	60	28	126
Bb	65	138	68	271
bb	35	67	30	132
Total	138	265	126	529

Solution. We get expected values from the formula (row total)(column total)/n. The details of computing chi-square are left as an exercise. Once we get χ^2, the question is: How many degrees of freedom do we have? The following skeleton table represents the situation schematically.

✓	✓	×	fixed
✓	✓	×	fixed
×	×	×	fixed
fixed	fixed	fixed	fixed

The checked cells can be freely filled in. Once these 4 cells are filled, the counts in the rest of the cells (marked ×) are fixed because of the fixed marginals. So there are 4 degrees of freedom.

Computation shows that $\chi^2_4 = 1.85$. This result suggests that the inheritances of the two attributes of peas (color and texture) is approximately independent. The probability of a poorer fit is 0.76.

In general, when we are testing for independence, *an* $r \times c$ *contingency table has* $(r - 1)(c - 1)$ *degrees of freedom.* Reason: We can fill in $c - 1$ cells in the first row, and then the cth cell is fixed because of the marginal total. This can be done for $r - 1$ rows. The remaining rth row is fixed because of the bottom marginals. We can thus freely fill in $c - 1$ cells in $r - 1$ rows. Hence we have $(r - 1)(c - 1)$ degrees of freedom.

Example 2. *Air Force accidents.* For one period of time, Table 11–1 gives the numbers of U.S. Air Force pilots involved in accidents, classified by type of service. Use the χ^2 statistic to assess the agreement in the proportions of pilots having accidents in the three types of service.

Solution. The bottom panel of Table 11–1 shows the expected values, if we assume that the number of accidents is proportional to the number of pilots. The chi-

Table 11–1

Numbers of U.S. Air Force pilots having accidents,
by type of service (Webb and Jones)

	Bomber	Transport	Liaison	Totals
Observed				
No accidents	26,307	36,244	8,183	70,734
Accidents	291	389	148	828
Totals	26,598	36,633	8,331	71,562
Expected				
No accidents	26,290	36,209	8,235	70,734
Accidents	308	424	96	828
Totals	26,598	36,633	8,331	71,562

square calculation is the sum of six terms. The three terms associated with cells having no accidents will make small contributions since their numerators are modest and denominators large; this is a common occurrence when we deal with rare events like accidents and one row contains nearly all the cases.

$$\chi^2 = 17^2 \left(\frac{1}{308} + \frac{1}{26,290} \right) + 35^2 \left(\frac{1}{424} + \frac{1}{36,209} \right) + 52^2 \left(\frac{1}{96} + \frac{1}{8,235} \right)$$

$$= 0.94 + 0.01 + 2.89 + 0.03 + 28.17 + 0.33 \approx 32.$$

This χ^2 has $(2 - 1)(3 - 1) = 2$ degrees of freedom, so either its critical ratio should be computed for a distribution with mean 2 and standard deviation 2, or the probability of a more extreme deviation should be looked up in a table. Obviously the result is enormous, largely owing to the liaison activity. The large-sized cells associated with no accidents made little contribution to the total. When numbers are as large as these, quite small departures from independence are readily detected. We conclude that type of service makes a difference in the fraction of pilots having accidents. The reason is still undiscovered: Are the pilot pools different, is the liaison duty more difficult, or does it offer more hours of exposure? The investigation has just begun.

EXERCISES FOR SECTION 11–4

1. Compute chi-square for the data in Example 1.

2. *Spinning warped pennies?* The children of R. E. Beckwith spun pennies that had been minted in different years and were in good condition, on a smooth surface to see if there had been a change in the probability of tails through the years. Each penny

was spun 100 times. Use the chi-square test to test for change. (The D stands for Denver mint.)

Penny	Tails	Heads	Total
1940	49	51	100
1955D	50	50	100
1961D	61	39	100
1962D	74	26	100
1963D	84	16	100
1964D	86	14	100

(For a few years, some pennies were dish-shaped.)

3. *Lobster catches* (H. Thomas). The following table gives numbers of lobsters caught by each of three types of creels, for 10 fishing days. Test for independence between fishing day and creel type. No matter how the test comes out, explain what such independence means.

					Day					
Type	10	11	12	13	14	15	16	17	18	19
Standard	2	2	4	2	3	5	2	4	4	3
Fine mesh cat-walk	2	2	3	3	2	10	1	6	7	5
Non-escape	2	4	2	0	4	4	4	6	3	6

4. *Status of women.* At Harvard University, numbers of women holding different types of posts on the faculty were tabulated for two different academic years. Obviously there were more women on the faculty in 1968–1969. But had the structure of the distribution changed? Use the chi-square test to find out. If the structures are significantly different, explain how they changed.

	1959–60	1968–69
Full Professors	4	3
Associate Professors	8	8
Assistant Professors	8	17
Instructors	52	69
Lecturers	28	50
Research Associates, etc.	107	244
Deans, etc.	18	41
Directors, etc.	68	142

Source: C. W. Bynum and J. M. Martin, *Radcliffe Quarterly*, June, 1970, Vol. 52, No. 2, p. 13.

Comment. Since the same persons may be holding full professorships, deanships, or directorships, at both dates, some independence may be lost between the individual observations making up the two columns.

5. *Taste preferences.* Twenty-five men and twenty-five women scored a new one-drink-meal according to a scale running from excellent ($+3$) to very poor (-3). After deciding on appropriate pooling, test for differences in preference between the sexes.

	$+3$	$+2$	$+1$	0	-1	-2	-3	Total
Males	2	5	10	7	1	0	0	25
Females	3	6	9	5	1	1	0	25
Total	5	11	19	12	2	1	0	50

Source: E. Street and M. Carroll. "Preliminary Evaluation of a New Food Product" in J. Tanur *et al.* (Eds.), *Statistics: A Guide to the Unknown*, Holden-Day, San Francisco, 1972.

6. *Survival by season.* The following table gives, by season, numbers of 100-day survivals and deaths after X-ray therapy for carcinoma of bronchus. Is the seasonal effect statistically significant at the 5% level?

	Survived	Died	Total
Spring (March, April)	24	10	34
Summer (May, June, July, August)	60	22	82
Autumn (September, October)	25	10	35
Winter (November, December, January, February)	30	30	60

7. (Continuation) In the data above, consider the numbers of months assigned to the seasons, and see if the total numbers being treated are proportional to the numbers of months? Use the χ^2 test with 0.05 significance level.

8. In the Swiss hail suppression experiment, various sections of the target area were randomly seeded or not seeded after a forecast that a storm worth seeding was at hand. Test whether, for hail days, the target section is independent of seeding.

	Number of hail days		
Section	Unseeded	Seeded	Total
A	7	8	15
B	14	15	29
C	5	14	19
D	9	19	28
	35	56	91

Source: Paul Schmid, "On 'Grossversuch III', a randomized hail suppression experiment in Switzerland," *5th Berkeley Symposium*, Vol. 5, p. 145.

9. *Blood pressure* (E. A. Thacker).   Students who on a first test had abnormally high or low blood pressure were tested later.   The numbers remaining the same or returning to normal at later rechecks are shown in Table 11–2.   Test for association between later state and initial state.   Interpret the result.

Table 11–2
Number of students in various categories

Later blood pressure	Initially high (above 150 mm Hg)	Initially low (below 104 mm Hg)
Remained unchanged	96	56
Returned to normal at first recheck	251	192
Returned to normal at second or later recheck	40	69
Totals	387	317

Source: E. A. Thacker, "A comparative study of normal and abnormal blood pressures among university students, including the cold-pressure test," *American Heart Journal*, Vol. 20: 1940, pp. 89–97 (p. 96).

11–5 ALTERNATIVE WAYS OF EXAMINING ASSUMPTIONS FOR SEVERAL BINOMIAL SAMPLES

Treating several large binomial samples. An investigation may seem to call for binomial methods when it does not call for them or when the assumption of binomiality needs scrutiny.

Example 1. *Perception experiment.* A psychological experiment in perception may have several subjects, each of whom has a large number of trials, say 1000. Suppose that 5 subjects have success counts of 800, 600, 300, 700, 400. Does the fraction of successes exceed $\frac{1}{2}$?

Solution. It is tempting to combine the samples into one big binomial and to test the pooled $\bar{p}$ against $\frac{1}{2}$ with variance

$$\sigma_{\bar{p}}^2 = pq/n = (\tfrac{1}{2})(\tfrac{1}{2})/1000n.$$

That test may be all right, but it may not.   For our 5 subjects, the foregoing approach would give

$$\bar{p} = \tfrac{2800}{5000} = 0.56, \quad \text{and} \quad \sigma_{\bar{p}} = \sqrt{(\tfrac{1}{2})(\tfrac{1}{2})/5000} \approx 0.007.$$

Therefore $\bar{p}$ is more than 8 standard deviations from $\frac{1}{2}$, a resoundingly significant result.   We are suspicious of the result because only 3 out of 5 subjects did better than $\frac{1}{2}$, and yet we are reporting a difference of 8 standard deviations. What is the trouble?

The trouble is that the responses contain much more than binomial variation. If we had binomials with $p = \frac{1}{2}$, the $\bar{p}$'s for individuals would be subject to the standard deviation

$$\sigma_{\bar{p}} = \sqrt{pq/n} = \sqrt{(\tfrac{1}{2})(\tfrac{1}{2})/1000} \approx 0.016;$$

and yet only 5 $\bar{p}$'s range from 0.3 to 0.8, a distance of 30 standard deviations! The subjects are performing differently.

If one treats the data as measurements and uses the t-test with 4 degrees of freedom, he gets $t_4 \approx 0.65$, a sensible result not far from 0, the null value. Our conclusion is that although subjects do not each have scores that vary binomially around $p = \frac{1}{2}$, their average performance is still compatible with an average $p = \frac{1}{2}$ based on only 5 subjects. A more telling test of the $p = \frac{1}{2}$ hypothesis will require more subjects if we want to know whether the average p is near $\frac{1}{2}$.

A genuine example where the binomiality may be nearly met follows.

Example 2. *Spock jurors* (Hans Zeisel). Dr. Benjamin Spock, author of a famous book on baby care, and others were initially convicted of conspiracy in connection with the draft during the Vietnam war. The defense appealed, one ground being the sex composition of the jury panel. The jury itself had no women, but chance and challenges could make that happen. Although the defense might have claimed that the jury lists (from which the jurors are chosen) should contain 55% women, as in the general population, they did not. Instead they complained that six judges in the court averaged 29% women in their jury lists, but the seventh judge, before whom Spock was tried, had fewer, not just on this occasion but systematically. The last 9 jury lists for that judge contained the following counts:

	Women	Men	Total	Proportion women
	8	42	50	0.16
	9	41	50	0.18
	7	43	50	0.14
	3	50	53	0.06
	9	41	50	0.18
	19	110	129	0.15
	11	59	70	0.16
	9	91	100	0.09
	11	34	45	0.24
Grand totals	86	511	597	0.144

How shall we test the claim?

Solution

a) *The sign test.* We take as the null hypothesis $p = 0.29$. We may think of each of the 9 results as one binomial trial with $P(\bar{p} \leq 0.29) \approx \frac{1}{2}$, where $\bar{p}$ is the proportion of women on a jury list. Since 9 jury lists out of 9 came out lower than 0.29, the average of the other judges, the one-sided significance level is about $(\frac{1}{2})^9 = 0.002$.

b) *The t-test.* Using the variability of the percentage of women in the jury lists, we assume as the null hypothesis that the mean percentage is $\mu_0 = 29\%$. The variability of the percentages around the sample mean $\bar{X} = 14.4\%$ gives us $s_{\bar{X}} = 1.76$. The *t*-test gives

$$t = \frac{\bar{X} - \mu_0}{s_{\bar{X}}} = \frac{14.4 - 29}{1.76} \approx -8.3,$$

with 8 degrees of freedom. From Table A–3, we find that

$$P(t_8 \geq 8.3) \approx 0.$$

This treatment uses the data a little more effectively than the sign test does, and it gives a more significant result

c) *The pooled binomial.* We can use the χ^2 test to see whether the variability in the proportions for the 9 lists is compatible with random drawings from the same binomial. Using the data shown in the table, we find a χ^2 value of 10.6. with 8 degrees of freedom. This is neither so large nor so small that we have reason to reject the binomiality. Consequently, it is reasonable to pool the binomial counts. For the sample in the table, we have a grand observed proportion of women $\bar{p} = 0.144$, and we estimate the standard deviation of $\bar{p}$ as

$$\sqrt{\bar{p}\bar{q}/n} = \sqrt{0.144(0.856)/597} \approx 0.0144.$$

Since the mean proportion for the other judges is 0.29, the departure is $0.144 - 0.29 = -0.146$, which is low by about 10 standard deviations. The deviation would have been slightly smaller if we had used the $p = 0.29$ of the other judges to estimate the standard deviation. But either way, the departure is large, and roughly compatible with the result of the *t*-test. All three methods point to the conclusion that the jury lists of the seventh judge regularly had a smaller proportion of women than the lists of the other judges. For this and other reasons the conviction was set aside.

11–6 THE ESTIMATED VALUES

In the simple problems we have dealt with, it has usually been obvious how to get the values of the E_i's for the chi-square test. When the E_i's are *estimated* expected values, trouble may arise. If poor estimates of the theoretical counts are used, then one would not be able to guarantee that the approximation of χ^2 still held even for large samples.

One good way to estimate the parameters is by the method of maximum likelihood, which we have described earlier and carried out for a few simple examples. This usually requires calculus or numerical methods, often with several variables and constraints. Under very general conditions, maximum likelihood estimates produce E_i's that give the appropriate approximate chi-square distribution.

Another method that may come in handy is to find the values of the E_i's under the model that minimize chi-square itself for the particular problem. This may sometimes be convenient as a numerical device with a computer when other approaches seem out of reach. It is not only a good way to get chi-square values but a good method of estimation.

We shall not pursue these matters very far here. The student needs to be aware of the problem and to know some solutions that lead to satisfactory results. The method of moments used in Section 10–2, Example 4, is not either of those recommended here, and indeed it may not be satisfactory. As a way out, the estimate of p given in Exercise 8 of that section would give a maximum likelihood estimate that would use the first of the solutions offered here. In the actual case, the estimates of p differ little— 0.414 versus 0.404; but we are concerned about what happens over the whole distribution. The minimum chi-square estimate is 0.405.

Example 1. *Three categories.* To determine the proportions (p_1, p_2, p_3), $\sum p_i = 1$, of three kinds of vehicles along a road, an investigator took three samples of 100 each. In the first sample he counted the number of type 1 vehicles, in the second he counted the number of type 2, and so on. In the first sample he found 22 of type 1, 78 of types 2 and 3. In the second sample he found 13 of type 2, 87 of types 1 and 3. In the third he found 57 of type 3, 43 of types 1 and 2. Estimate p_1, p_2, p_3.

Solution

a) *Maximum likelihood.* The likelihood for these three independent samples is

$$L = p_1^{22}(1 - p_1)^{78}p_2^{13}(1 - p_2)^{87}p_3^{57}(1 - p_3)^{43}.$$

We need to maximize L by finding values of (p_1, p_2, p_3) subject to the conditions $0 < p_i < 1$, $\sum p_i = 1$. The naive estimates would be $\bar{p}_1 = 0.22$, $\bar{p}_2 = 0.13$, $\bar{p}_3 = 0.57$. Unfortunately these add to 0.92 instead of 1.00. As a reasonable starting place for our hunt for a numerical maximum, we might divide these $\bar{p}_i$ by 0.92 and then explore values near the resulting triplet, (0.239, 0.141, 0.620). On a high-speed computer this is quick and easy, and we find the maximum likelihood estimate to be (0.247, 0.148, 0.605).

b) *Minimum chi-square.* Let $n = 100$, $\hat{p}_i$ be the estimates, and $\hat{q}_i = 1 - \hat{p}_i$. Then we have

$$O_1 = 22, \quad \bar{O}_1 = 78; \quad O_2 = 13, \quad \bar{O}_2 = 87; \quad O_3 = 57, \quad \bar{O}_3 = 43;$$

$$E_1 = n\hat{p}_1, \quad \bar{E}_1 = n\hat{q}_1; \quad E_2 = n\hat{p}_2, \quad \bar{E}_2 = n\hat{q}_2; \quad E_3 = n\hat{p}_3, \quad \bar{E}_3 = n\hat{q}_3;$$

$$\chi^2 = \sum_{i=1}^{3} \left[\frac{(O_i - E_i)^2}{E_i} + \frac{(\bar{O}_i - \bar{E}_i)^2}{\bar{E}_i} \right].$$

We fit numerically so that $\sum p_i = 1$, and so that χ^2 is a minimum. With 6 cells, as soon as one O_i is known in a pair, so is $\bar{O}_i$. Thus we have 3 degrees of freedom to begin with. We fit when we choose p_1 and p_2, because then p_3 is forced. We have fitted 2 parameters in addition to the 3 n's (each 100), and so have 1 degree of freedom. By numerical methods on a high-speed computer we find the minimizing triplet $(p_1, p_2, p_3) = (0.247, 0.149, 0.604)$, $\chi^2_{\min} = 1.16$, almost the same as the maximum likelihood triplet.

To minimize, one can choose a reasonable starting triplet and then explore values of p near it. For example, one could hunt first for the best triplet whose p's are whole hundredths. Then, having found the best of these, one could search in the neighborhood for the best triplet whose p's are whole thousandths.

CHAPTER 12

The Kruskal-Wallis Statistic

Foreword. For those who have studied the analysis of variance, it will be clear that the Kruskal-Wallis statistic is a one-way analysis of variance using ranks rather than the original measurements. For those who have not studied analysis of variance, this material will be an introduction. It is self-contained.

12–1 TESTING THREE INDEPENDENT SAMPLES

In Chapter 3, we used ranks and the Mann-Whitney test to see if two independent samples of measurements might have come from the same distribution. Since the Mann-Whitney test proved both effective and convenient, it is natural to expect that the use of ranks might help us to make decisions about three independent samples and their associated distributions. The following discussion suggests a a way of defining a statistic for testing three independent samples.

Suppose that we have the Scholastic Aptitude Test scores for three groups of students.

Group 1: 772 764 600 564
Group 2: 792 612 592
Group 3: 752 680 624 580 572

Assuming that these groups represent three independent samples, how might we test these samples to see if they could reasonably have come from the same distribution?

Let us rank the scores and get the rank sums of the three samples, as follows:

Score:	792	772	764	752	680	624	612	600	592	580	572	564
Rank:	1	2	3	4	5	6	7	8	9	10	11	12
Group:	2	1	1	3	3	3	2	1	2	3	3	1

If R_i is the rank sum of the ith sample, then

$$R_1 = 2 + 3 + 8 + 12 = 25,$$

$$R_2 = 1 + 7 + 9 = 17,$$

$$R_3 = 4 + 5 + 6 + 10 + 11 = 36.$$

As a second step, let us look at the mean rank for each sample, R_i/n_i ($n_1 = 4$, $n_2 = 3$, $n_3 = 5$), and see how far it is from the grand mean of all the ranks, $(N + 1)/2 = 13/2$. We get

$$\tfrac{25}{4} - \tfrac{13}{2}, \qquad \tfrac{17}{3} - \tfrac{13}{2}, \qquad \tfrac{36}{5} - \tfrac{13}{2},$$

or

$$-\tfrac{1}{4}, \qquad -\tfrac{5}{6}, \qquad \tfrac{7}{10}. \tag{1}$$

These three deviations provide some information, but they fail to give us what we want: a *single* measure of dispersion for the three sample means.

How about weighting each deviation in (1) with the number of measurements in the corresponding sample and taking the sum? We get

$$4(-1/4) + 3(-5/6) + 5(7/10).$$

It appears that we might have something here—until we notice that the sum is $-1 + (-5/2) + 7/2 = 0$. At first this result may seem fortuitous, but a little thought tells us that there is nothing fortuitous about it. We have just reminded ourselves of the fact that the sum of the deviations from a mean is zero.

The next suggestion is likely to be this: Take the weighted sum either of the absolute values or of the squared values. This has a familiar ring: We met a similar situation in defining variance. For that definition, we chose the squared values because absolute values are often intractable in computation, and we make the same choice here. We denote our single measure of dispersion by D, where

$$D = 4(-1/4)^2 + 3(-5/6)^2 + 5(7/10)^2$$

$$= 1/4 + 25/12 + 49/20 = 287/60.$$

Note that in terms of ranks the statistic D is *the sum of the squares of the deviations of the sample means from the grand mean, with each squared deviation weighted by the number of measurements in the sample.*

For the statistic D to help us make decisions about the distributions from which the samples are drawn, we need to know something about the distribution of D. We have a general notion that large values of D suggest that the three samples are not from the same distribution. But how large must D become in order to justify our rejecting the hypothesis of equal distributions at, say, the 5% level? The answers to such questions are given in the sections that follow. We shall use there the ideas of the foregoing discussion to define a statistic that measures dispersion for several independent samples, and we shall study its distribution.

EXERCISES FOR SECTION 12–1

Exercises 3 and 4 prepare the student for the next section.

Use the method of this section to compute the statistic D for the sets of three independent samples in Exercises 1 and 2.

1. Sample 1: 88, 74, 64, 61
   Sample 2: 78, 75, 70, 60
   Sample 3: 85, 80, 66, 52

2. Sample 1: 24.1, 20.2, 17.5, 16.2, 13.8
   Sample 2: 22.7, 21.2, 18.4, 15.2, 14.6
   Sample 3: 20.5, 19.8, 19.5, 15.6

3. Prove that $\Sigma\, n_i[R_i/n_i - (N + 1)/2] = 0$, when n_i is the number of measurements in sample i, $N = \Sigma\, n_i$, and R_i is the sum of the ranks of the n_i measurements in sample i.

4. Check Appendix VI–2 for the formula for the variance of the mean of a sample of size n_i drawn from a finite population of N elements. Use that formula to find the variance of R_i/n_i, the mean of the sum of the ranks in a random sample of n_i, drawn from the first N positive integers without replacement.

12–2 THE KRUSKAL-WALLIS STATISTIC FOR SEVERAL INDEPENDENT SAMPLES

Given three or more independent samples of measurements, we may wish to discover whether or not these samples could reasonably have come from distributions with the same location (such as mean or median). To get a test statistic for such an investigation, we shall generalize the findings of the last section.

To this end, we shall use the following notation:

N denotes the total number of measurements in C samples.

$n_1, n_2, \ldots, n_C$ denote, respectively, the number of measurements in the 1st, 2nd, $\ldots$, Cth samples, $\Sigma n_i = N$.

$R_1, R_2, \ldots, R_C$ denote, respectively, the rank sums of the 1st, 2nd, $\ldots$, Cth samples, when all N are ranked together.

We shall use these symbols to develop formulas for the special case in which $C = 3$. The algebra for the general case is identical but more laborious; and the result for the general case is evident from an inspection of the results obtained for $C = 3$.

Taking our cue from the discussion and results of Section 12–1, we shall define, as our measure of dispersion of sample means from the grand mean, the statistic

$$D = n_1\left(\frac{R_1}{n_1} - \frac{N+1}{2}\right)^2 + n_2\left(\frac{R_2}{n_2} - \frac{N+1}{2}\right)^2 + n_3\left(\frac{R_3}{n_3} - \frac{N+1}{2}\right)^2. \quad (1)$$

Note that this random variable D weights each squared deviation of a sample mean R_i/n_i from the grand mean $(N + 1)/2$ by the sample size n_i. Thus *the effect of a departure from the grand mean is measured* (a) *by the square of its size*, and (b) *by the number of ranks involved*. A squared deviation counts once for each measurement in its sample.

We can simplify expression (1) by squaring out and collecting terms:

$$
D = n_1 \cdot \frac{R_1^2}{n_1^2} + n_2 \cdot \frac{R_2^2}{n_2^2} + n_3 \cdot \frac{R_3^2}{n_3^2} - 2\left(\frac{N+1}{2}\right)\left(n_1 \cdot \frac{R_1}{n_1} + n_2 \cdot \frac{R_2}{n_2} + n_3 \cdot \frac{R_3}{n_3}\right)
$$

$$
+ (n_1 + n_2 + n_3)\left(\frac{N+1}{2}\right)^2
$$

$$
= \frac{R_1^2}{n_1} + \frac{R_2^2}{n_2} + \frac{R_3^2}{n_3} - (N+1)(R_1 + R_2 + R_3) + \frac{N(N+1)^2}{4}.
$$

Recalling that $R_1 + R_2 + R_3 = N(N + 1)/2$ (the sum of the first N integers), we get

$$
D = \frac{R_1^2}{n_1} + \frac{R_2^2}{n_2} + \frac{R_3^2}{n_3} - \frac{N(N+1)^2}{2} + \frac{N(N+1)^2}{4}.
$$

Finally, we get the simplified form for expression (1),

$$
D = \frac{R_1^2}{n_1} + \frac{R_2^2}{n_2} + \frac{R_3^2}{n_3} - \frac{N(N+1)^2}{4}. \tag{2}
$$

The mean of D. We now study the distribution of D under the null hypothesis of identically distributed populations. It is convenient to use form (1) to get the mean of D. Recall from the definition of a variance that the expected value of

$$
\left(\frac{R_i}{n_i} - \frac{N+1}{2}\right)^2
$$

is just the variance of the mean rank, R_i/n_i, since $(N + 1)/2$ is its mean. We know too from Appendix VI–2 that the variance of the mean of a sample of n_i, drawn without replacement from a finite population of size N, is

$$
\frac{\sigma^2}{n_i}\left(\frac{N - n_i}{N - 1}\right), \tag{3}
$$

where σ^2 is the variance of the set of individual measurements in the population. Since our population of measurements consists of the integers $1, 2, \ldots, N$, their variance is

$$
\sigma^2 = \frac{N^2 - 1}{12}. \tag{4}
$$

We can now use (1), (3), and (4) to get the mean of D:

$$E(D) = \sum_{i=1}^{3} n_i E \left(\frac{R_i}{n_i} - \frac{N+1}{2} \right)^2 = \sum_{i=1}^{3} n_i \frac{\sigma^2}{n_i} \left(\frac{N-n_i}{N-1} \right)$$

$$= \frac{(N^2-1)(3N-N)}{12(N-1)} = \frac{2N(N+1)}{12}. \tag{5}$$

We do not reduce the fraction because we want to exhibit the pattern for the generalization that follows.

If we had C samples instead of 3 samples, the same procedures would give the simplified form of D as

$$D = \sum_{i=1}^{C} \frac{R_i^2}{n_i} - \frac{N(N+1)^2}{4} \tag{6}$$

and the expected value

$$E(D) = \frac{(C-1)N(N+1)}{12}. \tag{7}$$

(See Exercise 10.)

Note 1. We are dealing with the distribution of D under the null hypothesis of identically distributed random variables.

Note 2. When $C = 3$, formula (7) reduces to formula (5), as might be expected.

Finally, to get a statistic whose mean is independent of the sample size, we need only define

$$H = \frac{D}{N(N+1)/12},$$

and this is precisely what we shall do.

Definition. *The Kruskal-Wallis Statistic.* We define the Kruskal-Wallis statistic H as follows:

$$H = \frac{12}{N(N+1)} \left(\sum_{i=1}^{C} \frac{R_i^2}{n_i} \right) - 3(N+1), \tag{8}$$

where C = the number of samples,

n_i = the number of measurements in the ith sample,

$N = \sum n_i$ = the total number of measurements,

R_i = the rank sum of the ith sample.

Note that under the null hypothesis

$$E(H) = \frac{12}{N(N + 1)} \cdot E(D) = \frac{12}{N(N + 1)} \cdot \frac{(C - 1)N(N + 1)}{12} = C - 1.$$

Ties. When ties occur, give them the average rank of the tied items, as usual.

EXERCISES FOR SECTION 12–2

1. Use formula (6) to check the value of D obtained in Section 12–1.

2. Use formula (6) to check the values of D found in Exercises 1 and 2 of Section 12–1.

   For each of the following sets of data, find H and $E(H)$.

3. The data used in the three samples of Section 12–1.

4. The data of Exercise 1, Section 12–1.

5. The data of Exercise 2, Section 12–1.

6. To test four different diets, 24 young turkeys are randomly divided into 4 groups of 6, and each group is fed a different diet. At the end of the experiment, the gain in weight for each turkey is recorded. Results are as follows:

Diet	Weight gained in pounds					
A	20.2	18.5	17.7	17.2	17.0	15.7
B	18.7	17.3	16.1	14.6	13.9	12.2
C	21.7	19.9	19.6	18.8	18.3	17.5
D	19.4	18.0	17.8	16.5	15.0	12.5

Compute H and $E(H)$ for these data.

7. The data of Table 12–1 give fuel oil consumption for 5 buses in 6 different summer

Table 12–1

Fuel consumption, gallons per 1000 miles

Month	Bus				
	1	2	3	4	5
May		114	111		
June		108	109	107	
July		110	112	106	105
August		110	110	116	98
September			114	111	102
October	118	110	115	102	

Source: A. P. Davies and A. W. Sears, "Some makeshift methods of analysis applied to complex experimental results," *Applied Statistics*, 1955, Vol. 4, p. 48.

months.   Disregarding possible differences between months, thereby treating the measurements as independent, compute the value of the Kruskal-Wallis statistic H, and compare it with $E(H)$. What conclusion do you draw about differences between buses in fuel consumption.

8. Carry out the steps leading to the generalization given by equation (7).

9. Derive the mean value of H by starting with formula (8) and taking expected values. [*Hint:* $E(X^2) = \sigma_X^2 + \mu_X^2$.]

12–3 RELATION OF H TO MANN-WHITNEY STATISTIC T; DISTRIBUTION OF H

To gain some insight into the distribution of H, we shall discuss the special case where $C = 2$. This discussion will give us an inkling of what to expect when $C > 2$.

If $C = 2$, we might expect a close relation between the Kruskal-Wallis statistic H and the Mann-Whitney statistic T (the sum of the ranks of a sample of size n, when ranked with another of size m). And there is. Recall that when two samples are randomly drawn from identically distributed populations and m and n are both large, the random variable T is approximately normally distributed, and the corresponding random variable with zero mean and unit variance is

$$Z = \frac{T - \left(\dfrac{N+1}{2}\right) n}{\sigma_T}, \tag{1}$$

where Z is approximately distributed according to the *standard* normal distribution, and

$$\sigma_T^2 = \frac{n(N-n)(N+1)}{12}.$$

In our Kruskal-Wallis notation, $T = R_1$, $n = n_1$, and N is the total number of measurements, as before. Therefore, we can write alternatively

$$Z = \frac{R_1 - \left(\dfrac{N+1}{2}\right) n_1}{\sqrt{\dfrac{n_1(N-n_1)(N+1)}{12}}}. \tag{2}$$

We shall now exhibit the relation of the Kruskal-Wallis statistic H to the standardized Mann-Whitney statistic Z. The idea is to use algebra to transform H so as to get an expression involving the right-hand side of (2). For $C = 2$, we have

$$H = \frac{12}{N(N+1)} \left(\frac{R_1^2}{n_1} + \frac{R_2^2}{n_2}\right) - 3(N+1).$$

By taking a common denominator on the right, we get

$$H = \frac{12}{n_1 n_2 N(N + 1)} \left[n_2 R_1^2 + n_1 R_2^2 - n_1 n_2 N \frac{(N + 1)^2}{4} \right]. \tag{3}$$

Since

$$R_2 = \frac{N(N + 1)}{2} - R_1,$$

the expression in the square brackets becomes

$$n_2 R_1^2 + n_1 \left(\frac{N(N + 1)}{2} - R_1 \right)^2 - n_1 n_2 N \frac{(N + 1)^2}{4}. \tag{4}$$

By squaring out and rearranging, we can transform expression (4) into

$$N \left(R_1 - n_1 \frac{N + 1}{2} \right)^2. \tag{5}$$

Now we replace the expression in the square brackets in (3) by the equivalent expression in (5) and get

$$H = \frac{12}{n_1 n_2 N(N + 1)} \left[N \left(R_1 - n_1 \frac{N + 1}{2} \right)^2 \right] = \frac{\left(R_1 - n_1 \frac{N + 1}{2} \right)^2}{n_1(N - n_1)(N + 1)/12}.$$

Therefore,

$$H = Z^2. \tag{6}$$

We have noted that, for large n_1 and $N - n_1$, the standardized Mann-Whitney statistic Z has approximately the standard normal distribution. It follows that, when $C = 2$, the Kruskal-Wallis statistic H has approximately the distribution of the *square* of a standard normal random variable. Hence, when $C = 2$, H is approximately χ_1^2, a chi-square random variable with $C - 1 = 1$ degree of freedom. (See Section 8–2.) For this special case, $E(H) = E(\chi_1^2) = 1$, which checks with the result obtained in Section 12–2.

The foregoing results for the special case $C = 2$ make plausible the following theorem, which we shall accept without proof.

Theorem. *Distribution of the Kruskal-Wallis Statistic H.* If C samples of sizes $n_1, n_2, \ldots, n_C$ are randomly drawn from identically distributed populations, and if the values of n_i are large, then the Kruskal-Wallis statistic,

$$H = \frac{12}{N(N + 1)} \sum_{i=1}^{C} \frac{R_i^2}{n_i} - 3(N + 1), \tag{7}$$

is approximately distributed as a chi-square random variable with $C - 1$ degrees of freedom.

The degree of freedom we lose comes from the linear constraint we discussed in Section 12–1: The weighted deviations add to zero. (See also Exercise 3 of Section 12–1.)

As a consequence of this theorem, we can get information about the distribution of H under the null hypothesis of identically distributed populations by consulting the chi-square tables. In practice, we may expect reasonably good approximations when all the n_i's are greater than or equal to 3, $C > 2$. We choose the number 3 because the sum of a random sample of 3 integers from the first $N \geq 9$ is approximately normal, whereas the sum for 1 is uniform and for 2 is triangular. For $C = 2$, one should refer to the Mann-Whitney chapter.

Example 1. The Kruskal-Wallis statistic H is computed for five independent samples, each of which contains six measurements. The result is $H = 10.2$. Find, under the null hypothesis, $P(H \geq 10.2)$.

Solution. H is approximately distributed as χ^2_4, a chi-square random variable with $5 - 1 = 4$ degrees of freedom. From the chi-square table, we find that the $P(H \geq 10.2)$ lies between 0.05 and 0.02. Interpolation gives $P(H \geq 10.2) = 0.04$.

Example 2. Twenty individuals with cancer of the same type and at the same stage are given four types of treatment, and the records of their progress are used to compare the treatments. The individuals are randomly divided into four groups of five, each group is given a different treatment, and the survival time of each individual in years is recorded.

Type of treatment	Number of years survived				
A	14.2	10.6	9.4	5.6	2.4
B	12.8	12.3	6.4	6.1	1.6
C	11.5	10.1	5.1	5.0	4.8
D	14.9	13.7	8.5	7.7	5.9

Compute H. Under the null hypothesis of identically distributed populations, find the probability of getting a value of H greater than the one computed.

Solution. We first rank the 20 measurements.

14.9	14.2	13.7	12.8	12.3	11.5	10.6	10.1	9.4	8.5
1	2	3	4	5	6	7	8	9	10
D	A	D	B	B	C	A	C	A	D

7.7	6.4	6.1	5.9	5.6	5.1	5.0	4.8	2.4	1.6
11	12	13	14	15	16	17	18	19	20
D	B	B	D	A	C	C	C	A	B

Hence

$$R_1 = 2 + 7 + 9 + 15 + 19 = 52; \qquad R_2 = 4 + 5 + 12 + 13 + 20 = 54;$$

$$R_3 = 6 + 8 + 16 + 17 + 18 = 65; \qquad R_4 = 1 + 3 + 10 + 11 + 14 = 39.$$

Then $\sum R_i^2 = 52^2 + 54^2 + 65^2 + 39^2 = 11{,}366$, and formula (7) gives, since all n_i are equal,

$$H = \frac{12}{(20)(21)} \cdot \frac{11{,}366}{5} - 3(21) = 1.9.$$

Since H is approximately a chi-square random variable with $4 - 1 = 3$ degrees of freedom, we find from the chi-square table that

$$P(H \geq 1.9) = 0.60, \quad \text{approximately.}$$

There is little evidence here of differences in the effects of treatments on survival.

EXERCISES FOR SECTION 12-3

1. Given six independent samples with $H = 9.2$, find $P(H \geq 9.2)$ if the samples are randomly drawn from identically distributed populations.

2. Rework Example 1, given that there are seven independent samples.

3. Complete the algebraic details of the transformation of expression (4) into expression (5).

Assuming that the samples are randomly drawn from identically distributed populations, find, in each of the following exercises, the probability that the determined value of H may be exceeded.

4. Exercise 2, Section 12-1.

5. Exercise 7, Section 12-2. What do you conclude about differences between buses?

6. Three chemical sprays for killing flies are tested, and the percentage of kills is recorded as follows:

Brand A:	72	65	67	75	62	73
Brand B:	55	59	68	70	53	50
Brand C:	64	74	61	58	51	69

Compute H. Under the null hypothesis of no differences between brands, find the probability that H may have a greater value than that computed? What do you conclude?

12-4 TABLES FOR THE KRUSKAL-WALLIS STATISTIC

For small values of C and of the n_i, tables of the Kruskal-Wallis statistic are available. But because there are so many sets of n_i for a given $C > 2$ and so many possible values of H, or equivalently, of $\sum R_i^2 / n_i$, we cannot afford very extensive tabulation.

Table 12-2

Exact distribution of Kruskal-Wallis statistic when null hypothesis is true, $C = 3$, $n_1 = n_2 = n_3 = 3$

h	$\sum R_i^2/n_i$	Frequency	$P(H \le h)$	$P(H > h)$
0.0	225.00	12	0.007	0.993
0.09	225.67	108	0.071	0.929
0.27	227.00	84	0.121	0.879
0.36	227.67	84	0.171	0.829
0.62	229.67	180	0.279	0.721
0.80	231.00	96	0.336	0.664
1.07	233.00	60	0.371	0.629
1.16	233.67	144	0.457	0.543
1.42	235.67	54	0.489	0.511
1.69	237.67	120	0.561	0.439
1.87	239.00	96	0.618	0.382
2.22	241.67	36	0.639	0.361
2.40	243.00	36	0.661	0.339
2.49	243.67	72	0.704	0.296
2.76	245.67	72	0.746	0.254
3.20	249.00	36	0.768	0.232
3.29	249.67	60	0.804	0.196
3.47	251.00	48	0.832	0.168
3.82	253.67	48	0.861	0.139
4.27	257.00	12	0.868	0.132
4.36	257.67	54	0.900	0.100
4.62	259.67	24	0.914	0.086
5.07	263.00	24	0.929	0.071
5.42	265.67	36	0.950	0.050
5.60	267.00	36	0.971	0.029
5.69	267.67	6	0.975	0.025
5.96	269.67	24	0.989	0.011
6.49	273.67	12	0.996	0.004
7.20	279.00	6	1.000	0.0

Total 1680

For $C = 3$, with $n_1 = n_2 = n_3 = 3$, Table 12-2 gives the exact distribution of H and of its equivalent, except for a linear transformation, $x = \sum R_i^2/n_i$. Since the total number of arrangements of 9 different things (the ranks) into 3 groups of three is given by the multinomial coefficient

$$\frac{9!}{3!\,3!\,3!} = 1,680,$$

1680 is the total frequency. Table 12–2 gives the exact frequencies and the cumulative relative frequencies in both directions. The table shows that to be at or above the upper 5% value, H must equal or exceed $h = 5.60$, or equivalently, x must equal or exceed 267.00.

No value of H is near a lower 5% point in Table 12–2: The choices there are 12.1%, 7.1%, or 0.7%.

Table A–14 at the back of the book gives additional tables for H. To save space, we have had to limit severely the tabled values. A complete distribution for just $C = 3$, $n_1 = 7$, $n_2 = 3$, $n_3 = 2$ would take 191 lines. In computing full tables for this statistic, we found our offices overflowing with computer output.

Example 1. Let us consider the following rankings and compute H, $P(H \leq h)$, and the corresponding results for χ^2.

A:	7	8	9	$R_1 = 24$	$R_1^2/3 = 192$
B:	1	3	5	$R_2 = 9$	$R_2^2/3 = 27$
C:	2	4	6	$R_3 = 12$	$R_3^2/3 = 48$

$$x = \text{Total} = 267$$

Solution. The probability of 267 or less from Table 12–2 is 0.971.

Substituting $x = 267$ into formula (8) of Section 12–2 gives

$$\chi_2^2 \approx H = \frac{12x}{N(N+1)} - 3(N+1) = \frac{12(267)}{9(10)} - 3(10) \approx 5.6.$$

Careful interpolation in a chi-square table gives $P(\chi^2 < 5.6) \approx 0.94$. Although 0.94 looks close to 0.97, it is well to compare the complements. One is 0.06, the other 0.03. Looked at in this way, the approximation is off by 100%. On the other hand, both tell you clearly that, if these numbers do come from the same population a rare event has occurred.

Figure 12–1 shows a comparison of the cumulatives of the Kruskal-Wallis H for 3 samples of 3 each with that for the chi-square distribution with two degrees of freedom.

EXERCISES FOR SECTION 12–4

1. Find 3 of the 12 sets of rankings (see Table 12–2) for 3 sets of 3 samples whose value of H is 0.

2. Repeat Exercise 1 for $H = 7.20$.

3. If $n_1 = 4$, $n_2 = 3$, $n_3 = 2$, $C = 3$, and $H = 6.75$, use Table A–14 at the back of the book to find $P(H \geq 6.75)$. Compare this result with that obtained from the chi-square approximation using Table A–4.

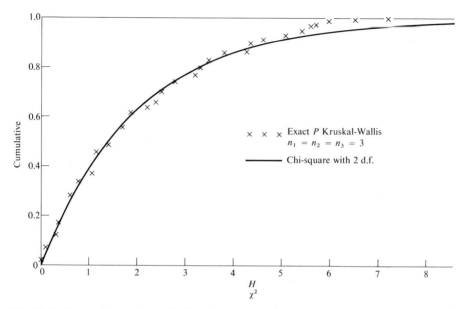

Fig. 12–1 Comparison of cumulative of exact Kruskal-Wallis statistic with that for χ^2 with 2 degrees of freedom.

4. With the help of Table A–14, for $n_1 = 2$, $n_2 = 2$, $n_3 = 1$, $C = 3$, find the exact distribution of H, and make a plot like that of Figure 12–1. What is the largest difference you observe on your graph?

Reading scores. (Data for Exercises 5, 6, and 7) The average reading test scores reported for Grade 7 by schools in 6 Manhattan districts for 1969 are tabulated below. (The national norm for Grade 7 is 7.7 for this test. The low scores are said to be partly due to a long teachers' strike.)

			District			
1	2	3	4	5	6
5.9	5.6	6.4	6.3	5.9	7.1
5.6	5.6	6.4	6.0	6.5	5.6
5.6	7.5	7.3	5.6		5.4
8.0					5.4

5. Use the Kruskal-Wallis test and Table A–14 to assess differences among districts 1, 2, and 3. Interpret the result.

6. Repeat Exercise 5 for districts 4, 5, and 6.

7. Use the Kruskal-Wallis test to assess differences among districts 1 through 6. [Compute H, and use Table A–4.] What conclusion do you draw?

8. In an analysis of the behavior of stocks of various companies, Benjamin F. King, Jr., estimated the percent of total variance of stock prices contributed by the behavior of the market. Use the Kruskal-Wallis tables to test whether there is industry-to-industry variation.

| | Percentage of variances | |
Oil	Light manufacturing	Heavy manufacturing
20.0	8.6	11.2
23.7	12.1	14.1
	7.1	31.9
	46.9	
	7.6	
	11.9	

9. The table below gives percentages of admissions to the Graduate School of Arts and Sciences at Harvard University in 1966–1967 for the 11 departments receiving the most applications. Use the tables for the Kruskal-Wallis test to decide whether there is substantial variation between the three divisions of the Graduate School in admission practices. Make any additional comment you think the data require.

| | Percentage admitted | |
Natural sciences	Social sciences	Humanities
34.5	23.0	25.4
36.0	26.6	24.6
25.0	20.3	18.8
28.2		58.3

CHAPTER 13

The Problem of *m* Rankings: The Friedman Index

13–1 THE PROBLEM OF MORE THAN TWO RANKINGS; A SPECIAL CASE

Just as the Kruskal-Wallis test generalizes the Mann-Whitney test from two samples to several, so the Friedman index extends the idea of rank correlation from two sets of data to several. To fix ideas, let us look at an instructive example.

Example 1. *Rankings of handwriting.* Five judges (Guilford, 1936, p. 247) ranked 12 samples of handwriting for readability. The results are shown in Table 13–1. Discuss the extent of agreement in the rankings.

Table 13–1
Results of rankings by 5 judges in Guilford's handwriting example

	A	*B*	*C*	*D*	*E*	*F*	*G*	*H*	*I*	*J*	*K*	*L*
1	12	8	11	10	9	6	5	4	7	1	2	3
2	11	12	10	9	7	8	6	4	1	3	5	2
3	10	11	12	9	7	8	6	5	3	4	1	2
4	9	12	11	10	8	7	4	3	5	6	2	1
5	12	10	11	9	4	7	8	1	6	2	5	3
Sum	54	53	55	47	35	36	29	17	22	16	15	11
Range	3	4	2	1	5	2	4	4	6	5	4	2

Source: *Psychometric Methods* by J. P. Guilford. Copyright 1936. Used with permission of McGraw-Hill Book Company.

Discussion. Naturally, since we have rankings and have developed the idea of the rank correlation between pairs of measurements, we would like to look at the rank correlations between all pairs of judges. These have been computed and are shown in Table 13–2.

Table 13–2
Rank correlations between pairs of judges

Judge	Judge			
	1	2	3	4
1	—	—	—	—
2	0.727	—	—	—
3	0.818	0.902	—	—
4	0.783	0.832	0.916	—
5	0.790	0.804	0.769	0.720

What we see in Tables 13–1 and 13–2 is a substantial amount of correlation between pairs of judges. No judge's ratings stand out as differing much from those of the others.

We have made the judgment that the judges are much alike. How about the handwriting items? One inexpensive way to get a systematic look at the items is to calculate the

$$\text{range} = \text{largest rank} - \text{smallest rank}$$

for each item. This number appears as the last entry in each column of Table 13–1. For item A, the range $= 12 - 9 = 3$.

The ranges run from 1 to 6, and none seems especially large. Some item might have stood out, possibly with a range of 10 or 11. Such an item probably should, if possible, be investigated or, if not, removed, and the rankings should be re-calculated. Two possible reasons for a wild item are: (1) clerical errors in rating, (2) the existence of a "hidden" dimension.

For example, a student once developed a most beautiful, regular handwriting. Seen at a distance, it inspired exclamations over its evenness and beauty. Seen close up, it looked like a sequence of humps from m's and n's. Few could decode it. If a judge in Example 1 accidentally rated such handwriting for beauty or regularity, he would be likely to give it a rank of 1 or 2; but if he rated it for read-ability, he would be likely to give it 12. Such writing clearly was not part of the general population of most handwriting examples. Unless the variable being rated was strictly adhered to, such an item could lead to much disagreement. The example illustrates why preliminary work and agreement on definitions is especially important in this sort of investigation. Sometimes rerankings are required when it is discovered that a judge has not used the variable agreed upon.

Since we cannot reinvestigate either judges or items, we wish to summarize the rankings. One good way is to use *the average of all possible rank correlations*. The average of the 10 correlations for these data is 0.806.

This numerical value can be compared with results for other groups of subjects or other groups of handwriting examples. We would say that the average correla-tion is "high," the agreement "good."

EXERCISES FOR SECTION 13–1

1. For Table 13–3, the average rank correlation of amounts of public debt held by various owners for pairs of months is 1. Without actually writing out the rankings or making any calculations, explain how you can tell that this statement is true.

Table 13–3
Ownership of the public debt, in billions of dollars

	Mutual banks	Insurance companies	Individuals	State and local governments
January 1969	3.6	7.9	75.9	27.8
February	3.6	7.8	76.2	28.4
March	3.6	7.7	77.0	28.1
April	3.5	7.6	76.6	28.7
May	3.7	7.9	76.8	28.1
June	3.5	7.7	76.4	27.3

Source: *Federal Reserve Bulletin*, December 1969, p. A42.

2. For Table 13–3, for each type of ownership rank the debt held for the 6 months given (from 1 for least to 6 for greatest). (a) By looking at the ranks, guess how high the average rank correlation is between pairs of ownership types. (b) By looking at the ranges of the ranks across ownerships, see whether any types of ownership seem to misbehave.

13–2   AVERAGE RANK CORRELATION COEFFICIENT FOR SEVERAL PAIRS OF JUDGES

Often we can inspect by eye a table like Table 13–1 and decide that it is satisfactory to compute the average intercorrelation. Calculating $m(m - 1)/2$ correlations can be a long job, but there is an easy formula that produces the average without calculating the correlation for each pair of judges.

Theorem. *Average rank correlation.* If the same I items are ranked m times, and the sum of the ranks for item i is R_i, $i = 1, 2, \ldots, I$, then the average rank correlation between the $m(m - 1)/2$ pairs of rankings is

$$\text{average } r_S = \frac{12 \sum R_i^2}{m(m - 1)I(I^2 - 1)} - \frac{1}{I - 1}\left[\frac{2(2I + 1)}{m - 1} + 3(I + 1)\right]. \qquad (1)$$

Discussion. The derivation of formula (1) is a long algebraic exercise that we shall not take space for here.

Example 1. Find the average rank correlation for the handwriting data of Section 13–1.

Solution. Applying formula (1) to the data of Example 1 of Section 13–1 gives

$$\text{average } r_S = \frac{12[54^2 + 53^2 + \cdots + 11^2]}{5(4)12(143)} - \frac{1}{11}\left[\frac{2(25)}{4} + 3(13)\right] = 0.806,$$

which agrees with the average calculated from the rank correlations of the 10 pairs of judges shown in Table 13–2.

Example 2. *Optional operations.* C. E. Lewis reports operative rates for 10 regions covering most of Kansas for six surgical operations whose performance is sometimes optional. The regional rates have been ranked from least to greatest for each operation, and the rankings are given in Table 13–4. Find the average intercorrelation between the ranks for pairs of operations, and interpret the result.

Table 13–4

Ranks of rates of surgical operations in 10 regions in Kansas

					Region					
	a	b	c	d	e	f	g	h	i	j
Tonsillectomy and adenoidectomy	4	1	5	6	9	8	10	7	3	2
Appendectomy	4	1	5	8	10	7	9	6	3	2
Hernia repair	3	2	5	9	8	6	10	7	4	1
Hemorrhoidectomy	4	1	2	7	5	9	8	10	6	3
Cholecystectomy	4	2	7	6	10	9	8	5	3	1
Varicose veins	2	3	7	6	5	9	8	4	10	1
	21	10	31	42	47	48	53	39	29	10

$m = 6, I = 10$

$$\text{Average } r_S = \frac{12 \sum R_i^2}{m(m-1)I(I^2-1)} - \frac{1}{I-1}\left[\frac{4I+2}{m-1} + 3(I+1)\right]$$

$$= \frac{12(13050)}{6(5)10(99)} - \frac{1}{9}\left[\frac{42}{5} + 3(11)\right]$$

$$= 0.67$$

Solution. The sums of ranks under the columns of Table 13–4, when substituted into equation (1), as shown at the bottom of the table, give an average rank correlation of 0.67. This high correlation suggests that the sorts of conditions and decisions that lead to high rates for one of these operations also apply to the others.

The particular operations that seem out of line are hemorrhoidectomy in regions *c*, *e*, and *h* and varicose veins in regions *e* and *i*. You can tell this by running your eye down each column to see if any rankings stand out far from the rest. One could make this procedure formal by taking the average rank for a column and subtracting it from each rank in the column to get a *residual*. A study of residuals is often a second step in a statistical analysis. Frequently, as here, visual scanning is good enough.

EXERCISES FOR SECTION 13–2

1. Find the average rank correlation in Table 13–3 between the pairs of ownership ranks of monthly holdings. Explain what the result means.

2. The three socioeconomic indices used in Table 13–5 are said to be highly correlated with one another. Consider how each should be ranked so as to bring this out (some can be ranked least to greatest, others greatest to least). After choosing the method for ranking them, use formula (1) to get the average rank intercorrelation.

Table 13–5
Three indices for 1961

Area	Socioeconomic classification	Persons per household (3.04 :100)	Household index 1961
London boroughs			
Bermondsey	33	95	64
Chelsea	211	77	144
Greenwich	84	100	92
Hackney	55	92	74
Lewisham	91	98	95
Shoreditch	31	97	64
Stoke Newington	70	92	81
Wandsworth	106	93	100

Source: W. E. Cox, "The estimation of incomes and expenditures in British towns," *Applied Statistics*, Vol. 17, (1968), p. 254.

3. (Continuation) Compute r_S for each pair of rankings for the first 5 boroughs for the data in Table 13–5, and show that the average of the three r_S's is exactly that given by the formula. (Ties may prevent exact agreement.)

4. *Research problem. Effects of choice of order.* If you are allowed to choose the order of the rankings, as in Exercise 2, there is some danger of "capitalizing on chance". To illustrate this danger, take 3 sets of 4 random numbers from Table A–22 at the back of the book in the order in which they are given. Rank the first

set of 4 from least to greatest. Now rank each of the other two columns twice, first from least to greatest and then in a separate table from greatest to least. You now have a choice of 4 sets of 3 rankings for these numbers. Compute the average rank correlation between the sets and find the size of the largest one. (Assuming that different students used different sets of random numbers, get the average of the maxima for the class.)

5. (Continuation) For sets of random numbers, if we increase the number of things ranked in a set from 4 to a large number n, what do you think will happen to the maximum average rank correlation for a fixed number of sets? Will it increase or decrease as n increases?

★6. *Research. Effects of optimizing rankings when data are random.* Consider all possible patterns of 3 rankings of 3 things. The first ranking might as well be regarded as in the order 1, 2, 3. For each ranking you may rerank from greatest to least instead of least to greatest so as to maximize the average rank correlation among the pairs of rankings in a pattern. Example:

	A	1	2	2	gives a higher average	1	2	2
Item	B	2	1	3	r_S if we rerank the	2	1	1
	C	3	3	1	third column	3	3	3

After getting the highest correlating rankings for each pattern, get the grand average of the average r_S. Compare it with 0, the average r_S if the numbers are randomly chosen.

13–3 TESTING FOR RANDOMNESS; THE FRIEDMAN STATISTIC

Ordinarily, when several judges rank items or when we obtain several rankings of the same sorts of items, *we expect correlations between the rankings.* And we want to know how big the correlation is, as in Section 13–2.

Occasionally, we are not at all confident that correlation exists, and we want to *test for its existence.* Then Friedman's statistic is what we wish to calculate. The Friedman statistic, which measures the disagreements between the rank sums of the judges, is very similar to the Kruskal-Wallis statistic in form and distribution. When the null hypothesis of random independent rankings holds, the Friedman statistic has approximately a chi-square distribution with $I - 1$ degrees of freedom.

Definition. *The Friedman statistic.* Given m rankings of I items, the Friedman statistic is

$$\chi_r^2 = \frac{12}{mI(I + 1)} \sum R_i^2 - 3m(I + 1),$$

where R_i is the sum of the m ranks assigned to item i, $i = 1, 2, \ldots, I$.

The subscript r on χ^2 refers to rank correlation rather than degrees of freedom.

Distribution. *Distribution of Friedman's χ_r^2 under random rankings.* If the rankings are random, then for large m and large I, Friedman's χ_r^2 is approximately distributed according to chi-square with $I - 1$ degrees of freedom.

Discussion. The motivating idea behind the Friedman statistic is that, under the null hypothesis, any R_i is approximately normally distributed with known mean $m(I + 1)/2$ and known variance $m(I^2 - 1)/12$. Consequently, the random variable

$$\frac{R_i - m(I + 1)/2}{\sqrt{m(I^2 - 1)/12}}$$

has approximately the *standard* normal distribution, and its square has approximately the chi-square distribution with 1 degree of freedom. The sum of these squares for all items would be distributed approximately chi-square with I degrees of freedom, were it not for the dependence among the R's resulting from the fact that $\sum R_i = mI(I + 1)/2$. This constraint reduces the degrees of freedom by one.

Another way of thinking of Friedman's formula is this: Consider the sum of the squares of deviations of rank sums from the mean. This sum is

$$T = \sum(R_i - m(I + 1)/2)^2.$$

We note that when the judges agree perfectly and give the same ranking to all items, this sum T gets as large as it possibly can. But T would be 0 if all R_i's were equal. Under randomness, it turns out that the expected value of T is $mI(I - 1)(I + 1)/12$. Therefore, to get a statistic with mean $I - 1$, we divide T by $mI(I + 1)/12$.

Since $\sum R_i^2$ is the only quantity depending on the actual values of the rankings in either average r_S or χ_r^2, it is natural that all three are linearly related.

EXERCISES FOR SECTION 13–3

1. Table 13–6 gives the logarithm of the concentration of three halogens in parts per million in the hair of 10 human subjects. It comes from a study of the application of medicine to crime. Rank the concentration for each halogen from least to greatest, and then test whether there is a significant average correlation among them.

2. Show that $\sum R_i = m(I)(I + 1)/2$.

3. Why is $T = 0$ when all R_i are equal?

   In Exercises 4, 5, 8, assume that the null hypothesis of random ranking holds.

4. Under the null hypothesis, show that
   a) $E(R_i) = m(I + 1)/2$;
   b) $\text{Var}(R_i) = m(I^2 - 1)/12$.

Table 13–6

Logarithm of concentration (parts per million) of halogens in hair of 10 human subjects

Subject	Cl	Trace element I	Br
1	8.268	2.945	3.893
2	7.460	3.642	3.133
3	8.439	1.631	3.040
4	7.637	2.339	3.533
5	7.740	4.488	2.855
6	5.858	2.589	5.705
7	6.005	2.355	2.350
8	8.069	2.594	3.763
9	7.571	2.174	2.847
10	8.804	2.104	3.470

Source: J. B. Parker and A. Halford, "Optimum test statistics with particular reference to a forensic science problem," *Applied Statistics*, Vol. 17, (1968), p. 246.

5. (Continuation) Show that

$$E(R_i^2) = \frac{m(I + 1)}{12} [I - 1 + 3m(I + 1)].$$

Recall that $E(R_i^2) = \mathrm{Var}\,(R_i) + [E(R_i)]^2.$]

6. (Continuation) Prove that

$$T = \sum R_i^2 - \frac{m^2 I(I + 1)^2}{4}.$$

7. Find how large T can get when all judges agree on all the rankings.

8. (Continuation) Prove that

$$E(T) = mI(I^2 - 1)/12.$$

9. (Continuation) Show that

$$\frac{12T}{mI(I + 1)}$$

is the Friedman statistic.

Table 13–7†

Major subdivisions of civics with rank order numbers as rated by teachers of various school types

	School type							
	Elem. school	Comp. school 7	Comp. school 8	Comp. school voc. stream 9	Comp. school acad. stream 9	Real-skola	Voc. real-skola	Girls' school
1. Home and family	1	1	5	2	9	9	8	8
2. Leisure	10.5	6	12	15	15	17	17	1.5
3. Schools, vocations, and their selection	2	2	1	1	1.5	2.5	13	12.5
4. Social behavior	4	3	3	7	10	18	11.5	17
5. Law and equity	3	4.5	2	3	7	7	4	6
6. Suffrage and elections	9	7	7	5	4	4	2	4
7. Community civics	5.5	10	10	6	5	2.5	3	3
8. The State	8	11	9	4	3	1	1	1.5
9. The Church	13	17	17	17	17	14	14.5	16
10. Domestic economy	7	4.5	8	11	11	12	11.5	12.5
11. Political economy	10.5	13	11	10	8	6	6	9
12. Social issues, social policy	5.5	8	4	9	6	5	5	5
13. Social movements	14	14	14	13	13.5	10	10	11
14. Defense and military	17	18	18	18	18	15	16	18
15. Democracy: meaning and problems	12	9	6	8	1.5	8	7	7
16. Scandinavian and international cooperation	16	15	16	12	12	11	9	10
17. Mass media	15	12	13	16	13.5	13	14.5	14
18. Psychological and sociological problems	18	16	15	14	16	16	18	15

† This table, with minor changes kindly provided by the authors, is reproduced with the permission of the publishers from Torsten Husén and Gunnar Boalt, *Educational Research and Educational Change: The Case of Sweden*, John Wiley and Sons, New York, Almqvist and Wiksell, Stockholm, 1968, p. 72.

★**REVIEW EXERCISE**

(T. Husén and G. Boalt) Civics

Table 13–7 shows the rankings of importance assigned to various parts of civics by teachers in different kinds of Swedish schools. Examine these rankings, and study the differences between the schools. Make such calculations as you regard appropriate and report your view of (1) the main groupings of the schools, and (2) the categories that cause most disagreement within groupings. Then give your summary of the results. (One would have to know a good bit about Swedish schools to do a complete interpretation, but much can be seen by immersing oneself in these data.)

CHAPTER 14

Order Statistics: Distributions of Probabilities; Confidence Limits, Tolerance Limits

14–1 WHAT ARE ORDER STATISTICS AND WHY STUDY THEM?

To test inexpensive light bulbs for length of life, the inspector ordinarily puts a large number of them into an electrical panel, turns them all on, and then records the elapsed times when they burn out. This method of gathering data is unusual in that *the observations arrive in the order of their sizes*, the smallest one first, then the next smallest, and so on. Naturally, in practice, a few bulbs burn out almost immediately or do not light at all. These are replaced at once, just as they would be in the retail store, and are not regarded as part of the basic distribution of well-made bulbs. A record is kept of their proportion.

Suppose that it is known, from considerable experience, that length of bulb life follows an approximately normal distribution, with mean 500 hours and standard deviation 100 hours. If 20 light bulbs are going to be tested, guess when the first bulb will burn out. Well under 500 hours, of course. The answer is 313 hours, on the average.

The smallest observation from such a test sample is known as the first order statistic. We shall learn from Table A–16 that the first order statistic in a sample of size 20 from a normal distribution averages 1.87σ less than the mean, μ, of the normal. Thus the expected value of the first order statistic in samples of size 20, drawn from our normal distribution, is $\mu - 1.87\sigma = 500 - 187 = 313$.

To see what the expected value in this instance means, it is well to review the idea of repeated sampling. Think of drawing a sample of 20 bulbs, finding the lengths of their bulb lives, ordering these lengths, and writing down the smallest of them. Then draw a new sample of 20 bulbs, and repeat the process. Imagine doing this many times. Then the average of these smallest values is close to the expected value of the first order statistic. Later, we shall consider this idea further.

In Table 14–1 this program has been carried out for 5 samples of size 20 from a standard normal distribution. The first line shows the 5 first order statistics and their average, -1.75. If we had infinitely many samples of size 20, the average would be -1.87.

Table 14-1
Five ordered samples of size 20, from the standard normal distribution

Order statistic	1	2	3	4	5	Mean	Range
1	−1.78	−1.58	−1.45	−2.51	−1.45	−1.75	1.06
2	−1.22	−1.21	−1.38	−1.35	−1.44		0.23
3	−1.22	−1.20	−0.95	−1.17	−1.27		0.32
4	−0.86	−1.03	−0.78	−0.66	−1.08		0.42
5	−0.74	−0.93	−0.52	−0.60	−1.06		0.54
6	−0.70	−0.85	−0.44	−0.52	−0.60		
7	−0.51	−0.78	−0.34	−0.43	−0.59		
8	−0.50	−0.58	−0.19	−0.15	−0.38		
9	−0.48	−0.20	−0.06	−0.15	−0.36		
10	−0.44	0.04	−0.03	−0.08	−0.35		
11	−0.21	0.08	0.10	0.01	−0.08		
12	−0.09	0.13	0.18	0.26	0.16		
13	−0.04	0.44	0.34	0.34	0.17		
14	0.07	0.47	0.43	0.38	0.19		
15	0.18	0.49	1.02	0.39	0.50		
16	0.26	0.68	1.61	0.53	0.50		1.35
17	0.28	0.89	1.62	0.70	0.70		1.34
18	0.78	1.42	1.79	1.12	0.89		1.01
19	0.99	1.60	1.98	1.35	1.07		0.99
20	1.20	1.81	2.50	1.55	1.47		1.30

Source: *A Million Random Digits with 100,000 Normal Deviates*, The RAND Corporation, Free Press, Glencoe, Ill., Table of Gaussian Deviates, page 3, lines 0105–0124, columns 1–5. Each column is one sample.

Think of this set of 20 light bulbs as a random sample drawn from an enormous batch of well-made bulbs. Let us abandon the normal distribution assumption and admit that we do not know the distribution of bulb lives in the batch. Suppose that after the third light bulb burns out at, say, 371 hours (the third order statistic), we are asked what fraction of the bulbs *in the batch* will burn out in less than 371 hours. Can we give any useful answer? Yes, and *we don't need to know the shape of the distribution*. We need only know that the distribution is continuous, as indeed we expect these bulb lifetimes to be. We shall see that on the average 3/21 (yes, 21, not 20) of the bulbs in the batch will burn out in less than 371 hours. The average is over *batches*. Again think of taking many samples of size 20, one sample from each batch. For each *batch*, find the proportion in that *batch* that burns out in a shorter time than the third member of the *sample*. The average of these proportions tends to 3/21. For the 5 samples in Table 14–1, the proportions to the left of the third order statistic can be found from the normal table to be 0.111, 0.115, 0.171, 0.121, 0.102. These proportions average 0.124, which is close to $3/21 \approx$ 0.143.

The order statistics are more directly related to the probability in the original distribution of measurements than the statistics we more commonly compute. It is important to understand order statistics for the following reasons. First, they have specific uses as statistics, as we shall see. Second, they offer quick and easy methods of estimation. Third, these methods often do not depend much on the shape of the distribution. Fourth, they give us a much deeper understanding of selection effects than we get in other statistical studies, and this may be even more important than the first three reasons.

EXERCISES FOR SECTION 14–1

1. Find the average value of the 20th order statistic in 5 samples of size 20 of Table 14–1. What should its value be with infinitely many samples of size 20?

2. For the standard normal distribution, why would the expected value of the first order statistic in samples of size n be the negative of the expected value of the nth order statistic?

3. What should be the average of the two expected values of the 10th and the 11th order statistics in the samples of size 20 from a standard normal? Use the data in Table 14–1 to get an empirical estimate.

4. Use the average area under the standard normal curve to the left of the 7th order statistic in the 5 samples of Table 14–1 to get an estimate of the theoretical value.

5. Take 10 samples of size 1 from a uniform distribution (take 4-digit numbers from your random number table, and put a decimal point in front of each). Obtain the average value. Compare it with the theoretical value.

6. Take 10 samples of size 2 from the uniform distribution. Find the 10 first order statistics for the 10 samples, and get their average.

7. Take 10 samples of size 2 from your tables of random normal deviates. Obtain the 10 first order statistics for the 10 samples, and get their average.

8. (Continuation) Obtain the range (largest minus smallest measurement) for each sample, and find the average range. Compare it with the theoretical value for $n = 2$ given in Table A–18.

9. (Continuation) For the first order statistics of Exercise 7, obtain the probability to the left from the normal tables. Get their average and compare with the theoretical value of $\frac{1}{3}$.

14–2 CONFIDENCE LIMITS FOR THE MEDIAN AND OTHER QUANTILES OF THE POPULATION

By applications of the binomial distribution, we can readily obtain confidence limits for the population median, which we denote by $x_{0.5}$.

Definition. *Population median.* For a continuous cumulative distribution function F, the *population median*, $x_{0.5}$, is a value of x that satisfies the equation

$$F(x) = 0.5.$$

For such F's, $x_{0.5}$ is a value of x having the probability to its left and the probability to its right both equal to $\frac{1}{2}$. Let us suppose that F has a density function f. When F is strictly monotonic near $x_{0.5}$, then $x_{0.5}$ is unique, as a graph of the density function f suggests,

but when f vanishes at and near $x_{0.5}$, the value is not unique.

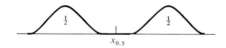

In what follows we shall regard $x_{0.5}$ as unique, but the results will hold for any $x_{0.5}$, whether unique or not.

Definition. *Order statistics.* A random sample of n measurements, $X_1, X_2, \ldots,$ X_n, is drawn from a distribution with continuous cumulative F. The measurements are ordered from least to greatest. Denote these ordered values of the measurements by $U_1, U_2, \ldots, U_n$ so that

$$U_1 \le U_2 \le \cdots \le U_n. \tag{1}$$

Then the U's are the order statistics in the sample of n.

In Section 14–1, Table 14–1 gives 5 sets of 20 U's drawn from a standard normal distribution.

Ties. Since F is continuous, the probability of ties is zero; and so hereafter we shall often omit the equals signs given in (1). As usual, we denote the specific numerical values of the random variables U_i by u_i.

Confidence limits for the median. To illustrate confidence limits for the median, we give the following example.

Example 1. *Strontium-90 concentration* (B. G. Greenberg). Ten samples of a test milk were measured for strontium-90 concentration in micromicrocuries per liter. Neither the smallest nor the largest measurement was given, but the rest in order of size were 7.1, 8.2, 8.4, 9.1, 9.8, 9.9, 10.5, 11.3. Without making assumptions about the shape of the distribution, set confidence limits for the population median $x_{0.5}$, and give the confidence level.

Solutions. Several symmetrical solutions of the form

$$U_i < x_{0.5} < U_{10-i+1}, \qquad i = 2, 3, 4, 5, \tag{2}$$

could be used. (We cannot set $i = 1$ because neither the smallest nor the largest measurement was reported.) Let us choose $i = 3$. To find the confidence level, we need the probability that

$$U_3 < x_{0.5} < U_8. \tag{3}$$

Inequality (3) is satisfied, provided that at least 3 measurements are to the left of the median and at least 3 are to the right. Expressed entirely in terms of left, (3) is satisfied if at least 3 and no more than 7 measurements are to the left of $x_{0.5}$. Let L be the number of measurements to the left of $x_{0.5}$. The confidence coefficient, C, is

$$C = P(3 \le L \le 7). \tag{4}$$

Its value can be computed from the binomial distribution because (1) each measurement X_i is either to the left or to the right of the median, (2) the probabilities are $P(X_i \le x_{0.5}) = \frac{1}{2} = P(X_i \ge x_{0.5})$, and (3) the X's are independent. Applying the binomial distribution, we have

$$C = P(3 \le L \le 7) = \sum_{i=3}^{7} \binom{10}{i}\left(\frac{1}{2}\right)^{10} = 1 - 2\sum_{i=0}^{2} \binom{10}{i}\left(\frac{1}{2}\right)^{10}.$$

From the cumulative binomial Tables A–5, we find

$$\sum_{i=0}^{2} \binom{10}{i}\left(\frac{1}{2}\right)^{10} = 0.055,$$

and so the confidence coefficient is $C = 1 - 2(0.055) = 0.89$. The numerical values of the lower and upper confidence limits in this example are $u_3 = 8.2$ and $u_8 = 10.5$ (remember that u_1 and u_{10} were not given). The numerical confidence interval is

$$8.2 < x_{0.5} < 10.5,$$

with confidence $C = 0.89$. Other coefficients could be obtained by using different U_i, U_{n-i+1} pairs.

Theorem 1. *Confidence limits for the median.* Given symmetrically chosen confidence limits u_i, u_{n-i+1} for the median, based on order statistics from a sample of size n, then the confidence coefficient for the inequality

$$u_i < x_{0.5} < u_{n-i+1}, \qquad 1 \le i < n - i + 1,$$

is given by

$$C = \sum_{k=i}^{n-i} \binom{n}{k} \left(\frac{1}{2}\right)^n = 1 - 2 \sum_{k=0}^{i-1} \binom{n}{k} \left(\frac{1}{2}\right)^n.$$

Proof. See Exercise 6.

The generalization to confidence limits for quantiles other than the median is immediate.

Definition. *pth quantile.* For a distribution with continuous cumulative F, the pth quantile, x_p, is a solution of the equation,

$$F(x) = p.$$

As in the case of the median, x_p may not be unique; but we shall discuss it as if it were. The value of x below which 25% of the distribution lies would have $p = 0.25$, and we could denote it by $x_{0.25}$.

Confidence limits for x_p. When $p \ne \frac{1}{2}$, there may be advantage in choosing the order statistics for the confidence limits asymmetrically. Consider, for example, the limits

$$U_i < x_p < U_{n-j+1}, \qquad 1 \le i < n - j + 1 \le n.$$

Theorem 2. *Confidence limits for x_p.* If the order statistics u_i and u_{n-j+1}, $1 \le i < n - j + 1 \le n$, are taken as lower and upper confidence limits for x_p, the confidence coefficient is

$$C = \sum_{k=i}^{n-j} \binom{n}{k} p^k (1 - p)^{n-k}.$$

Proof. See Exercise 9.

Example 2. Use the data of Example 1 to set lower and upper confidence limits for $x_{0.3}$.

Solution. Since we cannot use u_1 or u_{10}, let us use u_2 and u_8 (to give an example with asymmetry); then $i = 2, j = 3, n - j = 7$. If L is the number of order statistics to the left of $x_{0.3}$, then $P(2 \leq L \leq 7)$ is, from Table A–5

$$C = \sum_{k=2}^{10} b(k; 10, 0.3) - \sum_{k=8}^{10} b(k; 10, 0.3) = 0.851 - 0.002 = 0.849,$$

and the confidence interval is $7.1 < x_{0.3} < 10.5$.

Note that even had we been willing to assume normality for the distribution of strontium-90 measures, we still could not have used the t-distribution approach because of the missing values.

EXERCISES FOR SECTION 14–2

In the following exercises assume that the distribution F, from which the sample is drawn, is continuous.

1. Use the data of Example 1 to set u_2 and u_9 as lower and upper confidence limits for the median. Evaluate the confidence coefficient C.

2. In a sample of size n, if u_1 and u_n are taken as lower and upper confidence limits for $x_{0.5}$, find the confidence coefficient C.

3. In a sample of size n, where n is an even integer, find the probability that the population median is between the two middle measurements.

4. Use the data of Example 1 to set a one-sided confidence limit, $u_2 < x_{0.5}$, for the population median. Find the confidence coefficient C.

5. For a large sample of size n, explain how, without using binomial tables, and without making assumptions about the shape of F, you would choose i so that $u_i < x_{0.5} < u_{n-i+1}$ would give a confidence coefficient of about 0.95.

6. Prove Theorem 1.

7. Use the data of the first sample of Table 14–1, ignoring the knowledge that it comes from a normal distribution, to set approximate 90% confidence limits on $x_{0.2}$. [Find the u_i, u_{20-j+1} pair that most nearly approximates $C = 0.90$.]

8. Suppose that, as in Exercise 7, data are drawn from a normal and that confidence limits for the mean (which is also the median) are computed by using (1) the Student t-distribution approach and (2) the method of Theorem 1, with exactly the same confidence levels for both computations. Then, how do you think the lengths of the confidence intervals of the two methods will compare, on the average? Why?

9. Prove Theorem 2.

10. Ignoring the normality, use the total sample of 100 in Table 14–1 to find i and thus u_i so that $P(x_{0.01} < U_i) \approx 0.9$. Did the interval from $-\infty$ to u_i cover the true value? (Since we happen to know $x_{0.01}$ from normal tables in this special problem, we can answer the question, but of course, ordinarily one could not know.)

11. Ignoring the normality, use the total sample of 100 in Table 14–1 to set an approximate 95% confidence interval on the median. Does the interval include the true value?

14–3 DISTRIBUTION OF THE PROBABILITY TO THE LEFT OF THE *i*TH ORDER STATISTIC: STATISTICALLY EQUIVALENT BLOCKS

In this chapter and the next, we study the distributions of U_i, the *i*th order statistic, and Y_i, the probability to the left of U_i, namely $F(U_i)$. In this chapter, we emphasize Y_i for arbitrary distributions; and in Chapter 15, we emphasize U_i, especially for normal distributions.

In this section, we shall largely pass over derivations, but where it is easy to make the results plausible, we shall do so.

Samples of size $n = 1$. When we draw a random observation, X, from a distribution with continuous cumulative F, we know that $P(X < x_p) = F(x_p) = p$. Since $Y = F(X)$, $P(Y < p) = p$. In words, the distribution of the probability to the left of a random measurement is uniform on the interval $[0, 1]$. We have proved the following theorem.

> **Theorem 1.** For a random variable X with continuous cumulative distribution F, the probability $Y(=F(X))$ to the left of X is uniformly distributed on $[0, 1]$.

Among other things, this result means that when we draw an observation at random, it divides the distribution in such a way that the average probability to the left of the observation is $\frac{1}{2}$. A hasty reader might think that we are merely saying that when a total probability is broken into two parts, the average part must have half the probability. Although true, it is not our point. We could have a method of breaking probability into a left-hand part and a right-hand part so that the left-hand part averaged less than half the probability. Instead, when the observation is used to break the probability into two parts, the left-hand part *does* average to half.

Samples of size $n = 2$. When two measurements U, V are drawn at random from a continuous F, they divide the distribution into three parts.

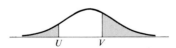

Let $U < V$, so that U and V are the order statistics of the sample. We can imagine that the average probability under the left tail might be the same as that under the right (the two shaded areas), but what about the middle? Although we shall not prove it, on the average the probability in the middle interval is the same as for each of the others. In other words, the order statistics divide the probability into thirds, on the average.

Sample of size n. Given the previous results, we are not surprised that for a sample of size n, the order statistics $U_1, U_2, \ldots, U_n$ divide the distribution into $n + 1$ parts, each of which contains on the average, probability $1/(n + 1)$. Consequently,

$$E(Y_i) = i/(n + 1).$$

The exact distribution of Y_i is known; its probability density function is

$$h(y) = \frac{n!}{(i - 1)! \, (n - i)!} \, y^{i-1}(1 - y)^{n-i}, \qquad 0 \le y \le 1. \qquad (1)$$

As a few examples of this distribution, we consider the following:

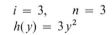

Case $i = 1$, $n = 1$.
$h(y) = 1$, $0 \le y \le 1$

gives the uniform distribution, as we mentioned above.

$i = 1$, $n = 2$
$h(y) = 2(1 - y)$

$i = 2$, $n = 2$
$h(y) = 2y$

$0 \le y \le 1$

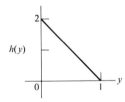

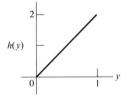

$i = 1$, $n = 3$
$h(y) = 3(1 - y)^2$

$i = 2$, $n = 3$
$h(y) = 6y(1 - y)$

$i = 3$, $n = 3$
$h(y) = 3y^2$

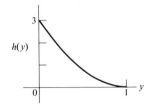

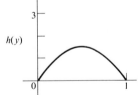

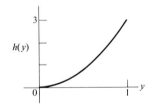

When both i and $n - i$ are large, the distribution is approximately normal. For example:

$$i = 6, \qquad n = 20$$

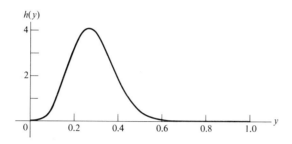

Theorem 2. *Distribution of* Y_i. In a sample of n drawn from a continuous distribution F, the probability density function of the random variable Y_i is given by equation (1), and its mean and variance are

$$\mu_{Y_i} = \frac{i}{n + 1} \qquad \text{Var } Y_i = \frac{i(n - i + 1)}{(n + 1)^2(n + 2)}. \qquad (2)$$

Note that, since i is the number of parts to the left of Y_i and $n + 1$ is the total number of parts, $i/(n + 1)$ is p, the proportion of parts to the left of Y_i. Similarly,

$$\frac{n - i + 1}{n + 1} = 1 - p,$$

the proportion of parts to the right of Y_i. Hence Var Y_i can be thought of as

$$\frac{p(1 - p)}{n + 2},$$

which is reminiscent of the binomial variance

$$\frac{p(1 - p)}{n}.$$

Theorem 3. *Normality*. When i and n grow large so that $0 < i/(n + 1) < 1$, Y_i is approximately normally distributed, with mean and variance given by expressions (2).

When n is large, it is attractive to make the comparison between the mean and variance of Y_i and those of the sample proportion of successes $\bar{p}$ in a binomial experiment of size $n + 2$.

If we think of the quantile x_p,

then the binomial experiment can be thought of as producing L, the number of measurements X_i that fall to the left of x_p in a binomial experiment of $n + 2$ trials. Then $\bar{p} = L/(n + 2)$, $E(\bar{p}) = i/(n + 1) = p$, and $\sigma_{\bar{p}}^2 = p(1 - p)/(n + 2) =$ Var Y_i. Thus, $\bar{p}$ is an unbiased estimate of p.

Corresponding to this, when we consider Y_i, we note that Y_i is the sum of the probabilities in the i intervals $(-\infty, U_1), [U_1, U_2), \ldots, [U_{i-1}, U_i)$ where $[a, b)$ means $a \le x < b$. (It is convenient to think of $U_0 = -\infty$, $U_{n+1} = +\infty$.) Let us call these probabilities $P_1, P_2, \ldots, P_i$, respectively.

$$Y_i = \sum_{i=1}^{i} P_i.$$

Each P_i has mean $1/(n + 1)$, and $E(Y_i) = i/(n + 1)$. If we knew Y_i, it would be an unbiased estimate of p.

Let u_i be the value of U_i in an experiment and p_i the value of P_i. Then the distribution would be divided as shown in the following figure.

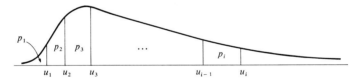

In the binomial experiment, L varies so that $\bar{p}$ is sometimes smaller than p, sometimes larger. In the order statistics, U_i varies so that it is sometimes smaller than x_p, sometimes larger, but in such a way that its related Y_i is correspondingly smaller or larger than p and averages to p. Y_i and $\bar{p}$ have the same variance.

We call the intervals $(-\infty, U_1), \ldots, [U_{i-1}, U_i), \ldots, [U_n, +\infty)$ *statistically equivalent blocks* partly because P_i, the probability in block i, $i = 1, 2, \ldots, n + 1$, has the same distribution for each block, and more generally because the sum of the probabilities in any collection of i mutually exclusive blocks has the same distribution as Y_i.

Theorem 4. The distribution of the sum of the probabilities of i mutually exclusive statistically equivalent blocks is that of Y_i.

We note that the *i* blocks must be specified without reference to the actual outcome. It is convenient to think of them as selected in advance of the experiment. Obviously, if one block has probability larger than $1/(n + 1)$, the others must be smaller on the average. The covariance between P_i and P_j is

$$\text{Cov} (P_i, P_j) = -1/(n + 1)^2(n + 2). \tag{3}$$

Example 1. *Two-sided tolerance limits.* To define the "normal" lower and upper limits for a particular physiological function, a physician may study well people or patients whose disease does not affect the physiological function. He wants to get values between which he is quite sure that a large percentage, say 95%, of measurements for such patients would lie. How can he set such a limit and be very sure, say 90% sure, that he has included at least 95% of the values for normally functioning patients?

Solution. He might use symmetrical limits (u_i, u_{n-i+1}) from a large random sample of patients whose disease does not affect the function. Then the expected proportion of all normal patients W in the interval (U_i, U_{n-i+1}) is $p = (n - 2i + 1)/(n + 1)$, because there are $n - 2i + 1$ statistically equivalent blocks.

Using formula (2), we have

$$\sigma_W^2 = \text{Var } Y_{n-2i+1}$$

$$= \frac{(n - 2i + 1)[n - (n - 2i + 1) + 1]}{(n + 1)^2(n + 2)}$$

$$= \frac{2i(n - 2i + 1)}{(n + 1)^2(n + 2)}$$

$$= p \left(\frac{2i}{n + 1} \right)\left(\frac{1}{n + 2} \right) = p(1 - p)/(n + 2).$$

Hence the standard deviation of W is

$$\sigma_W = \sqrt{p(1 - p)/(n + 2)}.$$

Let us set up and solve this problem in some generality. Suppose some large value of *n* is chosen, such as 1000; then we want to know *p* so as to get *i*. We use the following notation.

p^* = the desired proportion, 0.95 in the example.
$1 - \alpha$ = desired confidence, 0.90 in the example.
z_α = the α quantile for a standard normal distribution, -1.28 in the example.

Then using the normal approximation for W, we want to find p as a function of p^*, n, and z_α so that

$$\frac{p - p^*}{\sigma_W} \geq |z_\alpha| \tag{4}$$

When we substitute in (4) for p^*, σ_W, and z_α, we get a quadratic inequality in p. Instead of trying to solve this inequality by formula to find the smallest p satisfying (4), it may be more convenient to try values of p and interpolate. (*Warning:* the explicit formula for the solution of the quadratic is extremely sensitive to rounding errors.)

Suppose that $n = 998$ (about 1000). Let us try $p = 0.95$ and $p = 0.96$. Naturally $p = 0.95$ makes the left side of inequality (4) zero. For $p = 0.96$, the left side of (4) gives $0.01/\sqrt{0.04(0.96)/1000} = 1.61$. Since we want 1.28, and $1.28/1.61 \approx 0.8$, the desired p is about 0.958.

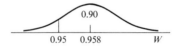

Consequently, since $(n - 2i + 1)/n = p$, we find that i is 21 or 22. The physician might use u_{21} and u_{978}.

EXERCISES FOR SECTION 14–3

1. When $i = 1$, $n = 2$, use the density given in equation (1) to show that $P(Y_i < \frac{1}{2}) = \frac{3}{4}$.

2. Graph the probability density function of Y_3 when $n = 5$. Note that the bell shape is beginning to emerge. Do this carefully on graph paper, using a large scale.

3. (Continuation) Find the mean and variance of Y_3.

4. (Continuation) In a sample of size $n = 5$, find the mean and variance of W_3, where $W_3 = P_1 + P_3 + P_5$.

5. (Continuation) From the graph of Exercise 2, find approximately $P(0.2 < Y_3 < 0.8)$.

6. From consideration of the probability density function in equation (1), show that when n is odd, the average probability to the left of the sample median is $\frac{1}{2}$.

7. Note that
$$P_1 + P_2 + \cdots + P_{n+1} = 1,$$
and so
$$\left(P_1 - \frac{1}{n+1}\right) + \left(P_2 - \frac{1}{n+1}\right) + \cdots + \left(P_{n+1} - \frac{1}{n+1}\right) = 0.$$

Assuming that all P_i's have identical distributions and that, as Theorem 4 implies,

$$\text{Var } P_i = \text{Var } Y_1,$$

prove that the covariance of P_i, P_j is that given by equation (3).

8. For a sample of 20 from a normal distribution, compare the probability to the left of the average first order statistic $\mu - 1.87\sigma$ with the average probability to the left, 1/21, discussed in this section. Explain why they need not be identical.

9. If $i = 10$, $n = 99$, find (a) the mean, (b) the approximate standard deviation of Y_i, and (c) the approximate probability that Y_i differs from 0.1 by more than 0.03.

The following information is used in Exercises 10–13. A quality-control man wants to know what values of X divide a large population of items into 6 groups of equal size. He takes a sample of 95 from the population and measures these carefully. Then he uses u_{16}, u_{32}, u_{48}, u_{64}, and u_{80} as the break points for forming the 6 groups.

10. Show that the intervals from $U_{16(k-1)}$ to U_{16k}, $k = 1, 2, 3, 4, 5, 6$, contain expected probability $\frac{1}{6}$.

11. Find the variance of the probabilities given in Exercise 10.

12. Find the mean and variance of the probability in the interval from U_{16} to U_{80}.

13. (Continuation) *Tolerance limits.* How confident can the man be that at least 60% of all items are in the interval given in Exercise 12.

14. *Guarantees.* A manufacturer wants to guarantee that at least 90% of his large inventory of a certain part have measurements between two values he will announce. If he takes a sample of 400 (you may use 398 for computational convenience), what values of i for u_i and u_{n-i+1} should he use to be 80% sure of his guarantee?

15. (Calculus) When $n = 5$, find exactly $P(0.2 < Y_3 < 0.8)$.

CHAPTER 15

Order Statistics: Point Estimation

15-1 IDEAS OF POINT ESTIMATION FOR MEANS AND MEDIANS

1. The mean μ of a normal distribution. We are very familiar with the sample mean $\overline{X}$ as an estimate of the population mean, μ. Note that each measurement in the sample has been given an equal weight, $1/n$, in this statistic,

$$\sum \left(\frac{X_i}{n} \right).$$

We know that if the population has a mean μ, then the sample mean is an unbiased estimate of μ. In the special case of the normal distribution, $\overline{X}$ happens also to be the maximum likelihood estimate of μ. It has further been shown, though the work is beyond the scope of this book, that for the purpose of estimating μ there is no information in the individual values of the sample measurements $X_1, X_2, \ldots, X_n$ that is not already contained in $\overline{X}$. This is a pleasant state of affairs for the family of normal distributions, but it does not hold in general.

Let us discuss estimation for a few other distributions, so as to shake ourselves loose from the idea of $\overline{X}$ as the only natural estimate of μ.

2. The uniform distribution. Let us consider the uniform distribution on the interval from a to b, $a < b$. Its probability density function has ordinate $1/(b - a)$ over this interval and 0 elsewhere. The mean of this distribution is $\mu = (a + b)/2$. It can be shown that all the information about μ is contained in the largest and smallest order statistics of the sample. Thus $(U_1 + U_n)/2$ is a natural estimate of μ for the uniform distribution, and it is a much better estimate than the sample mean. It happens also to be the maximum likelihood estimate.

3. The Laplace distribution. The normal distribution has the feature that its tail ordinates tend to zero like e^{-x^2}. The uniform has tails whose ordinates are exactly zero

Another distribution, called the Laplace distribution, has higher tails than either of these distributions. It has density

$$f(x) = \tfrac{1}{2}e^{-|x-\mu|}, \qquad -\infty < x < \infty.$$

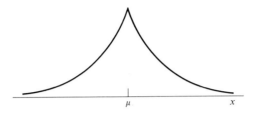

The tails of the Laplace distribution are higher than those of the normal because the ordinates tend to zero more slowly, like $e^{-|x|}$ rather than like e^{-x^2}. The maximum likelihood estimate of μ, the mean of the Laplace distribution, is the sample median. It satisfies the condition that $\sum |x_i - \mu|$ is a minimum.

Population median. In symmetrical distributions, the mean, μ, is identical with the median.

4. The Cauchy distribution. Finally, let us consider a distribution with very high tails, indeed. Some would say pathologically high. It is the Cauchy distribution, which has density

$$f(x) = \frac{1}{\pi[1 + (x - \mu)^2]}, \qquad -\infty < x < \infty.$$

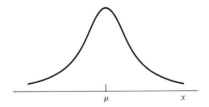

The Cauchy distribution is just a t-distribution with 1 degree of freedom, centered at μ instead of 0. And so, however pathological this distribution may be, it has practical import.

Although the Cauchy looks like the normal distribution, estimating its median or center of symmetry, μ, presents a problem different from that of a normal. The Cauchy has no mean and no variance. It turns out that $\overline{X}$ for a sample of size n is no better than the X of a sample of size 1 for estimating the median. The sample median of a Cauchy distribution can be used to estimate μ, however. And as n grows, the median will be certain to be close to μ.

The maximum likelihood estimate is a value of μ that satisfies the equation

$$\frac{x_1 - \mu}{1 + (x_1 - \mu)^2} + \frac{x_2 - \mu}{1 + (x_2 - \mu)^2} + \cdots + \frac{x_n - \mu}{1 + (x_n - \mu)^2} = 0.$$

This requirement differs from that for the sample mean because weights are being applied to the deviations $(x_i - \mu)$, and these weights are dependent on the observations themselves. Recall that the corresponding requirement for the sample mean is

$$(x_1 - \mu) + (x_2 - \mu) + \cdots + (x_n - \mu) = 0,$$

where equal weights are applied.

There may be more than one root to the equation. Then it is necessary to compute the value of the likelihood function at each root to see which value gives the largest product. Recall that the likelihood function for both discrete and continuous distributions with a density is the product of the ordinates at the values observed:

$$L = f(x_1)f(x_2) \cdots f(x_n).$$

The variety of these estimates shows that, depending on the form of the distribution, one may get different, good estimates of the center of a symmetric population. These estimates essentially apply different weights to the order statistics, and we have seen that these good weights run a gamut: We put all the weight on the middle measurement, the sample median, in the case of the Laplace; we put equal weights on the deviations from the sample mean for the normal distribution; and we put half the weight on each of the two end order statistics for the uniform.

Considerable research, partly theoretical but largely computational, has gone into finding out what weights might be applied to the order statistics in small and large samples to get a good estimate of the location of a distribution when one does not know which family of distributions the sample is likely to come from. This research has gone forward especially for symmetrical distributions, usually with one mode like the normal, the Cauchy, and the Laplace.

Next we discuss the distribution of the order statistics from the normal distribution; and then we return to the uses of order statistics for estimation when we do not know the form of the distribution from which the measurements are drawn.

EXERCISES FOR SECTION 15–1

1. In a sample of size $n = 3$, the measurements are -1, 1, 2. Find the maximum likelihood estimate of μ, the population median, when the distribution from which the observations are drawn is (a) normal, (b) uniform, (c) Laplace, (d) Cauchy. Comment on the relations among these estimates.

2. In a sample of size $n = 5$, the measurements are 1, 2, 3, 4, 90. Repeat Exercise 1 for these measurements.

15–2 DISTRIBUTION OF U_i

In Chapter 14, we discussed the probability to the left of an order statistic. What about the distribution of the order statistic itself? How is it distributed?

We shall not derive the approximate distribution of U_i, but we state the result. Recall that F and f are the cumulative and density of the original measurements.

> *Approximate normality of U_i.* As n grows large, U_i is approximately normally distributed with approximate mean μ_i and approximate variance σ_i^2, where
>
> $$F(\mu_i) = i/(n + 1) \tag{1}$$
>
> and
>
> $$\sigma_i^2 = \frac{F(\mu_i)[1 - F(\mu_i)]}{n[f(\mu_i)]^2}, \tag{2}$$
>
> on condition that
> a) $i/(n + 1)$ tends to a value not near 0 or 1,
> b) $f(\mu_i) > 0$.

The purposes of the conditions are as follows:

First, as n grows large, the ith order statistic will be constrained to a narrower and narrower distribution from equation (2), while the main body of its density stays near a particular location, as implied by equation (1). This will happen if $i/(n + 1)$ stays near some given value, say $\frac{1}{4}$. But if i is 1 or some other fixed number, then as n grows large, the order statistic will drift to the left, and the limiting distribution may not have the properties given. If $i = 1$, then $i/(n + 1)$ tends to 0. Similar problems arise when i is near n, say $n - 2$. Then $(n - 2)/(n + 1)$ tends to 1 as n grows large, and again condition (a) fails. The purpose of condition (a) is *to keep the main body of the density of U_i away from the far tails of the distribution F and concentrated near μ_i.*

Second, if f is zero at and near μ_i, then the density may look like the two-humped figure below. The result is that the variance of U_i may remain large even though n grows large.

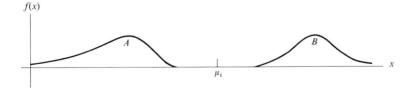

Whereas, when $f > 0$ for values near μ_i, as n grows large, the distribution of U_i becomes dense near μ_i. In our two-humped figure, it cannot, no matter how large n is.

Third, the continuity of F guarantees (with probability 1) that there are no ties, and the differentiability, together with $f(\mu_i) > 0$, ensures that f will be positive in a neighborhood of μ_i, not just at μ_i.

Probability density function for normal distributions. The values of the ordinates of distributions from the family of normal distributions are given by

$$f(x) = \frac{1}{\sqrt{2\pi}\sigma} e^{-(x-\mu)^2/2\sigma^2}, \qquad -\infty < x < +\infty,$$

where $e = 2.718\ldots$, $\pi = 3.1415\ldots$, μ is the mean, σ the standard deviation.

Example 1. *Sample median from a normal distribution.* If a large sample of size n, n odd, is drawn from a *normal* distribution with mean μ and variance σ^2, find the mean and approximate variance of the sample median $U_{(n+1)/2}$.

Solution. From the symmetry of the normal distribution, the sample median has mean μ. This is true for all symmetrical distributions that have means.

From (2), we have approximately that

$$\text{Var}\left[U_{(n+1)/2}\right] = \frac{\frac{1}{2}(\frac{1}{2})}{n\left(\dfrac{1}{\sqrt{2\pi}\sigma}\right)^2} = \frac{\pi\sigma^2}{2n},$$

since $f(x)$, the density of the normal at μ, is $1/\sqrt{2\pi}\sigma$.

Definition. *Efficiency.* If two statistics S and T are normally distributed about the same mean μ, then if $\sigma_S^2 < \sigma_T^2$, then

$$e = \sigma_S^2/\sigma_T^2$$

is called the relative efficiency of T with respect to S for estimating their common mean.

The term *efficiency* is sometimes used also when S and T are not assumed to be normally distributed.

Example 2. *Efficiency of the sample median compared with the sample mean.* Let $T = $ sample median and $S = \bar{X}$, and find the efficiency of T with respect to S for a large sample of size n, n odd.

Solution. Recall that the sample mean $\bar{X}$ has variance σ^2/n. Since, from Example 1, the sample median has variance $\pi/2$ times as large as that for $\bar{X}$, the efficiency of the sample median for estimating the normal population mean μ, relative to the sample mean, is

$$\text{Var } \overline{X}/\text{Var } U_{(n+1)/2} \approx \frac{\sigma^2/n}{\pi\sigma^2/2n} = \frac{2}{\pi} \approx 0.64.$$

From the normal distribution, the sample mean of a sample of size $0.64n$ gives as good an estimate of the population mean μ as the sample median from a sample of size n. And so the sample mean has an advantage. On the other hand, it does not protect against wild observations as the median would.

We shall not prove the following theorem, but it is worth knowing.

Theorem. *Median of the distribution of the sample median.* For samples of size n, n odd, from an arbitrary continuous distribution F, the median of the distribution of the sample median $U_{(n+1)/2}$ is $x_{0.5}$, the population median.

This remarkable result for sample medians is in a sense more general than the corresponding result for $\overline{X}$ and μ, since μ may not exist, whereas $x_{0.5}$ must. We can say, then, that the sample median, $\tilde{x}$, is a median-unbiased estimate of the population median $x_{0.50}$.

When the sample size is even, it is practical to take the average of the two middle values $U_{n/2}$, $U_{n/2+1}$ as the median so as to have a unique median. But this choice is arbitrary.

The exact distribution of the *i*th order statistic. Exact rather than approximate results for the distribution of the *i*th order statistic are given by a theorem which we shall not prove.

Theorem. *Distribution of U_i.* If a sample $X_1, X_2, \ldots, X_n$ is drawn from a distribution with cumulative F and density f, then the probability density function g of the *i*th order statistic U_i is given by

$$g(u) = \frac{n!}{(i-1)!\,(n-i)!}\, F(u)^{i-1}[1 - F(u)]^{n-i}f(u). \tag{3}$$

Example 3. *Uniform distribution.* If n X's are drawn from a uniform distribution on the interval $(0, 1)$, find the distribution of U_i.

Solution. Since $f(x) = 1$ on $(0, 1)$ and zero elsewhere, and $F(u) = u$ on $(0, 1)$, we can write

$$g(u) = \frac{n!}{(i-1)!\,(n-i)!}\, u^{i-1}(1-u)^{n-i}.$$

Thus we have proved a corollary to the theorem.

> **Corollary.** The distribution of the ith order statistic from a sample of n drawn from the uniform distribution on $(0, 1)$ is identical with that of Y_i, the probability to the left of the ith order statistic in samples of n drawn from an arbitrary distribution.

Example 4. *Exponential distribution.* Find the distribution of U_5 in a sample of size 10 drawn from the exponential distribution for which

$$f(x) = e^{-x}, \qquad 0 \le x.$$

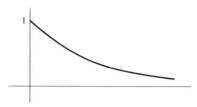

Solution. It takes calculus to show that $F(u) = 1 - e^{-u}$ $(= \int_0^u e^{-x}\, dx)$. Then we can substitute in expression (3) to get the exact distribution

$$g(u) = \frac{10!}{4!\,5!}(1 - e^{-u})^4(e^{-u})^5 e^{-u}, \qquad 0 \le u.$$

If we rewrite this by letting $p = 1 - e^{-u}$, we can recast it in the form of a multiple of an individual term of a binomial distribution,

$$g(u) = 6 \binom{10}{4} p^4 (1 - p)^6$$

$$= 6b(4;\, 10,\, p).$$

Now using a table of individual terms for a binomial distribution, we can compute ordinates and make a graph of the probability density function. The result is exhibited in Fig. 15–1. The parent exponential distribution is shown with the graph of g superimposed. Note that the graph of g is shaped rather similarly to a normal, even though f looked nothing like a normal distribution.

EXERCISES FOR SECTION 15–2

1. A sample of 9 is drawn from the standard normal distribution. Find approximately the mean and the variance of U_3. What if the normal has mean 3 and standard deviation 5?

2. A sample of 25 is drawn from a normal distribution with mean μ and standard deviation σ. Find approximate values for the mean and the variance of the sample median.

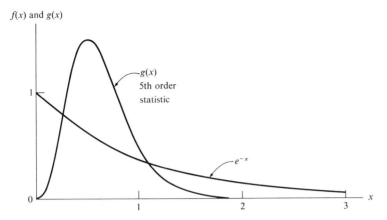

$f(x)$ and $g(x)$

g(x)
5th order
statistic

e^{-x}

Fig. 15–1 Graph of the exponential $f(x) = e^{-x}$ and of the density g of the fifth order statistic, U_5, from a sample of size $n = 10$ drawn from it.

3. Use the large sample results from Section 15–2 to find the approximate probability that at least 50% of the original measurement distribution is to the left of the 60th measurement in a sample of 99 drawn from a uniform distribution. Compare this with the result from Theorem 3 of Section 14–3.

4. Eleven rods are randomly drawn from a large inventory and their inside diameters measured. This dimension is approximately normally distributed with standard deviation $\sigma = 0.002$ inches. If the sample median is 0.320 inches instead of the standard 0.318, is there reason to suppose that much of the inventory has too large an inside diameter?

5. Two batches of material have samples of sizes n_1 and n_2, respectively (n_i odd). The characteristic being measured in each batch is approximately normally distributed with unknown means μ_1 and μ_2, respectively, and common standard deviation σ known. Devise a test for the equality of the means based on the difference of the sample medians, $\tilde{x}_1$ and $\tilde{x}_2$.

6. In Exercise 5, let $\tilde{x}_1 = 1.01$, $\tilde{x}_2 = 0.87$, $\sigma = 0.15$, $n_1 = 17$, $n_2 = 35$. Test for a difference between μ_1 and μ_2 at the 5% level.

7. Find the exact probability density function of the largest order statistic in a sample of n from the exponential distribution with $f(x) = e^{-x}$. The distribution e^{-x} is a good representation of the distribution of lifetimes of objects that fail or die by accident rather than by wearing out and have an average life of one unit. Use this model to interpret the meaning of the largest order statistic.

8. (Continuation) Make a graph of the density g found in Exercise 7, for $n = 10$.

9. (Calculus) Find the exact distribution of the median for a sample of size $2n + 1$ drawn from a standard normal distribution. Let $F(u) = \int_{-\infty}^{u} f(x)dx$, since the integration cannot actually be carried out in closed form.

10. Graph the results for Exercise 9 for the case of $2n + 1 = 3$ and compare it with the corresponding graph for the sample mean.

15–3 TABLES OF EXPECTED VALUES FOR THE ORDER STATISTICS OF THE STANDARD NORMAL DISTRIBUTION

The integrals one has to evaluate to get expected values for order statistics (other than those of the uniform distribution) are usually difficult or impossible to get in closed form. That is, they require numerical integration. Getting the expected values for the order statistics of the normal, except for a few (like the median), requires a serious effort. Fortunately, the results have been tabulated, and we give values for $n = 1$ to $n = 100$ in Table A–16 at the back of the book. These have been drawn from the valuable and much more extensive tables of H. Leon Harter.

The symmetry of the normal gives us a few useful relations. First, when n is odd, the expected value of the median $U_{(n+1)/2}$ is the population mean μ, as we have seen.

Second, the symmetry of the normal also guarantees that

$$\mu - \mu_i = \mu_{n-i+1} - \mu, \tag{1}$$

where

$$\mu_i = E(U_i), \qquad i = 1, 2, \ldots, n.$$

Note that equation (1) says that the ith order statistic and the $(n - i + 1)$st order statistic have means equidistant from μ.

We can compare the probability to the left of the mean of the ith order statistic with $i/(n + 1)$ for given i, n. For this purpose, let us choose a few values of n and i that are especially easy to check. We can pick n's ending in 9: 9, 19, 29, and so on, and values of i that will make $i/(n + 1)$ take values shown at the head of Table 15–1. For example, choose $n = 19$ and $i = 2$ so that $i/(n + 1) = 0.1$. From Table A–16, for $n = 19$ we get $\mu_2 = -1.38$. The normal table gives $P(Z < -1.38) = 0.084$, as shown in Table 15–1. Looking down the columns of

Table 15–1

Probabilities to the left of the expected values of selected order statistics from the normal

n	\multicolumn{6}{c}{$i/(n + 1)$}
	0.05	0.1	0.2	0.3	0.4	0.5
9	—	0.069	0.176	0.284	0.392	0.500
19	0.033	0.084	0.188	0.292	0.396	0.500
29	—	0.089	0.192	0.295	0.397	0.500
39	0.041	0.092	0.194	0.296	0.398	0.500
49	—	0.093	0.195	0.297	0.398	0.500
59	0.044	0.095	0.196	0.297	0.399	0.500
69	—	0.095	0.197	0.298	0.399	0.500
79	0.045	0.096	0.197	0.298	0.399	0.500
89	—	0.096	0.197	0.298	0.399	0.500
99	0.046	0.097	0.198	0.298	0.399	0.500

Table 15–1, one sees that as n grows larger, the probability to the left of the mean, $E(U_i)$, comes closer to the fraction $i/(n + 1)$ given at the head of the columns.

We can use Table A–16, as n grows, to see how fast the mean value of the extreme order statistics move away from the mean μ. For example, for $n = 10$, 20, 30, the largest order statistics are, on the average, respectively, 1.54, 1.84, 2.04 standard deviations from the mean. Naturally, as n increases, the extreme order statistics move further away from μ, on the average. Those order statistics adjacent to the sample median, on the other hand, move toward the mean. Thus, if $n = 11$, $|\mu_5| = \mu_7 = 0.22$; if $n = 21$, $|\mu_{10}| = \mu_{12} = 0.12$.

Figure 15–2 shows the mean value of the largest order statistic plotted against $\sqrt{\log n}$. Note the approximate linearity. That $\sqrt{\log n}$ would produce such near linearity is not at all obvious.

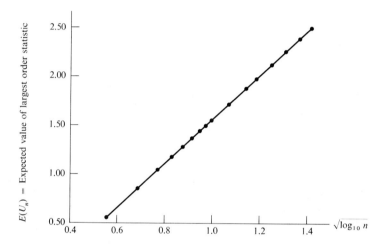

Fig. 15–2 Plot of $E(U_n)$ against $\sqrt{\log_{10} n}$ showing approximate linearity.

Figure 15–3 shows, for odd values of n, the distance between the expected value of the median and that of the next order statistic, plotted against $1/n$. Note the approximate linearity. We might expect this because of the flatness of the normal in the neighborhood of its mean.

Example 1. *Selectivity.* The mean and standard deviation of CEEB scores are $\mu = 500$ and $\sigma = 100$. If a sample of 5 CEEB scores is drawn, how big is the largest score on the average?

Solution. In Table A–16, we find that the expected value of the largest of 5 *standard* normal observations is 1.16, which means that the expected value of the largest is 1.16 standard deviations *above the mean*. Consequently, the mean of the largest of 5 CEEB scores is

$$500 + (1.16)100 = 616.$$

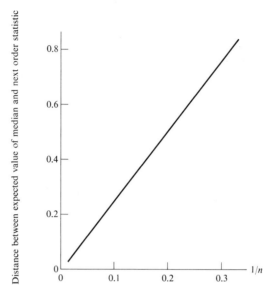

Fig. 15–3 Distance between expected value of median and expected value of next order statistic (from normal) plotted against $1/n$ for odd values of n, where n is sample size.

Example 2. Suppose employees are hired in groups of 5 drawn randomly from the pool of acceptable applicants. An employee's time with the company is approximately 6 years on the average with a standard deviation of 2 years, and it is approximately normally distributed. In the groups of 5, the longest-staying employees stay how many more years, on the average, than the second-longest-staying employees?

Solution. For normally distributed measurements with $n = 5$, Table A–16 shows $E(U_5) = 1.16\sigma$, $E(U_4) = 0.50\sigma$. The difference in expected values is $1.16\sigma - 0.50\sigma = 0.66\sigma$. Since $\sigma = 2$, the difference is about 1.32 years.

Example 3. In a wholesale warehouse, self-service is allowed. Certain bags of merchandise worth $1 a pound have a mean of 100 pounds with standard deviation 1 pound. A purchaser can weigh as many bags as he likes before choosing one, but his time is valuable, and it costs $0.15 per weighing of a bag. A prospective purchaser plans to buy 2 bags and plans to weigh according to this strategy: Weigh n bags and choose the best 2. He wants to know how big n should be to maximize his profit on the transaction.

Solution. If $n = 2$, he would not weigh any bags, and his net gain would be zero. If he weighs $n = 3$ and picks the best 2, he gains $0.85 - 0.45 = $0.40 because $E(U_3) + E(U_2) = 0.85\sigma = 0.85$ in dollars; and the weighing cost is $3 \times 0.15 = 0.45$ in dollars. We make a table of further results.

n	Total excess pounds	Cost	Net gain in dollars
3	0.85	0.45	0.40
4	1.33	0.60	0.73
5	1.66	0.75	0.91
6	1.91	0.90	1.01
7	2.11	1.05	1.06
8	2.27	1.20	1.07†
9	2.42	1.35	1.07†
10	2.54	1.50	1.04

† Apparently the maximum gain occurs when 8 or 9 bags are weighed.

EXERCISES FOR SECTION 15-3

1. For samples of size $n = 9, 19, 29, 39$, and 49 from a standard normal distribution, compute the range M_n of the expected values of the middle 5 observations, and the range H_n of the expected values of the largest 5 observations. Then for each n, compute the ratio M_n/H_n, and plot it against n. Explain what it shows.

2. If weights of pears are approximately normally distributed, and you always choose the largest from a box of a dozen, how many standard deviations larger than the mean does your choice average?

3. At a cafeteria, helpings of ice cream have a standard deviation in weight of 0.5 oz and are approximately normally distributed. At any moment, there are about 20 servings to choose from. Every day a fat man and his thin friend eat ice cream. The fat man always chooses the biggest serving, the thin man the smallest. Over a working year (200 days), how much more ice cream does the fat man eat than the thin man?

4. Find the expected difference between the 8th and 4th order statistics in a sample of 10 from a normal distribution with standard deviation σ.

5. Refer to Example 3. With conditions the same, what is the optimum n for the purchase of 1 bag?

6. In Example 3, if the purchaser plans to weigh n bags and buy the best 3, what is his optimum n?

★7. (Continuation) *Research problem.* If one had the expected values on a computer, it would be interesting to see how the optimum n in Example 3 changes with the purchase of r bags, for various costs. [Cost can be thought of as a fraction of the price of one standard deviation. Choose a set of r's and a set of costs.]

8. In a sample of size 6 from an approximately normal distribution, the smallest measurement is 6.4, the largest 12.1. Use Table A-16 to estimate σ.

9. *Quasi-range.* In Example 1 of Section 14-2, the smallest and largest measurements are missing. If the strontium-90 concentrations are approximately normally distributed, use $u_9 - u_2$ together with Table A-16 to estimate σ. (The random variable $U_{n-i+1} - U_i$ is called a quasi-range.)

15–4 VARIANCES AND COVARIANCES OF THE ORDER STATISTICS FROM A NORMAL DISTRIBUTION

For the *standard normal* distribution, Table A–17 at the back of the book gives variances and covariances of the order statistics for samples up to size 10. Recall that Cov $(X, X) = $ Var (X). Hence entries on the main diagonal of those tables are variances. From the symmetry of the normal distribution, it follows that

$$\text{Var } (U_i) = \text{Var } (U_{n-i+1}) \quad \text{and} \quad \text{Cov } (U_i, U_j) = \text{Cov } (U_{n-i+1}, U_{n-j+1}).$$

Contaminated normal distribution. Among distributions that have been studied for purposes of estimation, one is called "the contaminated normal". This distribution is formed as the weighted sum of two normals with the same mean, μ, but different variances. For example, one normal may have 9 times the variance of the other. A weight of 0.95 might be assigned to the normal with small variance, and 0.05 to the normal with large variance. This is one model for a measuring situation where there is a 0.05 chance of getting a measurement from a process with a much larger variance than that of the primary process under study. It is a model for a distribution plagued with wild observations.

Remarkable results have been found. For example, for the purpose of estimating the common mean of the normal distributions, if the normal with the smaller variance is contaminated as much as ten percent by a normal with a variance 9 times as large, then the sample mean is no better than the sample median.

Part of the reason for intensive investigations like those mentioned in Section 15–1 is that we do not often know the form of our distributions in practice. Even when we have a good idea, as in the case of many physical measurements which approximately follow the normal law, there are usually "wild observations" or gross errors that we cannot eliminate on the basis of evidence in hand. The question then arises: How do we estimate the population mean or median so as to have good efficiency and at the same time not assume very much about the form of the distribution? Merely choosing the median seems to sacrifice too much of the structural information in the kinds of distributions that we usually have.

Since the work is largely based on lengthy computations, we shall have to take the results on faith, but for a large family of approximately symmetrical, unimodal distributions, good efficiency is achieved by using the trimmed mean. We need again the notation $[M]$, which means the greatest integer contained in M. Examples: $[3] = [3.1] = [\pi] = 3$.

Definition. *Trimmed mean.* In a sample size of n, the trimmed mean with trimming coefficient α is

$$\frac{1}{n - 2[\alpha n]} (U_{[\alpha n]+1} + U_{[\alpha n]+2} + \cdots + U_{n-[\alpha n]}).$$

Example 1. Find the trimmed mean of the measurements 10, 24, 25, 23, 25, 137 with trimming coefficient 20%.

Solution. Since $n = 6$, $0.2n = 1.2$. We eliminate $[1.2] = 1$ measurement from each end of the list of order statistics, leaving only 23, 24, 25, 25. The trimmed mean is their average, 24.25.

Although in large samples trimming coefficients of $\alpha = 0.25$ to 0.275 seem useful for the kinds of distributions often found in practice, in samples up to size $n = 20$, $\alpha = 0.20$ seems effective both at preserving efficiency and defending against high tails in distributions.

Example 2. *Estimation.* As insurance against wild outlying values in a sample of 5, you plan to estimate a normal mean μ by a "trimmed mean"

$$T = \tfrac{1}{3}(U_2 + U_3 + U_4),$$

which is the average of the 3 middle measurements. If there are no outliers, what is the efficiency of this estimate compared with the sample mean, $\overline{X}$?

Solution. The variance of $\overline{X}$ is $\sigma^2/5 = 0.2\sigma^2$. For the trimmed mean, the variance is

$$\sigma_T^2 = \tfrac{1}{9}[\sigma_{U_2}^2 + \sigma_{U_3}^2 + \sigma_{U_4}^2 + 2\,\mathrm{Cov}\,(U_2, U_3)$$
$$+ 2\,\mathrm{Cov}\,(U_2, U_4) + 2\,\mathrm{Cov}\,(U_3, U_4)].$$

(See Appendix V–2.) The symmetry gives $\sigma_{U_2}^2 = \sigma_{U_4}^2$ and $\mathrm{Cov}\,(U_2, U_3) = \mathrm{Cov}\,(U_3, U_4)$ since U_3 is the median. From Table A–17, we find

$$\sigma_T^2 = \frac{\sigma^2}{9}\,[0.312 + 0.287 + 0.312 + 2(0.208 + 0.150 + 0.208)]$$

$$= \frac{2.043\sigma^2}{9} = 0.227\sigma^2.$$

Thus the efficiency is $\sigma_{\overline{X}}^2/\sigma_T^2 = 0.2/0.227 = 0.88$. So, although we have tossed out 40% of the measurements, we have lost only 12% of the information. Since trimming end values can reduce the bad effects of wild observations, it is often worth paying the price.

Example 3. As an alternative to the trimmed mean, we might ask whether other weights we could apply to the 3 middle measurements in a sample of size 5 would reduce the variance of the estimate of μ, when the sample is actually normal. Let us try weights proportional to 1, 2, 1.

Solution. We want the variance of T_w, where

$$T_w = \tfrac{1}{4}(U_2 + 2U_3 + U_4).$$

$$\mathrm{Var}\,T_w = \frac{\sigma^2}{16}\,(0.312 + 4(0.287) + 0.312 + 2[2(0.208) + 0.150 + 2(0.208)])$$

$$= 0.233\sigma^2.$$

The efficiency is compared with T: $\sigma_T^2/\sigma_{T_w}^2 = 0.227/0.233 \approx 0.97$, and so equal weights were better than 1, 2, 1. But see Exercise 12.

Part of the point of Example 3 is that we can compute means and variances of weighted sums of order statistics from the normal for small samples using Tables A–16 and A–17.

EXERCISES FOR SECTION 15–4

Refer to Tables A–16 and A–17 at the back of the book.

1. If a sample of 10 is drawn from a standard normal distribution, find (a) Cov (U_4, U_5), (b) Var (U_4).

2. If a sample of 7 is drawn from a standard normal distribution, find (a) Cov (U_1, U_4), (b) Var (U_4).

3. In Exercise 2, check approximately the tabulated value of Var (U_4) by using formula (2) of Section 15–2.

4. Compute the correlation between the two most central order statistics for $n = 2, 6, 10$. Graph the resulting correlations against $1/n$ or $1/\sqrt{n}$, whichever comes nearer to giving a straight line relation. What do you think will happen as n grows large?

5. Compute the correlation between the two most extreme order statistics in samples of size 2, 6, 10. What do you think will happen as n grows large?

6. What is the smallest sample size for which the α trimmed mean eliminates observations?

7. Apply the trimmed mean with trimming coefficient 20% to the first sample in Table 14–1.

8. Apply the trimmed mean with trimming coefficient 20% to the data of Example 1 of Section 14–2.

9. For a sample of size 6 drawn from the normal distribution, compute the variance of the trimmed mean with trimming coefficient 20% and compare it with that of $\bar{X}$. Thus get the efficiency of this trimmed mean.

10. Repeat Exercise 9 for trimming coefficient 40%. Which trimmed mean is to be preferred for the normal?

11. For a sample of size 4, from a normal distribution, (a) find d_4 so that

$$E\left(\frac{U_4 - U_1}{d_4}\right) = \sigma,$$

and (b) find the standard deviation of $(U_4 - U_1)/d_4$ as an estimate of σ based on the range.

12. In a sample of size 5, to estimate the mean μ of a normal distribution, the weights w, $1 - 2w$, and w are attached to u_2, u_3, u_4, respectively. In terms of random variables, the estimate is

$$T_w = wU_2 + (1 - 2w)U_3 + wU_4.$$

By either calculus or numerical methods, find the value of w that minimizes Var T_w.

13. For large n, $i < j$, and large i and $n - j$, the approximate covariance between U_i and U_j is

$$\text{Cov}\,(U_i,\,U_j) \approx \frac{F(x_p)[1 - F(x_r)]}{nf(x_p)f(x_r)},$$

where

$$p = \frac{i}{n + 1}, \qquad r = \frac{j}{n + 1}.$$

Use this fact to devise a large sample test to see whether the interval between $x_{0.25}$ and $x_{0.75}$ is longer or shorter than a known standard L, if the observations are supposedly drawn from an approximately normal distribution with known standard deviation σ.

15–5 USE OF RANGES TO ESTIMATE σ

As a quick, inexpensive *estimator of standard deviations of roughly normal distributions*, the rule

$$\frac{\text{range}}{\text{divisor}}$$

can often be recommended. The divisor depends on the sample size n.

1. Divisor for small samples. If $n \le 15$, the range divided by $\sqrt{n}$ estimates the standard deviation well enough for rough purposes. For more precise estimation, Table A–18 lists the divisor d_n which gives an unbiased estimate of σ.

Frequently the ranges of several small samples of size n are averaged and then divided by d_n to estimate the standard deviation. This procedure gives a more efficient range estimate than that obtained from pooling the small samples and making one estimate from the grand range. The reason is that the range loses efficiency compared with s as n increases. At $n = 2$ they are identical.

Example 1. For the 5 observations 1, 1, 2, 2, 4, estimate σ (a) using the divisor $\sqrt{n}$, (b) using the divisor d_n in Table A–18.

Solution

a) Since the range is $4 - 1 = 3$ and $\sqrt{5} = 2.24$, we get

$$\hat{\sigma} \approx 3/\sqrt{5} \approx 1.34.$$

b) Using the divisor $d_n = 2.33$, we get

$$\hat{\sigma} \approx 3/2.33 \approx 1.29.$$

It may be of interest to compare $\hat{\sigma}$ with the actual sample standard deviation $\sqrt{\sum(x - \bar{x})^2/(n - 1)} \approx 1.225$.

2. Divisor for large samples. If $n \geq 16$, we can obtain the correct divisor d_n from Table A–18 provided $n \leq 1000$.

Example 2. For a sample of 50 having range 30, the estimate for σ is

$$30/4.50 = 6.67.$$

A substitute s for finding a substitute t. Recall that, in using the t-test, we evaluate the random variable

$$t = \frac{\overline{X} - \mu}{s/\sqrt{n}}.$$

The column in Table A–18 headed "divisor to get substitute s" gives a better divisor than d_n for finding a range-based value of s for use in a t-test. It has the merit of giving a quick test and a rough check on the t-calculation.

Example 3. *Substitute s for t-test.* The 5 observations 1, 1, 2, 2, 4 are to be used to test the deviation of their population mean from 0. Compare the usual t-value with that obtained by using the substitute s.

Solution. The usual t-value with 4 degrees of freedom, based on $\overline{x} = 2$ and $s^2 = \Sigma(x_i - \overline{x})^2/(n-1) = 6/4$, is

$$t = \frac{\overline{x} - \mu_0}{s/\sqrt{n}} = \frac{2-0}{\sqrt{\frac{6}{4 \times 5}}} = 3.65.$$

If we replace s by the estimate (range/2.48), where the 2.48 is taken from the fourth column in Table A–18, we get

$$\text{"substitute } s\text{"} = (4-1)/2.48 = 1.21.$$

Then our substitute t is

$$t' = 2\sqrt{5}/1.21 = 3.70,$$

which is close to t.

EXERCISES FOR SECTION 15–5

1. An observer records 9 observations as follows:

$$2, 2, 2, 2, 3, 3, 3, 4, 6.$$

Find a range estimate for σ (a) using the divisor $\sqrt{n}$, (b) using the divisor d_n in Table A–18.

2. A sample of 100 measurements has a range of 22. Find a range estimate for σ.

3. The 9 observations in Exercise 1 are used to test the deviation of their population

mean from 0. Compare the usual *t*-value with that obtained by using the divisor for substitute *s* in Table A-18.

4. The log concentration of parts per million of mercury in the hair of 10 subjects is 1.53, 1.02, 0.75, 1.71, 1.29, 1.25, 1.52, 1.81, 1.41, 0.51. Control subjects averaged 1.63 (which we shall take to be based on a very large sample). (a) Using the range, estimate the standard deviation for the population whence the 10 subjects were drawn. (b) Use the quick test to compare the mean of the population whence the 10 subjects were drawn with the control mean. (c) Use the 20% trimmed mean to estimate the population mean.

5. The following data (H. A. K. Rowland) give rise in hemoglobin concentration in grams per 100 ml during the first week of oral iron treatment for 6 patients: 0.06, 0.99, 1.29, 0.66, 1.09, 1.48. (a) Use the range to estimate the standard deviation. (b) Use the quick test to check whether the average rise is more than 0.40 grams per ml. (c) Set symmetrical 95% confidence limits on the average rise.

REFERENCES

1. H. Leon Harter. *Order Statistics and Their Use in Testing and Estimation*, Vols. 1 and 2. Aerospace Research Laboratories, United States Air Force, Washington, D.C.: U.S. Government Printing Office, Superintendent of Documents, 1969.

2. D. B. Owen. *Handbook of Statistical Tables.* Reading, Mass.: Addison-Wesley, 1962.

3. A. E. Sarhan and B. G. Greenberg. Estimation of location and scale parameters by order statistics from singly and doubly censored samples. Part I. The normal distribution up to samples of size 10. *Annals of Mathematical Statistics*, Vol. 27, 1956, pp. 427–451.

4. L. H. C. Tippett. On the extreme individuals and the range of samples taken from a normal population. *Biometrika*, Vol. 17, 1925, pp. 364–387.

CHAPTER 16

Designing Your Own Statistical Test or Measure

16–1 FIRST STEPS: TRY TO AVOID INVENTION

It is popular to say that every problem should have special statistical techniques tailored to its needs. Although there is something to this, much can be done with minor adjustments of standard devices. The practical investigator wants to spend his time primarily on the design, execution, and analysis of his study, and he doesn't want to build new statistics, unless forced, any more than he wants to make his own pencils, slide rules, or computers.

In a specific new problem, it is likely that you may come across a special situation that seems to you to need a new statistic. What should you do? There are several steps to take first:

1. Discuss the problem with others knowledgeable about the field and the particular application to find out whether anyone has already solved this problem to your satisfaction.
2. Review the literature. Various statistical bibliographies exist, and the *Statistical Theory and Method Abstracts* may supply the information you need. Standard texts may have your problem tucked away somewhere.
3. Find a statistical consultant who will help you design a new statistic or help you by proving that an available one is satisfactory.

After taking these first steps, you may decide to design your own test. If you do invent a new statistic, then you want some reason for supposing that, for your purpose, it is as good as, or better than, the ready-made statistics already available. What steps can you take to check this?

If you are designing a test, then you have two key objectives:

1. To define statistics that may be sensitive to the departures from the null hypothesis that you are especially interested in detecting.
2. To set up specific kinds of likely alternative hypotheses that you want to detect.

Possibly you may be able to compute the distribution of your statistic under the null hypothesis, but it will be unusual if you are able to do so for the alternative hypotheses. It is usually hard to compute the power of a proposed test. This is likely to be especially so if you have been driven to inventing your own test.

Nevertheless, you will want to make such calculations because otherwise you are not likely to know for certain that your new technique is valuable for your problem.

We have seen calculations of power earlier, but we shall take up further examples in the next two sections to drive home the general idea.

16-2 RUNS UP AND DOWN; COMPARING TESTS FOR TREND

Example 1. *Runs up and down.* Suppose that stock market prices on successive weeks closed with the following Dow-Jones averages:

631	633	642	645	624	622	630	636	660
	+	+	+	−	−	+	+	+

Below the sequence are the signs of the differences between the successive Dow-Jones averages. This sequence of signs has three runs: two runs of plus signs of length 3, called *runs up*; and one run of minus signs of length 2, called a *run down*. We might report (a) the total number of runs, (b) the length of the longest run up, or (c) some other statistic. The example suffices to define the notions of runs up and runs down and their length. Under the null hypothesis that all permutations of the original series are equally likely, we can compute or approximate distributions of such statistics, and thus provide a basis for tests of significance.

Although we have in this and previous sections mainly illustrated the power of various tests when the same test is compared for different sample sizes, or for different significance levels, that is only one use of the study of power.

The practitioner may be interested in the absolute probability of detecting departures from the null hypothesis, because if that probability is too small, he knows there is little use in carrying out the investigation. But he may also use the study of power as a way to decide which of several possible tests of a null hypothesis against an alternative will produce the strongest results—will be best at detecting specific departures from the null hypothesis.

In this section we illustrate this idea, using various statistics of runs up and down to test the null hypothesis of randomness against the alternative of a trend. First, we must specify the null hypothesis and the alternatives very precisely, including the sampling model. To get a trend, we assign to n successive measurements X_i normal distributions with mean $(i - 1)\delta$, $i = 1, 2, \ldots, n$, and standard deviation $\sigma = 1$. (Taking $\sigma = 1$ loses no generality since δ can be given arbitrarily large values.) Then when $\delta = 0$, the null hypothesis of n standard normal deviates applies, but when $\delta \neq 0$, we get a trend. We study the runs up and down for various values of δ which would give a positive trend.

To test for this trend using the signals of runs up and down, we have considered three statistics:

Statistic 1: the number of runs of either kind. These should be fewer as δ gets further from 0.

Statistic 2: the length of the longest run up. The more positive the trend, the more likely should be long runs up.

Statistic 3: the shortness of the longest run down. The more positive the trend, the shorter should be the longest run down.

The thought may occur to the reader that the rank correlation coefficient between i and X_i should be a good statistic to try here, and of course it is. Similarly, one could make a Mann-Whitney two-sample test, using the first half of the measurements against the last half, as we did in an example. Computing the power for these statistics and comparing the results with those for statistics 1, 2, 3 above are left as exercises.

It is not at all obvious which of statistics 1, 2, 3 would be preferable. But by simulating the experiment many times for different values of δ, we can get a good idea of the answer.

On the basis of the distribution theory of runs (not given here) under the null hypothesis, for a sample size of $n = 10$, we chose:

for Statistic 1, the rejection criterion 5 or more runs, all told;
for Statistic 2, the rejection criterion 3 or more for the length of the longest run up; and
for Statistic 3, the rejection criterion 0 or 1 as the length of the longest run down.

All three of these criteria give a probability near 0.2 for the probability of rejection when $\delta = 0$; thus the significance level in each instance is about 0.2. It is important that the tests have comparable significance levels, since, if a test starts ahead at $\delta = 0$, it is likely to stay ahead.

To create the required random numbers, we use

$$X_i = Z_i + (i - 1)\delta,$$

where Z_i is a standard random normal deviate from Table A–23.

Figure 16–1 shows the power for these tests based on the simulations. Two of the measures produce nearly a dead tie in the simulations, and the third lags considerably behind. The total number of runs does not have the power of the length of the longest run up or that of the longest run down. The message is this: Among the three statistics considered, statistics 2 and 3, no doubt highly correlated, are to be preferred when one is testing for positive trend in this model.

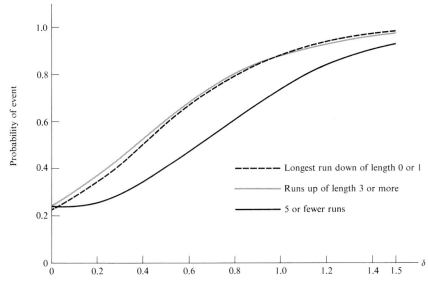

Fig. 16–1 (a) Solid curve gives probability of 5 or fewer runs up and down for $n = 10$, and various values of δ, the grid spacing of the means of the distributions. (b) Light solid curve gives probability of runs up of length 3 or more. (c) Dashed curve gives probability of longest run down of length 0 or 1.

EXERCISES FOR SECTION 16–2

Ideally, the following exercises will be done on a high-speed computer. If such an arrangement is not possible, each member of a class can do a few simulations, which can then be pooled, and perhaps power can be computed for only one point, say $\delta = 0.8$, in addition to the significance level.

1. *Rank correlation coefficient.* Using a significance level near 0.2, find the power of the rank correlation coefficient for $n = 10$ in testing for a trend. Compare the result with those of the runs tests, and comment.

2. *Mann-Whitney test.* Using a significance level of about 0.2, follow the suggestion in the text of forming a Mann-Whitney two-sample test for $n = 10$. Get the power function, and compare it with the results in Fig. 16–1. Comment on which test is to be preferred.

16–3 THE COMPARATIVE STUDY OF THE POWER OF SEVERAL TESTS FOR SLIPPAGE OR SHIFT IN LOCATION

As a second illustration of the comparison of several tests of the same hypothesis, we consider three tests for slippage in the problem of two samples from separate populations. As our standard example, we shall use sample size $m = 3$ for the left population and $n = 2$ for the right one, and we shall compute power.

For this illustrative purpose we consider three possible situations:

1. Two normal distributions with unknown means but identical known standard deviations. This leads to the standard normal test.

2. Two normal distributions with equal and unknown standard deviations, using the Mann-Whitney test. This test has no way of using the information about normality.

3. Same distributions as in 2, but we use the number of runs to make the test. Again, this test has no way of using the normality.

In this section, when we speak of runs, we are talking of runs of the same kinds of elements. If $X\,X\,X\,O\,O\,X\,O\,O\,O$ is a sequence of 9 elements, it has 4 runs: 1 of length 1 of X's, 1 of length 3 of X's, 1 of length 2 of O's, 1 of length 3 of O's. We use the 20% level, which, in the case of runs, means rejecting the null hypothesis when either $X\,X\,X\,O\,O$ or $O\,O\,X\,X\,X$ occurs, since there are 10 possible patterns of X's and O's. The X's represent numbers from one sample, the O's from the other. Their order is determined by their pooled ranks.

The simulations can be run in the usual way, as they were for the Mann-Whitney earlier. The powers are shown in Table 16–1 rather than by a figure, because the numbers are fairly close together for some of the situations. We would expect the test using statistic 1 to be the most powerful, because it uses the normality and the information about the known standard deviation. (Second best would be the t-test if we had it, because it would use the normality and the fact that the two standard deviations are known to be equal.) The question, then, is how well the Mann-Whitney test and the runs test do against the normal and the t-test.

The answer is that the Mann–Whitney holds fairly close to the normal test, and so it gets good marks. The runs test is not so good. Why not?

Table 16–1
One-sided, two known normals,
$n_1 = 3, n_2 = 2, \sigma_1 = \sigma_2 = \sigma = 1$

		Situation	
		2	
	1	Mann-	3
δ	Normal	Whitney	Runs
0	0.20	0.20	0.19
0.5	0.39	0.38	0.24
1.0	0.60	0.57	0.38
1.5	0.79	0.74	0.55
2.0	0.91	0.87	0.70
2.5	0.97	0.94	0.82
3.0	0.993	0.98	0.92

The runs test rejects when the sample of size two has the two biggest observations or the two smallest observations. These situations produce exactly two runs, the minimum. The result is that, as the mean for the population of the small sample gets larger, we mainly reject only for the pattern $X\ X\ X\ O\ O$, whereas the Mann-Whitney for the same significance level is rejecting for both $X\ X\ X\ O\ O$ and $X\ X\ O\ X\ O$.

The conclusion in this instance is that the Mann-Whitney is not losing much power compared with the normal test; since it is not making strong assumptions about normality or equality of standard deviations, it has much to be said for it.

EXERCISES FOR SECTION 16–3

1. Create another test for slippage, simulate its performance to get its power, and compare with the results of the tests discussed in this section.

2. From Table 16–1 you can estimate about what the power of the t-test would be if it were carried out. Explain.

Appendixes

The following Appendixes are intended to provide a concise review of some important definitions, theorems, and formulas used in the main body of the text. For those wishing a fuller treatment with further examples and exercises, one or more references to *PWSA*† are given at the head of each Appendix. Other sources of reference may be found among the texts listed in the bibliography.

---

† F. Mosteller, R. E. K. Rourke, and G. B. Thomas, Jr., *Probability with Statistical Applications*, Addison-Wesley, 2nd edition, 1970.

Discrete Distribution of a Random Variable

[References: *PWSA* Sections 5–1, 5–2, 6–1, 7–3]

I–1 WHAT IS A RANDOM VARIABLE?

The idea of an experiment is familiar, and we shall use it to lead us to the definition of a random variable.

Experiment. Three fair coins are tossed.

Query. How many of the coins will fall "heads"?

Discussion. The answer is determined by the outcome of the experiment. The number may be 0, 1, 2, or 3. Although we cannot predict the outcome exactly, we can list the possibilities and the probabilities as follows:

Possible outcomes:	HHH	HHT	HTH	THH	HTT	THT	TTH	TTT
Number of heads:	3	2	2	2	1	1	1	0
Probability:	$\frac{1}{8}$	$\frac{1}{8}$	$\frac{1}{8}$	$\frac{1}{8}$	$\frac{1}{8}$	$\frac{1}{8}$	$\frac{1}{8}$	$\frac{1}{8}$

If we let X denote the number of heads, then the foregoing table discloses three facts:

1. X has a set of possible values: 0, 1, 2, 3.
2. The particular value taken by X depends on the outcome of an experiment.
3. For each possible value of X, there is an associated probability. For example, the probability that $X = 2$ is $\frac{3}{8}$, and the probability that $X = 0$ is $\frac{1}{8}$, and so on.

This X that we have been discussing is an example of a random variable.

> **Definition.** *Random variable.* A variable whose value is a number determined by the outcome of an experiment is called a *random variable.*

Note that a random variable is like any other variable except that we may know more about the random variable, namely the probability that it takes any one of its possible values.

Notation. In this book, unless the contrary is explicitly stated, *we shall use capital letters* (such as X, Y, Z, W, and so on) *to denote random variables.*

We often wish to distinguish between a random variable and one or more of its values. To help us make this distinction, *we shall use small letters* (with or without subscripts) *to denote values of a random variable.* Here are some examples: the random variable X has possible values x_i, where $i = 1, 2, 3, \ldots, n$; the random variable Y has possible values y_i, where $i = 1, 2, 3, \ldots, t$; and so on. If we merely wish to denote an unspecified member of the set of possible values of a random variable, we do not need subscripts. Thus x stands for one of the possible values of X, and $P(X = x)$ stands for the probability that the random variable X takes an unspecified value x. The symbol "x" is just holding a place for a number yet to be named. Thus, in the foregoing experiment, if we let $x = 3$, we have

$$P(X = x) = P(X = 3) = \tfrac{1}{8}.$$

I–2 PROBABILITY FUNCTION OF A DISCRETE RANDOM VARIABLE

A random variable has a *discrete* probability distribution if and only if

1. the number of its possible values is finite or countably infinite;
2. the values are numbers (in advanced work, vectors);
3. the probability associated with each possible value is strictly positive or zero; and
4. the sum of the probabilities is 1.

Example 1. *A discrete distribution.* A sample of two is randomly chosen from a box of four light bulbs, two of which are defective. If X is the number of defective bulbs in the sample, find the probability table of X, and draw its graph.

Solution. There are six possible samples of two: one with no defectives, one with two defectives, and four with one defective. Since all samples are equally likely, we get the table and graph in Fig. I–1.

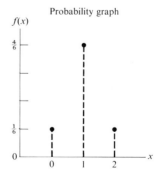

Probability table

x:	0	1	2
$P(X = x)$:	$\tfrac{1}{6}$	$\tfrac{4}{6}$	$\tfrac{1}{6}$

Figure I–1

We have positive probability for each of the three possible values of X. On the graph, the probability for each value of X is represented by *the height of the dot above it*.

Definition. *The probability function of a discrete random variable.* If, for a given discrete random variable X,

$$f(x_i) = P(X = x_i),$$

then the set of ordered pairs $(x_i, f(x_i))$, $i = 1, 2, \ldots, n$, is called *the probability function of X*.

I–3 THE CUMULATIVE DISTRIBUTION

When students receive their test scores from the College Entrance Examination Board, they may get something like Percentile 84, Score 600. This percentile means essentially that 84% of the students taking the test scored at or below 600. Indeed, the Board provides a table that gives the percentiles corresponding to various test scores. Such a table describes the *cumulative distribution* of the test scores.

Definition. *The cumulative distribution.* A cumulative distribution is a function F that gives, for each value of a random variable, the probability that a measurement randomly drawn from the parent population falls at or below that value.

Or we might use the following equivalent definition:

If X is a random variable, then for real values of x the equation

$$P(X \leq x) = F(x), \qquad -\infty \leq x \leq +\infty, \tag{1}$$

defines a set of ordered pairs $(x, F(x))$, which is the cumulative distribution function F of X.

The function F gives the proportion $F(x)$ of the population whose X-measurement falls at or below x, for real values of x.

The graph of the cumulative function for the random variable X in Example 1, Section I–2, is shown in Fig. I–2.

For discrete distributions, by noting the changes in the cumulative F at the jumps, we can get the probabilities $f(x_i)$ at the jump points x_i. (See Figs. I–1 and

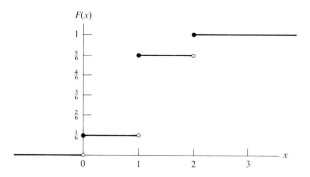

Fig. I–2 Cumulative for a discrete distribution.

I–2.) If the x_i's are ordered so that $x_{i-1} < x_i$, and there is a first jump point x_1, then

$$F(x_i) - F(x_{i-1}) = f(x_i), \qquad i = 2, 3, 4, \ldots, n;$$

$$F(x_1) = f(x_1).$$

For $x < x_1$, $F(x) = 0$, for $x \geq x_n$, $F(x) = 1$. Thus, in Fig. I–2

$$f(2) = F(2) - F(1) = 1 - \tfrac{5}{6} = \tfrac{1}{6}.$$

I–4 THE MEAN AND THE VARIANCE

For discrete probability distributions, we noted in Section I–1 that, if x_1, x_2, $x_3, \ldots, x_n$ are possible values of the random variable X with probabilities $f(x_i) = p_i$, then

$$\sum_{i=1}^{n} f(x_i) = \sum_{i=1}^{n} p_i = 1. \tag{1}$$

The mean and variance of X are defined as follows:

Definitions. *Mean and variance for discrete distributions.* If a discrete random variable X has possible values $x_1, x_2, \ldots, x_n$, where $P(X = x_i) = f(x_i) = p_i$, then the mean value of X is defined as

$$\mu = \sum_{i=1}^{n} x_i f(x_i) = \sum_{i=1}^{n} x_i p_i = E(X); \tag{2}$$

and the variance of X is defined as

$$\sigma^2 = \sum_{i=1}^{n} (x_i - \mu)^2 f(x_i) = \sum_{i=1}^{n} (x_i - \mu)^2 p_i = E[(X - \mu)^2]. \tag{3}$$

An alternative form of (3), which is much used for computation by hand, is

$$\sigma^2 = \sum_{i=1}^{n} x_i^2 f(x_i) - \mu^2 = E(X^2) - [E(X)]^2. \tag{4}$$

Notes

1. The expression $E(X)$ is "the expected value of the random variable X". The notation refers not to a function but to an operation on the values of X.

2. Often we attach a subscript to μ or to σ^2 to show explicitly the random variable whose mean or variance is being computed. As examples, μ_X, $\mu_{\bar{X}}$, $\mu_{\Sigma X}$, and μ_{T^2}, refer respectively to the mean of the random variable X; the mean of the sample mean $\bar{X}$; the mean of the sum of the values of the variables $X_1, X_2, \ldots, X_n$ (but the subscripts have been suppressed); and the mean of the square of T. Note that $\mu_{T^2} = E(T^2)$.

3. In formulas (1) through (4), the upper limit n may be $+\infty$, since n is either finite or countably infinite in a discrete distribution. In principle, the lower limit could be minus infinity. We can imagine a doubly infinite sequence of x_i's, such as

$$\ldots x_{-3}, x_{-2}, x_{-1}, x_0, x_1, x_2, x_3, \ldots.$$

A simple example is the set of all integers, positive, zero, and negative. It would then be convenient to write

$$\mu = \sum_{i=-\infty}^{\infty} x_i f(x_i).$$

In the countably infinite case, the mean and variance may or may not exist. When the mean exists, the variance may not.

Example 2. Find the mean and the variance for the random variable X in Example 1, Section I–2.

Solution. Using the probability table in Section I–2, we get

$$\mu_X = \sum_{i=1}^{n} x_i p_i = 0(\tfrac{1}{6}) + 1(\tfrac{4}{6}) + 2(\tfrac{1}{6}) = 1,$$

and

$$\sigma_X^2 = E(X^2) - [E(X)]^2$$
$$= [0^2(\tfrac{1}{6}) + 1^2(\tfrac{4}{6}) + 2^2(\tfrac{1}{6})] - 1^2$$
$$= \tfrac{8}{6} - 1 = \tfrac{1}{3}.$$

I-5 SOME PROPERTIES OF MEANS AND VARIANCES

The following theorems can be proved.

Theorem 1. Given a random variable X with mean μ and variance σ^2. If a and b are constants, then

$$E(aX + b) = aE(X) + b = a\mu + b, \tag{1}$$

and

$$\text{Var}\,(aX + b) = a^2\,\text{Var}\,(X) = a^2\sigma^2. \tag{2}$$

Theorem 2. Given n random variables $X_1, X_2, \ldots, X_n$ with means $\mu_1, \mu_2, \ldots, \mu_n$, respectively. Then

$$E(\textstyle\sum X_i) = \sum[E(X_i)] = \sum \mu_i. \tag{3}$$

In words: The mean of the sum of random variables equals the sum of their means.

Theorem 3. Given n *independent* random variables $X_1, X_2, \ldots, X_n$ with variances $\sigma_1^2, \sigma_2^2, \ldots, \sigma_n^2$, respectively. Then

$$\text{Var}\,(\textstyle\sum X_i) = \sum[\text{Var}\,(X_i)] = \sum \sigma_i^2. \tag{4}$$

In words: The variance of the sum of *independent* random variables equals the sum of their variances. Theorem 3 also holds if the condition "independent" is replaced by the weaker condition "uncorrelated".

Means and variances of *weighted* sums are discussed in Appendix V-4.

Continuous Distribution of a Random Variable

[References: *PWSA* Sections 7–2, 7–3, 7–4, 8–3]

II–1 PROBABILITIES REPRESENTED BY AREAS

If a random variable X denotes a physical measurement, we think of the values of X as being continuous, which means, roughly speaking, that there are no gaps or jumps in the set of admissible values. For example, measurements of length and time can, theoretically, take every real value in an interval, even though we can never attain complete precision in practice. In dealing with such situations, we must assign the probability smoothly, rather than in chunks.

Example 1. A blindfolded experimenter throws a switch that stops an electric clock. What is the probability that the clock stops at *exactly* 20 seconds after a full minute?

Discussion. By "exactly" we mean 20.00000 . . . seconds after the minute, where the zeros continue without end. The desired probability is obviously zero: there is no chance at all.

Similarly, for every real number of seconds between 0 and 60, the probability is zero. But the probability that the clock stops is 1, for it must stop somewhere. Thus it appears that, for continuous distributions, we cannot get the total probability by adding probabilities at points. We are confronted by a paradox of the infinite: The sum of zeros cannot give 1, but the total probability must be 1. The difficulty is not new. We may think of a line segment of unit length as composed of an infinite set of points. Each point has zero length, and we cannot get the length of the segment by adding the lengths of its points.

We get around this problem by assigning probabilities to intervals rather than to single points. *For continuous distributions, we represent probabilities as areas above intervals.* The method is illustrated in the following example.

Example 2. *The stopped clock.* If a clock stops at a random time, what is the probability that the hour hand stops between the numerals 1 and 5?

Solution. Since the arc between the numerals 1 and 5 is $\frac{4}{12}$ of the circumference, it is natural to say that the required probability is $\frac{1}{3}$. How shall we represent this result graphically?

We represent the probabilities as *areas* and get an *area probability graph* as follows:

We consider what is wanted: (a) The total area of the graph must be 1. (b) The base of the graph must extend from 0 to 12 on the time axis. (c) Equal probabilities must be assigned to equal time intervals, since the clock stops at a random point between 0 and 12. All three demands are met if we represent the total probability by a rectangle with base on the interval $[0, 12]$ of the time axis and with altitude $\frac{1}{12}$ unit, as shown in the following figure. The shaded area, which is $(5 - 1)/12 = \frac{4}{12}$ units, represents the probability that the clock stops between 1 and 5.

II–2 THE PROBABILITY DENSITY FUNCTION

We have seen that the probabilities associated with a continuous random variable X are represented by areas above intervals. What forms the upper boundary of such an area? In Example 2 of Section II–1, the upper boundary was a straight line. In general, the upper boundary is a curve that may, for example, look like one of the curves in Fig. II–1. Each of these curves is the graph of a function f defined by a set of ordered pairs $(x, f(x))$. But any arbitrarily chosen function f will not do. Acceptable functions f enjoy special properties and are given a special name, as set forth in the following definition.

Definition. *Probability density function* (pdf). The function f is the probability density function of the continuous random variable X if and only if

(1) the total area between the graph of f and the x-axis is 1 unit; and

(2) the area (shaded) bounded by $x = a$, $x = b$, the x-axis, and the graph of f equals

$$P(a < X < b).$$

(3) $f(x) \geq 0$ for all $-\infty \leq x \leq \infty$.

It is common practice also to refer to $f(x)$, the formula of f, as the probability density function of X or, more simply, as the *density* of X. The graph of f is called the *density curve* or, more carelessly, the density function of X, or even "the density".

How are the areas under density curves computed? For some functions, calculus provides the answer; for others, approximation methods may be used. In this book, we usually do the job with tables, as illustrated in Section 1–4.

In a manner similar to that used for discrete distributions, the cumulative function F of a continuous random variable X is defined by $F(x)$, where

$$F(x) = P(X \leq x)$$

= area under density curve to the left of x.

An ellipsis. Theoretically, a continuous random variable may take as a value any real number within a certain interval or set of intervals. Nevertheless, it is possible for a continuous random variable to have a discrete distribution. For example, if we are measuring lengths of rods, then the length of a rod, X, is a continuous random variable; but if all lengths turn out to be one foot or two feet, then the distribution of X is discrete.

In this book, when we use the phrase "continuous random variable", we shall mean "a random variable with a continuous cumulative distribution" unless the contrary is explicitly stated.

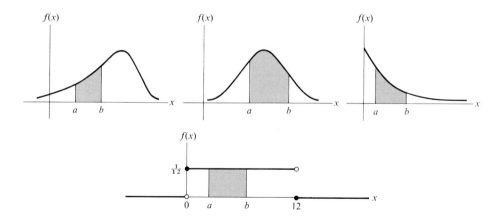

Fig. II–1 $P(a < X < b)$.

II–3 MEAN AND VARIANCE OF A CONTINUOUS DISTRIBUTION

The definitions given in Appendix I for the mean and variance of a discrete distribution must be modified to accommodate continuous distributions. To define the mean and the variance of a continuous probability distribution, we need some calculus. Essentially, we drop the subscripts used in Appendix I, replace the summation sign by the definite integral (recall that the definite integral may be regarded as the limit of a sum), and use limits $-\infty$ and $+\infty$ so as to cover the whole real axis.

Definitions. *Mean and variance for continuous distributions.* If X is a continuous random variable with a density function f that is continuous for real x, then the mean value of X is defined as

$$\mu = \int_{-\infty}^{\infty} x f(x) \, dx; \tag{1}$$

and the variance of X is defined as

$$\sigma^2 = \int_{-\infty}^{\infty} (x - \mu)^2 f(x) \, dx. \tag{2}$$

For convenience in computation, we have the alternative formula

$$\sigma^2 = \int_{-\infty}^{\infty} x^2 f(x) \, dx - \mu^2 = E(X^2) - [E(X)]^2. \tag{3}$$

It is possible for f not to have a mean or a variance under certain circumstances. Nonexistence of the mean or the variance means that the integrals (1) or (2) or both do not converge.

Similar troubles can arise in dealing with discrete distributions. These troubles are real and often of practical importance, but we need not bombard the reader with them in this book.

The computation of μ and σ^2. When the total density is restricted to an interval rather than spread out over the whole axis, we need not integrate over the whole axis. For example, if the probability is confined to the interval $[a, b]$, then

$$\int_{-\infty}^{a} f(x) \, dx = \int_{b}^{\infty} f(x) \, dx = 0,$$

and we can integrate just from a to b. This idea is illustrated in the following example.

Example 3. The pdf of a continuous random variable X is defined by $f(x) = 3x^2$, where $0 \leq x \leq 1$, and $f(x) = 0$ for all other values of x. Verify that $P(0 < X < 1) = 1$, and compute μ_X and σ_X^2.

Solution. From the definition of pdf, we have

$$P(0 < X < 1) = \int_0^1 3x^2 \, dx = x^3 \Big]_0^1 = 1 \therefore$$

Since all of the probability lies between 0 and 1, we get

$$\mu_X = \int_0^1 x f(x) \, dx = \int_0^1 3x^3 \, dx = \frac{3x^4}{4} \Big]_0^1 = \tfrac{3}{4}.$$

Also, since

$$E(X^2) = \int_0^1 x^2 f(x) \, dx = \int_0^1 3x^4 \, dx = \frac{3x^5}{5} \Big]_0^1 = \tfrac{3}{5},$$

we have

$$\sigma_X^2 = E(X^2) - \mu_X^2 = \tfrac{3}{5} - \tfrac{9}{16} = \tfrac{3}{80}.$$

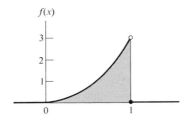

What happens if we are presented with integrands such as no mortal man can integrate? Unfortunately, this situation is extremely common. How do we attack it? Essentially, we go back to the definition of an integral as the limit of a sum. We break up the x-axis into tiny units, we evaluate the ordinate at the middle of each unit, we multiply this value by the width of the unit, and we add up all these tiny rectangular areas to get an approximation.

In short, we really return to the discrete case. (Naturally, more sophisticated numerical methods exist, but they all feature a discrete grid. Today high-speed computers help a lot with such integrations, but many fine extensive tables were built without them.) Since we shall not need to use such a method in this book, we shall not illustrate it. (For an example, see *PWSA*, pp. 264–265.)

Most introductory calculus texts explain the foregoing method of evaluating an integral, in the course of defining the definite integral of *f*.

Note. The theorems about means and variances listed in Section I–5 also hold for continuous distributions.

Normal Approximations: Central Limit Theorems

[References: *PWSA* Section 8–3; pp. 351–352]

III–1 THE VALUE OF APPROXIMATIONS

The most important property of a random variable is its distribution. Since exact distributions are often hard to get, we welcome approximations that bridge the gap. A quite general set of theorems, called Central Limit Theorems, deal with the limiting distributions of sums of random variables, and in our problems these limiting distributions are ordinarily normal distributions. The Central Limit Theorems tell us how to use the normal tables to get probabilities for a sum of random variables without knowing the exact distribution of this sum.

For example, if we think of the number of successes on each trial of a binomial experiment as a random variable, then the total number of successes in n trials, X, is a random variable that is the sum of n independent random variables. And there is a Central Limit Theorem that gives us a nice approximation to the probabilities for the binomial random variable X when n is large.

III–2 STANDARDIZED RANDOM VARIABLES

Normal approximations are effected by using the probability tables for the *standard* normal random variable Z, which has mean 0 and variance 1. In using a Central Limit Theorem to approximate the probabilities for a random variable X, we first transform X to a new, standardized random variable having mean 0 and variance 1. This operation is quite simple: The standardized form for a random variable X is

$$\frac{X - \mu_X}{\sigma_X}.$$

Since

$$E(aX + b) = aE(X) + b,$$

and

$$\text{Var}\ (aX + b) = a^2\ \text{Var}\ (X),$$

we have

$$E\left[\frac{X - \mu_X}{\sigma_X}\right] = \frac{E(X)}{\sigma_X} - \frac{\mu_X}{\sigma_X} = 0,$$

and

$$\text{Var}\left[\frac{X - \mu_X}{\sigma_X}\right] = \text{Var}(X)/\sigma_X^2 = 1.$$

III–3 THE CENTRAL LIMIT THEOREM FOR THE BINOMIAL

Consider the following example.

Example 1. A manufacturer produces electric fuses that are perfect 90% of the time. What is the probability of getting 95 or more perfect fuses in a randomly selected 100?

Discussion. The exact probability is given by the binomial formula as

$$\sum_{r=95}^{100} \binom{100}{r} (0.90)^r (0.10)^{100-r}.$$

Direct computation would add little to the gaiety of nations, and tables may well fall short. But the following theorem would see us through.

Theorem. *The De Moivre-Laplace theorem.* Let $X_1, X_2, \ldots, X_n$ be a sequence of binomial random variables, where X_n is the number of successes in a binomial experiment of n trials, each with $P(\text{success}) = p, 0 < p < 1$. Let Z_n, $n = 1, 2, 3, \ldots$, be the corresponding sequence of standardized random variables where

$$Z_n = \frac{X_n - np}{\sqrt{npq}}.$$

Then, as n becomes infinite,

$$P(Z_n \geq z) \to P(Z \geq z),$$

where Z is the *standard* normal random variable and z is a constant.

How do we use this theorem in practice? Suppose that, for a binomial experiment, we want to find $P(X_n \geq x)$. We proceed as follows:

1. *We make the continuity correction.* (See Section 2–6.) This step improves the approximation because we are using a continuous distribution to approximate a discrete distribution. We get

$$P(X_n \geq x) = P(X_n \geq x - \tfrac{1}{2}).$$

2. *We standardize the random variable.* Thus,

$$P(X_n \geq x) = P(X_n \geq x - \tfrac{1}{2})$$

$$= P\left(\frac{X_n - np}{\sqrt{npq}} \geq \frac{x - \tfrac{1}{2} - np}{\sqrt{npq}}\right).$$

3. *We apply the De Moivre-Laplace theorem.* Let

$$z = \frac{x - \frac{1}{2} - np}{\sqrt{npq}}.$$

Then the sequences can be standardized:

$$P(X_n \geq x) = P\left(Z_n \geq \frac{x - \frac{1}{2} - np}{\sqrt{npq}}\right) \approx P(Z \geq z),$$

where Z is the standard normal random variable and z is a value of Z. The normal tables finish the job.

We can now solve the problem in Example 1. Since

$$z = \frac{95 - \frac{1}{2} - 100(0.98)}{\sqrt{100(0.90)(0.10)}} = 1.5,$$

the normal table gives

$$P(X \geq 95) \approx P(Z \geq 1.5) = 0.067,$$

which agrees fairly well with the true value 0.058.

Note. The De Moivre-Laplace theorem may give a good approximation for the binomial even for modest values of n. If the mean μ is at least 3σ from both 0 and n, empirical investigations suggest that the maximum error is 0.011 in evaluating single terms, and 0.025 in evaluating the cumulative.

III–4 MORE GENERAL CENTRAL LIMIT THEOREMS

The De Moivre-Laplace theorem for binomial counts is a special case of the following more general theorem.

Theorem. *Central Limit Theorem.* Let $X_1, X_2, \ldots, X_n$ be a sequence of *identically distributed, independent* random variables, each with mean μ and variance σ^2. Let

$$T_n = X_1 + X_2 + \cdots + X_n.$$

Then, for each fixed value z, as n becomes infinite,

$$P\left(\frac{T_n - n\mu}{\sigma\sqrt{n}} \geq z\right) \rightarrow P(Z \geq z),$$

where Z is the standard normal random variable.

Notes

1. The mean of T_n is $n\mu$ because the mean of a sum is the sum of the means; and Var $(T_n) = n\sigma^2$ because, for *independent* random variables, the variance of a sum is the sum of the variances.

2. Since

$$\frac{T_n - n\mu}{\sigma\sqrt{n}} = \frac{(T_n/n) - \mu}{\sigma/\sqrt{n}} = \frac{\bar{X} - \mu}{\sigma_{\bar{X}}},$$

the Central Limit Theorem may take the form used in Section 8–1:

> Given n independent measurements drawn from a normal distribution with mean μ and standard deviation σ. If we adjust the sample mean $\bar{X}$ by subtracting the *population* mean μ and dividing by $\sigma_{\bar{X}}$ ($= \sigma/\sqrt{n}$), we get a random variable
>
> $$\frac{\bar{X} - \mu}{\sigma_{\bar{X}}} = \frac{\bar{X} - \mu}{\sigma/\sqrt{n}}$$
>
> with the *standard* normal distribution.

3. The Central Limit Theorem justifies our using a normal approximation for chi-square in Section 8–4 and elsewhere when the number of degrees of freedom is large

The foregoing Central Limit Theorem can be made even more general. We may give up the demand for *identical* distributions. Under broad conditions, for large values of n, if $X_1, X_2, \ldots, X_n$ are *independent* random variables with means

$$\mu_1, \mu_2, \ldots, \mu_n$$

and variances

$$\sigma_1^2, \sigma_2^2, \ldots, \sigma_n^2,$$

respectively, and $T_n = \Sigma X_i$, then the random variable

$$\frac{T_n - \Sigma\mu_i}{\Sigma\sigma_i^2}$$

has approximately the standard normal distribution.

Dependent Random Variables: Covariance and Correlation

[Reference: *PWSA* Section 10–4]

IV–1 COVARIANCE

If random variables are *independent*, the variance of their sum equals the sum of their variances:

$$\text{Var} \left(\sum X_i \right) = \sum (\text{Var } X_i).$$

What sort of relation, if any, holds when the random variables are dependent? To answer this question, we introduce the idea of covariance.

Consider two dependent random variables X and Y. Since these variables are dependent, we would like to get some measure of the extent to which their paired values are linearly related. Do the paired values of these variables increase and decrease together? Or does one of the paired values get larger as the other gets smaller? The measure that we use to provide answers to such questions is *the mean of the product of the paired deviations of X and Y from their respective means*. This measure is called the covariance between X and Y, and it is denoted by Cov (X, Y).

Definition. *Covariance.* The covariance between X and Y is

$$\text{Cov } (X, Y) = E[(X - \mu_X)(Y - \mu_Y)]. \tag{1}$$

Note that, if paired values of X and Y become large or small together, the two factors on the right side of (1) are usually both positive or both negative, and the covariance is usually positive. And if one of the paired values of X and Y gets large as the other gets small, one of the factors on the right side of (1) is usually positive and the other negative. Hence the covariance is usually negative.

IV–2 PROPERTIES OF COVARIANCE

For purposes of computation, formula (1) can be transformed into

$$\text{Cov } (X, Y) = E(XY) - \mu_X \mu_Y. \tag{2}$$

288

It is evident from either (1) or (2) that

$$\text{Cov} (X, Y) = \text{Cov} (Y, X); \tag{3}$$

and that

$$\text{Cov} (X, X) = E(X^2) - \mu_X^2 = \text{Var} (X) = \sigma_X^2. \tag{4}$$

Hence *the covariance of a random variable with itself is the variance.*
    From formula (1), we get

$$\text{Cov} (X + a, Y + b) = E[X + a - (\mu_X + a)][Y + b - (\mu_Y + b)]$$
$$= \text{Cov} (X, Y). \tag{5}$$

And from formula (2), we get

$$\text{Cov} (aX, bY) = E(abXY) - ab\mu_X\mu_Y = ab \text{ Cov} (X, Y). \tag{6}$$

Thus the covariance is unchanged when arbitrary constants are *added* to the variables; but the covariance is changed when the variables are *multiplied* by arbitrary constants.

Example 1. Let X and Y have the following joint probability function. Find $\text{Var} (X)$, $\text{Var} (Y)$, and $\text{Cov} (X, Y)$.

x	y	$P(x, y)$
1	1	0.2
1	-1	0.4
-1	1	0.1
-1	-1	0.3

Solution. We have

$$\mu_X = 1(0.6) + (-1)(0.4) = 0.2,$$
$$\mu_Y = 1(0.3) + (-1)(0.7) = -0.4,$$
$$\sigma_X^2 = [1^2(0.6) + (-1)^2(0.4)] - (0.2)^2 = 0.96,$$
$$\sigma_Y^2 = [1^2(0.3) + (-1)^2(0.7)] - (-0.4)^2 = 0.84,$$
$$\text{Cov} (X, Y) = E(XY) - \mu_X\mu_Y$$
$$= [(1)(1)(0.2) + (1)(-1)(0.4) + (-1)(1)(0.1) + (-1)(-1)(0.3)]$$
$$- (0.2)(-0.4)$$
$$= 0.08.$$

IV–3 CORRELATION

Since a change of units of measurements of X and Y (for example, changing from feet to inches) is equivalent to multiplying X and Y by constants, formula (6) shows that the covariance depends on the units of measurement. We want a measure of linear relationship between X and Y that is independent of the units of measurement. Such a measure is obtained if we take the covariance between standardized X and standardized Y.

> **Definition.** *Correlation.* The *correlation* between X and Y is
>
> $$\rho = \text{Cov}\,(X', Y'). \tag{7}$$

The *standardized* values of X and Y are

$$X' = \frac{X - \mu_X}{\sigma_X} \quad \text{and} \quad Y' = \frac{Y - \mu_Y}{\sigma_Y}.$$

Applying the results of (5) and (6) to (7), we get

$$\rho = \text{Cov}\,(X', Y') = \frac{\text{Cov}\,(X, Y)}{\sigma_X \sigma_Y}, \quad \text{or} \quad \text{Cov}\,(X, Y) = \rho \sigma_X \sigma_Y. \tag{8}$$

The correlation between aX and bY, where a and b are arbitrary nonzero constants, is

$$\frac{\text{Cov}\,(aX, bY)}{\sigma_{aX} \sigma_{bY}} = \frac{ab\,\text{Cov}\,(X, Y)}{|ab|\sigma_X \sigma_Y} = \pm \rho,$$

where ρ is the correlation between X and Y. Hence, except possibly for sign, $a, b \neq 0$, ρ is indeed independent of the units of measurement of X and Y.

If X and Y are independent, $E(XY) = \mu_X \mu_Y$ and $\text{Cov}\,(X, Y) = 0$ by formula (2). Hence $\rho = 0$. But the converse is not true: $\rho = 0$ *does not necessarily imply that X and Y are independent.* If $\rho = 0$, we say that X and Y are *uncorrelated.*

For purposes of computation, we often use the form

$$\rho = \frac{E(XY) - \mu_X \mu_Y}{\sigma_X \sigma_Y}. \tag{9}$$

If $\sigma_{X'} = 0$ or $\sigma_Y = 0$, ρ is undefined.

Example 2. Using the data of Example 1, find the correlation between the random variables X and Y.

Solution. Since Cov (X, Y) has been computed, the most convenient formula is

$$\rho = \frac{\text{Cov}(X, Y)}{\sigma_X \sigma_Y} = \frac{0.08}{\sqrt{(0.96)(0.84)}} \approx 0.09.$$

The range of ρ. It can be proved that

$$-1 \leq \rho \leq 1.$$

These limits are intuitively satisfying because, going back to the meaning of correlation, we might expect the correlation between X and Y to be a maximum when $Y = X$ and a minimum when $Y = -X$. Since

$$\text{Cov}(X, X) = \sigma_X^2 \qquad \text{and} \qquad \text{Cov}(X, -X) = -\sigma_X^2,$$

our hunch plus formula (8) leads to the correct limits for ρ. A proof would use a Schwarz inequality.

Sample correlation coefficients. Given a set of pairs of measurements (x_i, y_i), $i = 1, 2, \ldots, n$, we make the following definition, by analogy with ρ.

Definition. The *sample correlation coefficient* is

$$r = \frac{\sum(x_i - \bar{x})(y_i - \bar{y})}{\sqrt{\sum(x_i - \bar{x})^2 \, \sum(y_i - \bar{y})^2}}.$$

We also call r the *Pearsonian or the product-moment correlation coefficient*. It can be used as an estimate of ρ.

For purposes of computation, we may use the equivalent formula

$$r = \frac{\sum x_i y_i - n\bar{x}\bar{y}}{\sqrt{\sum(x_i - \bar{x})^2 \, \sum(y_i - \bar{y})^2}}.$$

Variances of Sums of Random Variables

[Reference: *PWSA* Section 10–4]

Equation (11) of Section V–3 below gives the variance of a weighted sum of possibly correlated random variables. This formula contains the basic information of this Appendix. The rest is motivation and important special cases.

V–1 VARIANCE OF $X + Y$ AND OF $X - Y$

Using the definitions of variance and covariance, we can prove that

$$\sigma^2_{X+Y} = \sigma^2_X + \sigma^2_Y + 2 \text{ Cov } (X, Y). \tag{1}$$

Since

$$\sigma^2_{-Y} = (-1)^2\sigma^2_Y = \sigma^2_Y,$$

and

$$\text{Cov } (X, -Y) = -\text{Cov } (X, Y),$$

we can replace Y by $-Y$ in (1) and get

$$\sigma^2_{X-Y} = \sigma^2_X + \sigma^2_Y - 2 \text{ Cov } (X, Y). \tag{2}$$

Using formula (8) of Appendix IV, we can write (1) and (2) as follows:

$$\sigma^2_{X+Y} = \sigma^2_X + \sigma^2_Y + 2\rho\sigma_X\sigma_Y, \tag{3}$$

$$\sigma^2_{X-Y} = \sigma^2_X + \sigma^2_Y - 2\rho\sigma_X\sigma_Y. \tag{4}$$

If X and Y are *independent or even uncorrelated*, $\rho = 0$ and we get the somewhat surprising result:

$$\sigma^2_{X+Y} = \sigma^2_X + \sigma^2_Y = \sigma^2_{X-Y}. \tag{5}$$

Example 1. For the data in Example 1, Appendix IV, compute σ^2_{X+Y} and σ^2_{X-Y}.

Solution. Using the variances and covariances computed in Example 1, Appendix IV, we have

$$\sigma_{X+Y}^2 = \sigma_X^2 + \sigma_Y^2 + 2 \operatorname{Cov}(X, Y)$$

$$= 0.96 + 0.84 + 2(0.08) = 1.96.$$

The covariance has increased $\operatorname{Var}(X + Y)$ to about 9% above the value $\sigma_X^2 + \sigma_Y^2 = 1.80$ that would have resulted if X and Y were independent. Similarly, we find $\sigma_{X-Y}^2 = 1.80 - 0.16 = 1.64$.

V-2 VARIANCE OF THE SUM OF n RANDOM VARIABLES

If X_1, X_2, X_3 are random variables with variances σ_1^2, σ_2^2, σ_3^2, respectively, and if

$$Z = X_1 + X_2 + X_3,$$

it may be no surprise to learn that formula (1) can be extended to give

$$\sigma_Z^2 = \sigma_1^2 + \sigma_2^2 + \sigma_3^2 + 2 \operatorname{Cov}(X_1, X_2) + 2 \operatorname{Cov}(X_2, X_3) + 2 \operatorname{Cov}(X_1, X_3). \quad (6)$$

In words: *The variance of a sum of several random variables is the sum of their variances plus twice the sum of the covariances of all possible selections of pairs of the random variables.*

The reason for the 2's which multiply the covariance is that, when we square a sum, $(X_1 + X_2 + X_3)^2$, the crossproduct terms have a 2 multiplying them. One way to think of this is as arising from the 2 orders, say $X_1 X_2$ and $X_2 X_1$. Then it is convenient to think of the variance of a sum as being composed of terms that form a square which is symmetrical about its main diagonal:

i \ j	1	2	3
1	σ_1^2	$\operatorname{Cov}(X_1, X_2)$	$\operatorname{Cov}(X_1, X_3)$
2	$\operatorname{Cov}(X_2, X_1)$	σ_2^2	$\operatorname{Cov}(X_2, X_3)$
3	$\operatorname{Cov}(X_3, X_1)$	$\operatorname{Cov}(X_3, X_2)$	σ_3^2

Even the main diagonal terms could be written $\operatorname{Cov}(X_i, X_i)$. The whole square is called a covariance matrix, or sometimes a variance-covariance matrix. Note that when we add these terms, since $\operatorname{Cov}(X_i, X_j) = \operatorname{Cov}(X_j, X_i)$, each covariance produces a coefficient of 2 for formula (6), or more generally for formula (7) below. Thinking about the square makes it easy to remember that every variable is being associated with every other variable in the variance of a sum. The square is a more orderly way to look at matters than formula (6) or (7).

The foregoing statement can be generalized to deal with n random variables.

Theorem. *Variance of a sum.* Let the random variables $X_1, X_2, \ldots, X_n$ have variances $\sigma_1^2, \sigma_2^2, \ldots, \sigma_n^2$, respectively, and covariances Cov (X_i, X_j), $i \neq j$, and let

$$Z = X_1 + X_2 + \cdots + X_n.$$

Then

$$\sigma_Z^2 = \sum_{i=1}^{n} \sigma_i^2 + 2 \sum_{i<j} \text{Cov } (X_i, X_j), \tag{7}$$

$i = 1, 2, \ldots, n$, and $j = 1, 2, \ldots, n$.

Note that the number of covariances to be added is $\binom{n}{2} = n(n-1)/2$, which is the number of selections of n variables, taken 2 at a time without counting order.

If we denote the correlation between X_i and X_j by ρ_{ij}, we can use formula (8) of Appendix IV and write

$$\sigma_Z^2 = \sum_{i=1}^{n} \sigma_i^2 + 2 \sum_{i<j} \rho_{ij} \sigma_i \sigma_j, \tag{8}$$

$i = 1, 2, \ldots, n$, and $j = 1, 2, \ldots, n$. Again, note that this formula could have its terms arranged in a square, because $\rho_{ij} = \rho_{ji}$.

If the X_i's are *independent or even uncorrelated*, then all of the ρ_{ij}'s are zero, and we get

$$\sigma_Z^2 = \sum_{i=1}^{n} \sigma_i^2. \tag{9}$$

In words: *The variance of the sum of n independent or uncorrelated random variables is the sum of their variances.* Recall that independence implies zero correlation, but zero correlation does not imply independence.

Example 2. *Simplified Army alpha test.* A test is composed of four parts: (1) arithmetic, (2) number completion, (3) word analogies, and (4) information. Let X_1 stand for the score on part 1, X_2 for the score on part 2, and so on. Scores on each part have mean 50 and standard deviation 10, and the variables are correlated as follows (after J. P. Guilford, *Psychometric Methods*, McGraw-Hill, 1936, p. 491):

	X_1	X_2	X_3	X_4	Sum
X_1		0.49	0.09	0.31	0.89
X_2			0.40	0.16	0.56
X_3				0.29	0.29
				Total	1.74

Find the variance of Z, the sum of the scores on the four parts.

Solution. Since we have the correlations ρ_{ij}, $i < j$, formula (8) is convenient. All σ_i's are 10, and we have

$$\sigma_Z^2 = \sum_{i=1}^{n} \sigma_i^2 + 2 \sum_{i<j} \rho_{ij}\sigma_i\sigma_j$$

$$= 4(100) + 2(100) \sum_{i<j} \rho_{ij}$$

$$= 400 + 200(1.74) = 748.$$

Because the mean of Z is $4(50) = 200$ and $\sigma = \sqrt{748} = 27.3$, and because the scores on such tests are usually approximately normally distributed, we conclude that about $\frac{2}{3}$ of the scores are within one standard deviation of the mean, that is, between 173 and 227.

A special case. If all of the ρ_{ij}'s equal ρ, and all of the σ_i's equal σ, then formula (8) becomes

$$\sigma_Z^2 = n\sigma^2 + 2\sigma^2[n(n-1)/2]\rho$$

$$= n\sigma^2[1 + (n-1)\rho]. \tag{10}$$

V–3 VARIANCE OF WEIGHTED SUMS

Given n random variables $X_1, X_2, \ldots, X_n$ with variances $\sigma_1^2, \sigma_2^2, \ldots, \sigma_n^2$, respectively. If these random variables are given constant weights $w_1, w_2, \ldots, w_n$, respectively, then we can use formula (7) to get the variance of the weighted sum

$$W = w_1 X_1 + w_2 X_2 + \cdots + w_n X_n.$$

We need only replace X_i with $w_i X_i$, and recall that

$$\sigma_{w_i X_i}^2 = w_i^2 \sigma_{X_i}^2 = w_i^2 \sigma_i^2,$$

and that

$$\text{Cov}(w_i X_i, w_j X_j) = w_i w_j \, \text{Cov}(X_i, X_j).$$

[See Appendix IV–2, equation (6).]

Substituting in formula (7), we get

$$\sigma_W^2 = \sum w_i^2 \sigma_i^2 + 2 \sum_{i<j} w_i w_j \, \text{Cov} \, (X_i, X_j). \tag{11}$$

If the X_i's are *independent or even uncorrelated*, all covariances vanish, and the variance of the weighted sum becomes

$$\sigma_W^2 = \sum w_i^2 \sigma_i^2. \tag{12}$$

Note that the terms of the right-hand side of formula (11) can form a square, the entry in the ith row and jth column being $w_i w_j \, \text{Cov} \, (X_i, X_j)$.

Example 3. Using the data of Example 1, Appendix IV, evaluate σ_W^2, where $W = 2X + 3Y$.

Solution. Using formula (11) along with the information in Example 1, Appendix IV, we get

$$\sigma_W^2 = \sum_{i=1}^{2} w_i^2 \sigma_i^2 + 2 \sum_{i<j} w_i w_j \, \text{Cov} \, (X_i, X_j)$$

$$= [2^2(0.96) + 3^2(0.84)] + 2[(2)(3)(0.080)]$$

$$= 12.36.$$

V–4 SUMS AND DIFFERENCES OF INDEPENDENT NORMAL RANDOM VARIABLES

The following facts concerning *independent normal* random variables are of frequent use.

1. The sum and the difference of two independent normal random variables are normally distributed.

2. Given the independent normal random variables X_i, $i = 1, 2, \ldots, n$, having, respectively, means μ_i and variances σ_i^2. Then:
   a) the *sum* $Z = \sum X_i$ is normally distributed with

$$\mu_Z = \sum \mu_i \quad \text{and} \quad \sigma_Z^2 = \sum \sigma_i^2;$$

   and

   b) the *weighted sum* $W = \sum w_i X_i$ is normally distributed with

$$\mu_W = \sum w_i \mu_i \quad \text{and} \quad \sigma_W^2 = \sum w_i^2 \sigma_i^2.$$

The idea of putting the terms in a square still works, but the off-diagonal terms are all zero.

V–5 SUMS AND DIFFERENCES OF
CORRELATED NORMAL RANDOM VARIABLES

Frequently normally distributed random variables are jointly distributed according to a distribution called a multivariate normal. Without defining it, we mention that, for this case, the distribution of a weighted sum of these variables is again a one-variable normal. This fact plus the Central Limit Theorem has the effect of allowing us frequently to treat sums of random variables rather cavalierly as approximately normal.

Sampling Theory

[References: *PWSA* pp. 353–356 and 372–377]

VI–1 SAMPLING WITH REPLACEMENT

To go from the sum of random variables to their mean is but a short step. If $X_1, X_2, \ldots, X_n$ represent n uncorrelated random variables, and

$$Z = X_1 + X_2 + \cdots + X_n,$$

then the mean of the X_i's is

$$\overline{X} = Z/n.$$

Hence

$$\mu_{\overline{X}} = \mu_Z/n \quad \text{and} \quad \sigma_{\overline{X}}^2 = \sigma_Z^2/n^2.$$

These facts coupled with the findings in Appendix V enable us to get the mean and the variance of *the distribution of sample averages*.

Theorem. *Means and variances of sample averages.* Given n uncorrelated random variables $X_1, X_2, \ldots, X_n$, having, respectively, means $\mu_1, \mu_2, \ldots, \mu_n$ and variances $\sigma_1^2, \sigma_2^2, \ldots, \sigma_n^2$. If $\overline{X}$ is the average of the n random variables, then

$$\mu_{\overline{X}} = \frac{1}{n}(\mu_1 + \mu_2 + \cdots + \mu_n) \tag{1}$$

and

$$\sigma_{\overline{X}}^2 = \frac{1}{n^2}(\sigma_1^2 + \sigma_2^2 + \cdots + \sigma_n^2). \tag{2}$$

Proof. If we let $Z = \Sigma X_i$, then

$$\mu_{\overline{X}} = \mu_Z/n = (\mu_1 + \mu_2 + \cdots + \mu_n)/n$$

because the mean of a sum equals the sum of the means. Also, by formula (9) of Appendix V, we have

$$\sigma_{\overline{X}}^2 = \sigma_Z^2/n^2 = (\sigma_1^2 + \sigma_2^2 + \cdots + \sigma_n^2)/n^2.$$

298

Special case: identically distributed random variables. If the n random variables X_i all have identical means, μ, and identical variances, σ^2, then (1) becomes

$$\mu_{\bar{X}} = (n\mu)/n = \mu, \tag{3}$$

and (2) becomes

$$\sigma_{\bar{X}}^2 = (n\sigma^2)/n^2 = \sigma^2/n. \tag{4}$$

Hence the standard deviation becomes

$$\sigma_{\bar{X}} = \sigma/\sqrt{n}.$$

Formulas (3) and (4) are important because they apply to:

1. *sampling with replacement from a finite population*, and

2. *sampling without replacement from an infinite population*.

If X_i is the measurement associated with the ith element of the sample, where $i = 1, 2, 3, \ldots, n$, then in sampling with replacement the X_i's all have the same probability function: The X_i's are identically distributed.   As the size of the population increases relative to the size of the sample, this situation is approximated with increasing accuracy in sampling *without* replacement.

In words: When uncorrelated random variables are identically distributed, *the expected value of the average of n measurements is the population mean, μ*; and *the standard deviation of the averages is the population standard deviation, σ, divided by $\sqrt{n}$.*

Thus the standard deviation of the averages is inversely proportional to the square root of the number of measurements in the sample.

Example 1. When a single die is tossed, the number of dots showing on the top face has $\mu = 3.5$ and $\sigma^2 = 35/12$. Find the mean and the variance of the sample average when 100 dice are tossed.

Solution. Let $\bar{X}$ be the average of the number of dots showing on the top faces of the 100 dice. Then

$$\mu_{\bar{X}} = \mu = 3.5,$$

and

$$\sigma_{\bar{X}}^2 = \sigma^2/n = 35/1200 = 0.0292.$$

Average proportion of successes for a binomial. Consider a binomial experiment of n trials with P (success) $= p$ and the total number of successes, X. For a binomial, we have

$$\mu_X = np \qquad \text{and} \qquad \sigma_X^2 = npq.$$

Let $\bar{p} = X/n$ be the proportion of successes.

Since $X = \sum X_i$, where X_i is the number of successes on the ith trial, the foregoing theory holds. We get

$$\mu_{\bar{p}} = \mu_X/n = np/n = p, \tag{5}$$

and

$$\sigma_{\bar{p}}^2 = \sigma_X^2/n^2 = npq/n^2 = pq/n. \tag{6}$$

Example 2. *Voting.* If 60% of a large population vote Republican, find the mean and the standard deviation of the proportion of Republican voters in a random sample of 100.

Solution. Since the population is large, we may regard the sampling as a binomial experiment with $p = 0.60$ and $n = 100$. Let $\bar{p}$ denote the proportion of Republican votes. Then

$$\mu_{\bar{p}} = \mu = 0.60,$$

and

$$\sigma_{\bar{p}}^2 = pq/n = (0.6)(0.4)/100 = 0.0024.$$

Therefore,

$$\sigma_{\bar{p}} = \sqrt{0.0024} = 0.049.$$

VI–2 SAMPLING WITHOUT REPLACEMENT

In many practical applications, sampling is done *without* replacement. What can be said about the mean and variance of the distribution of sample averages in such cases? For example, if we randomly select without replacement the CEEB scores of 10 students from a class of 50, what accuracy can we expect in drawing conclusions about the class from the sample?

Consider a population of N elements with mean μ and variance σ. A sample of n is randomly drawn from this population. Let X_1 be the measurement obtained from the first member of the sample, X_2 the measurement obtained from the second, ..., and, in general, X_i be the measurement obtained from the ith, $i = 1, 2, 3, \ldots, n$. Also, let

$$Z = X_1 + X_2 + \cdots + X_n.$$

We want to know the mean and variance of the sample average

$$\bar{X} = Z/n.$$

Note that our system of notation involves *order*. The *first* member of the sample corresponds to X_1, the second to X_2, and so on. Since the members are drawn without replacement, one might assume that the X_i's all have different means and variances. Such an assumption is false, for it can be proved that

1. The means of the X_i's are all equal and

$$E(\bar{X}) = \mu. \tag{1}$$

2. The variances of the X_i's are all equal to σ^2.

3. The correlations between pairs X_i, X_j are all equal to $-1/(N-1)$. Therefore, we can use formula (10) of Appendix V and get

$$\sigma_{\bar{X}}^2 = \frac{1}{n^2}\sigma_Z^2 = \frac{1}{n^2}\left[n\sigma^2\left(1 - \frac{n-1}{N-1}\right)\right],$$

whence

$$\sigma_{\bar{X}}^2 = \frac{\sigma^2}{n}\left(\frac{N-n}{N-1}\right). \tag{2}$$

The following special cases reenforce our confidence in formula (2).

1. If $n = N$, $\sigma_{\bar{X}}^2 = 0$. This is as it should be, for if the sample consists of the whole population, there is no variation in $\bar{X}$ from one sample to another: $\bar{X} = \mu$ for all samples, and the samples are all alike.

2. If $n = 1$, $\sigma_{\bar{X}}^2 = \sigma^2$. If the sample size is 1, then we are dealing with the population of individuals for which the variance is σ^2.

3. If N is large compared with n, then $N \approx N - 1$,

$$\frac{N-n}{N-1} \approx 1 - \frac{n}{N} \approx 1,$$

and

$$\sigma_{\bar{X}}^2 \approx \sigma^2/n,$$

which is the variance for sampling *with* replacement. This result is consistent with the statement that sampling *without* replacement from a very large population may be treated as sampling *with* replacement to yield a very good approximation. Since

$$\frac{N-n}{N-1} \approx 1 - \frac{n}{N},$$

authors often call n/N the sampling fraction f and approximate (2) by

$$\sigma_{\bar{X}}^2 \approx \frac{\sigma^2}{n}(1 - f).$$

This formula shows directly the effect of the size of the sampling fraction.

Example 3. The CEEB scores of a class of 50 students have mean 600 and standard deviation 50. Find the mean and standard deviation of the sample averages for a sample of 10 scores chosen at random.

Solution. We have

$$\mu_{\bar{x}} = \mu = 600$$

and

$$\sigma_{\bar{x}}^2 = \frac{\sigma^2}{n}\left(\frac{N-n}{N-1}\right) = \frac{2500}{10}\left(\frac{50-10}{50-1}\right) = 10{,}000/49.$$

Hence

$$\sigma_{\bar{x}}^2 = 100/7 = 14.3,$$

instead of $\sqrt{250} = 15.8$ for sampling with replacement.

Confidence Limits and Significance Testing

[References: *PWSA* Sections 9–4, 12–1, 12–2, 12–3, 12–4]

VII–1 THE GENERAL IDEA OF STATISTICAL INFERENCE

In the science of statistical inference, we use samples as a basis for making general statements about populations. For example, we use the distribution of the sample mean $\bar{X}$ to make statements about the population mean μ. In this book, we do this in two ways, sketched by the following descriptions.

1. By giving an interval $[\bar{x} - k, \bar{x} + k]$ within which we think μ is likely to lie. This technique is called *interval estimation*, and the method of *confidence limits* is one practical way to get such an interval.

2. By assuming that μ has the hypothetical value M. To decide how good our assumption is, we find how many standard deviations M is from $\bar{x}$. If $|\bar{x} - M|$ is more than a pre-chosen number of standard deviations, we "reject" the hypothesis that $M = \mu$; otherwise we "accept" the hypothesis. This is the method of *significance testing*.

Note that the words "accept" and "reject" are technical words as used in this connection: They are neutral labels indicating courses of action, with the exact courses not stated. In the foregoing discussion, a practical interpretation is that to "accept" the hypothesis means that we shall continue to believe that M is near μ; to "reject" the hypothesis means that we believe that it is unlikely that M is near μ.

Let us be more specific in outlining these two methods for several specific problems that arise in sampling from normal distributions.

Confidence limits for the population mean, known σ. The Central Limit Theorem guarantees that, for several independent measurements drawn from a continuous population with mean μ and standard deviation σ, the distribution of the sample mean $\overline{X}$ is approximately normal with $E(\overline{X}) = \mu$ and $\sigma_{\overline{X}}^2 = \sigma^2/n$. (See Appendix III–4.) Then

$$Z = (\overline{X} - \mu)/\sigma_{\overline{X}}$$

is approximately a *standard* normal random variable. Since, from Table A–1

$$P(|Z| \leq 1.96) = 0.95,$$

we have approximately

$$P(|(\overline{X} - \mu)/\sigma_{\overline{X}}| \leq 1.96) = 0.95.$$

An algebraic transformation of this inequality gives this fact:

$$\overline{X} - 1.96\sigma_{\overline{X}} \leq \mu \leq \overline{X} + 1.96\sigma_{\overline{X}} \qquad (1)$$

with probability 0.95.

The *confidence level* is the probability that the *confidence interval* $[\overline{X} - 1.96\sigma_{\overline{X}}, \overline{X} + 1.96\sigma_{\overline{X}}]$ includes the true mean. In the foregoing instance, the confidence level is 0.95. Other confidence levels may be obtained by replacing the multiplier 1.96; for example, for a confidence level of 0.99, we use the multiplier 2.32 because $P(|Z| \leq 2.32) = 0.99$, as shown in Table A–1. Note that as the confidence level gets higher, the confidence interval gets wider.

Significance testing. Suppose that we have a hypothetical mean M for the population. If we assume that $M = \mu$, then, for the distribution of sample means, we can say that $(\overline{X} - M)/\sigma_{\overline{X}}$ is approximately a standard normal random variable. And as before, the probability is about 0.95 that

$$\overline{X} - 1.96\sigma_{\overline{X}} \leq M \leq \overline{X} + 1.96\sigma_{\overline{X}}.$$

In other words, it is very likely that M is within 1.96 standard deviations of a value $\bar{x}$. If we substitute values of $\overline{X}$ in the inequality for sample after sample, 95% of the resulting statements will be true. Consequently, if M falls outside this confidence interval for a specific value $\bar{x}$, we are inclined to believe that M is not the true mean, and we reject the hypothesis that $M = \mu$. This is an example of significance testing.

The *significance level* of the test is

$$1 - \text{the confidence level,}$$

which is, in this case, $1 - 0.95 = 0.05$ or 5%. *The significance level is the probability of rejecting the hypothetical mean when it is the true mean.* When a hypothesis is set up for the purpose of seeing if the data force us to reject it, this hypothesis is called a *null hypothesis*. The word "null" conveys the idea "no difference". Essentially, the *significance level of a test is the probability of rejecting a true null hypothesis.*

In the rest of this Appendix, we shall illustrate standard uses of confidence limits and significance tests with some examples. Even though these important methods of statistical inference are widely used in their own right, we emphasize that *they are introduced here to exhibit ideas needed in this book*. Our purpose is to offer a brief review rather than a complete treatment of the methods. The reader who wishes a fuller discussion of the ideas should consult the references at the beginning of this Appendix or one of the texts listed in the bibliography.

VII–2 CONFIDENCE LIMITS AND SIGNIFICANCE TESTS ON μ; KNOWN σ

Example 1. *Confidence limits* (Machine setup). An operator sets his machine to make rods of a particular diameter. Different rods produced by this machine have different diameters. These diameters are approximately normally and independently distributed about their mean μ with standard deviation $\sigma = 0.0020$ inch. The first 9 rods made after a new machine setup have sample mean $\bar{x} = 1.2000$ inches. Set 95% confidence limits on the mean diameter of the population for this setup.

Solution. Let the random variable $\bar{X}$ denote the mean diameter of a sample. Since the items are independent and the measurements approximately normally distributed, we know from the Central Limit Theorem (Appendix III–4) that $\bar{X}$ is approximately normally distributed, and that the random variable $Z = (\bar{X} - \mu)/\sigma_{\bar{X}}$ is approximately *standard* normal. Hence the probability is about 0.95 that

$$\bar{X} - 1.96\sigma_{\bar{X}} \leq \mu \leq \bar{X} + 1.96\sigma_{\bar{X}}, \tag{1}$$

as in (1) of Section VII–1.

To apply (1) to our particular sample, recall (Appendix VI) that

$$\sigma_{\bar{X}} = \sigma/\sqrt{n} = 0.002/3 \approx 0.00067.$$

Substituting for $\bar{X}$ and $\sigma_{\bar{X}}$ in (1), we get the 95% confidence interval

$$1.1987 \leq \mu \leq 1.2013, \tag{2}$$

and our 95% confidence limits are $[1.1987, 1.2013]$. These limits give the operator a good notion of where his setting actually is.

Example 2. *Significance testing*. Suppose that in Example 1 the operator wanted to set the machine so that $\mu = 1.2008$ inches. Use the data of Example 1 to test, at the 5% level, the hypothesis that 1.2008 is the actual mean.

Solution 1. Since 1.2008 falls inside the 95% confidence interval (2), we do not reject the hypothesis that $\mu = 1.2008$ at the 5% level. Indeed, we could not reject any other value of μ in the interval (2) at the 5% level.

Solution 2. Alternatively, we may compute $z = (\bar{x} - \mu)/\sigma_{\bar{x}}$ and see if it falls between -1.96 and 1.96. We get

$$(1.2000 - 1.2008)/0.00067 = -1.2.$$

Since -1.2 is inside the 95% confidence interval for Z, we accept the null hypothesis at the 5% significance level.

VII–3 CONFIDENCE LIMITS AND SIGNIFICANCE TESTS ON μ; UNKNOWN σ, LARGE SAMPLES

If σ is unknown (and it usually is), we get an estimate of $\sigma_{\bar{x}}$ by using, instead of σ^2, the sample variance

$$s^2 = \frac{\Sigma(x_i - \bar{x})^2}{(n - 1)}.$$

We get

$$\text{estimate of } \sigma_{\bar{x}}^2 = s^2/n = \frac{\Sigma(x_i - \bar{x})^2}{n(n - 1)} = s_{\bar{x}}^2.$$

Although it is ridiculous to suppose that there is some sample size n above which samples are "large" and below which samples are "small", it is convenient to speak this way.

For *"large"* samples (for example, $n \geq 30$), it is very likely that $s_{\bar{x}}$ is "close" to $\sigma_{\bar{x}}$. Hence

$$(\bar{x} - \mu)/\sigma_{\bar{x}} \approx (\bar{x} - \mu)/s_{\bar{x}},$$

and we may use the normal tables as before to get the approximate 95% confidence interval

$$\bar{X} - 1.96s_{\bar{x}} \leq \mu \leq \bar{X} + 1.96s_{\bar{x}}. \tag{1}$$

What happens if $n < 30$ and we have a "small" sample? We discuss this case in the next section.

VII–4 CONFIDENCE LIMITS AND SIGNIFICANCE TESTS WITH SMALL SAMPLES; THE t-DISTRIBUTIONS

For the discussion of Section VII–3, suppose that the samples are small (say less than 30). Then the sample variance s^2, which varies from sample to sample, may not be close to the population variance σ^2. Hence we may no longer assume that the random variable

$$T = (\bar{X} - \mu)/s_{\bar{x}}$$

is approximately distributed like Z, the standard normal random variable. What, then, is the distribution of T for small samples?

W. S. Gosset, an investigator who used the pseudonym "Student", discovered the following facts about the distribution of T for samples drawn from a population that is normal.

1. *The distribution of T is not normal*; but it approaches normality more and more closely as *n* gets large. When *n* = 30, the normal approximation is close.

2. *The distribution of T is symmetric about t* = 0.

3. Even though *T* has a different distribution for each sample size *n*, *these distributions do not depend on μ or σ.* This property makes it easy to get confidence limits or tests of significance once we have a probability table like Table A–3.

We have denoted $(\overline{X} - \mu)/s_{\overline{x}}$ by *T* to emphasize the fact that it is a random variable. In practice, the lower-case *t* is used to denote both the random variable and one of its values. We shall conform to this custom, thus making an exception to our usual practice. We shall call these distributions *Student's t-distributions*, or just *t-distributions*.

The tabulated *t*-distributions are indexed under *n* − 1, where *n* is the sample size. For reasons that we shall not go into here, this index *n* − 1 is called the *number of degrees of freedom of the t-distribution.* For example, for sample size 2, we find the *t*-distribution indexed under 1 degree of freedom (d.f.). (See Table A–3.)

Figure VII–1 shows examples of *t*-distributions for *n* = 2, 3, and infinity, the latter being identical with the standard normal distribution. For small values of *n*, the tails of the *t*-distributions beyond 2σ are higher than those of the standard normal distribution. This has the effect of lengthening confidence intervals, as we shall see.

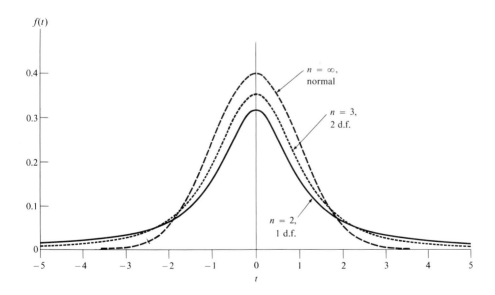

Fig. VII–1 Collection of *t*-distributions (d.f. = degrees of freedom).

The role of the t-distribution for small samples parallels that of the standard normal distribution for large samples. The procedure in setting confidence limits on μ for small samples, drawn from an approximately normal population with unknown σ, is analogous to that used in Section VII–3 for large samples. The one difference is this: Instead of using the random variable $Z = (\overline{X} - \mu)/s_{\overline{x}}$ and the standard normal distribution, we use the random variable $t = (\overline{X} - \mu)/s_{\overline{x}}$ and the t-distribution with the proper degrees of freedom.

For example, to set 95% confidence limits for a sample of size 4, we go to Table A–3 and find that $P(|t| \leq 3.18) = 0.95$. (This is sometimes written $t_{3,0.95} = 3.18$, where 3 is the number of d.f.). Hence the probability is 0.95 that

$$|(\overline{X} - \mu)/s_{\overline{x}}| \leq 3.18.$$

Rearrange this inequality as before and get

$$\overline{X} - 3.18s_{\overline{x}} \leq \mu \leq \overline{X} + 3.18s_{\overline{x}}, \tag{1}$$

which, apart from the multiplier 3.18, is identical with inequality (1) in Section VII–2. For different confidence levels, we use different multipliers for $s_{\overline{x}}$. These are obtained from Table A–3.

Example 3. *Confidence limits for μ; unknown σ, small sample.* In Example 1 of Section VII–2 (Machine setup), suppose that $\bar{x} = 1.2000$ as before, and that the observed value of $s_{\overline{x}}$ in a sample of 10 is 0.0008. Use the data to set symmetric 95% confidence limits on μ.

Solution. For $n = 10$ and 95% confidence limits, Table A–3 gives the multiplier 2.26. Therefore, from (1), we get

$$1.2000 - 2.26(0.0008) \leq \mu \leq 1.2000 + 2.26(0.0008),$$

or

$$1.1982 \leq \mu \leq 1.2018.$$

Then the confidence limits are $[1.1982, 1.2018]$, which enclose a slightly wider interval than that obtained in Section VII–2.

VII–5 DIFFERENCES BETWEEN MEANS; TWO SAMPLES, KNOWN σ's

Suppose that we have two *independent* samples drawn from *normal* populations with *known variances*. We may want an answer to this question: Does the difference in the sample means suggest that the samples are likely to come from populations with unequal means, or is this difference likely to stem from chance?

Let $\overline{X}$ and $\overline{Y}$ denote the means of two samples of sizes n_X and n_Y, independently drawn from normal populations with means μ_X and μ_Y and variances σ_X^2 and σ_Y^2. We use two facts:

1. The difference $\overline{X} - \overline{Y}$ is normally distributed. (Appendix V–4.)

2. $$\sigma_{\overline{X}-\overline{Y}} = \sigma_{\overline{X}}^2 + \sigma_{\overline{Y}}^2 = \frac{\sigma_X^2}{n_X} + \frac{\sigma_Y^2}{n_Y}. \qquad \text{(Appendixes V–1 and VI–1.)}$$

To set 95% confidence limits on $\mu_X - \mu_Y$, we use the normal random variable $\overline{X} - \overline{Y}$, just as we used $\overline{X}$ in Section VII–2, and compute

$$\overline{x} - \overline{y} \pm 1.96\sigma_{\overline{x}-\overline{y}}.$$

Example 4. *Lactose tolerance.* To study comparative tolerance of adults of different races to lactose, physicians computed peak rise in blood-reducing substances for 20 Orientals and 20 Caucasians after they had been given a dose of lactose (essentially milk). Larger rises imply more tolerance. The sample means for Caucasian and Orientals are $\overline{x} = 40$ and $\overline{y} = 15$, respectively, and the variances are $\sigma_X^2 = 100$ and $\sigma_Y^2 = 25$. Set 95% confidence limits on the difference between the population means.

Solution. We have

$$\sigma_{\overline{x}-\overline{y}} = \sqrt{\frac{\sigma_X^2}{n_X} + \frac{\sigma_Y^2}{n_Y}} = \sqrt{\frac{100}{20} + \frac{25}{20}} = \frac{5}{2}.$$

Hence the 95% confidence limits are

$$\overline{x} - \overline{y} \pm 1.96(\tfrac{5}{2}) = 40 - 15 \pm 4.9 \approx [20, 30].$$

The result shows a substantial difference in tolerance between the two groups of people. Such results may be important in deciding what forms of food are useful for undernourished adults in various parts of the world.

VII–6 DIFFERENCES BETWEEN MEANS; TWO SAMPLES, UNKNOWN σ's

We shall confine ourselves to the case in which the unknown σ's are equal. Using the notation introduced in Section VII–5, we find ourselves without a value for $\sigma_{\overline{x}-\overline{y}}^2$. It seems natural once more to use the sample to provide an estimate for the variance of the difference of means, and we do so. For an estimate of $\sigma_{\overline{x}-\overline{y}}^2$, we use

$$s_{\overline{x}-\overline{y}}^2 = \frac{\sum(x_i - \overline{x})^2 + \sum(y_i - \overline{y})^2}{n_X + n_Y - 2}\left(\frac{1}{n_X} + \frac{1}{n_Y}\right). \qquad (1)$$

Then the random variable $\overline{X} - \overline{Y}$ turns out to have a t-distribution with $n_X + n_Y - 2$ degrees of freedom. The confidence limits are

$$\overline{x} - \overline{y} \pm ks_{\overline{x}-\overline{y}},$$

where k is obtained from Table A–3 once the number of degrees of freedom and the confidence level is known.

Example 5. *Ambiguity and speech.* To test the hypothesis that people with less information about a job talk more about it, Hayes, Meltzer, and Lundberg gave 5 groups a clear diagram of a complicated Tinkertoy® to be built, and they gave 5 groups an ambiguous diagram that was smaller and shaded. Each group had to tell a silent builder, who had no diagram, how to make the Tinkertoy. The groups were scored for the amount of talking done. The groups with the clear diagram had a mean speech score $\overline{x} = 22.8$, with $\Sigma(x_i - \overline{x})^2 = 707.6$; and the groups with the ambiguous diagram had a mean speech score $\overline{y} = 40.8$, with $\Sigma(y_i - \overline{y})^2 = 1281.6$. Assuming that the population variances of the speech scores of the two groups are equal, set 90% confidence limits on $\mu_X - \mu_Y$ and test the hypothesis mentioned.

Solution. We use (1) to get an estimate of the standard deviation of the difference in means, thus:

$$s_{\overline{x}-\overline{y}}^2 = \frac{707.6 + 1281.6}{8}\left(\frac{1}{5} + \frac{1}{5}\right) = 99.5,$$

whence

$$s_{\overline{x}-\overline{y}} = 9.97.$$

The 90% symmetric confidence level for t with 8 degrees of freedom (Table A–3) is 1.86. Hence the 90% confidence limits are

$$\overline{x} - \overline{y} \pm 1.86s_{\overline{x}-\overline{y}} = 22.8 - 40.8 \pm 1.86(9.97)$$

$$= [-36.6, +0.6].$$

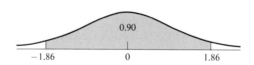

Then the difference in means, although large and in a direction that favors the hypothesis, has quite wide confidence limits. Note that 0 is contained in the confidence interval, so we would not reject the hypothesis of equal means at the 10% level of significance. Nevertheless, the fact that 0 is so near the upper confidence limit depresses us a little about the truth of the hypothesis that $\mu_Y = \mu_X$.

® Registered trademark

Tables

The authors hereby acknowledge permission to adapt and use material from the following sources as noted.

Table	Source
A–1	F. Mosteller, R. E. K. Rourke, and G. B. Thomas, Jr., *Probability with Statistical Applications*, Second edition. Reading, Mass.: Addison-Wesley, 1970, p. 473.
A–2	*Ibid.*, p. 493.
A–3	*Ibid.*, p. 492. Rounded from D. B. Owen, *Handbook of Statistical Tables*. Courtesy of the Atomic Energy Commission. Reading, Mass.: Addison-Wesley, 1962.
A–4	(1) H. L. Harter, *New Tables of the Incomplete Gamma-function Ratio and of Percentage Points of the Chi-square and Beta Distributions*. Aerospace Research Laboratories, Office of Aerospace Research, United States Air Force. Washington: U.S. Government Printing Office, 1964. (2) Owen, *Handbook*, pp. 49–55.
A–5	*PWSA*, pp. 484–489.
A–13	Owen, *Handbook*, pp. 400–406.
A–15	*Ibid.*, pp. 407–419.
A–16	H. L. Harter, *Order Statistics and Their Use in Testing and Estimation*, Volume 2: *Estimates Based on Order Statistics of Samples from Various Populations*. Aerospace Research Laboratories, Office of Aerospace Research, United States Air Force. Washington: U.S. Government Printing Office, pp. 425–455.
A–17	Owen, *Handbook*, pp. 163–164.
A–20	E. V. Huntington, *Four Place Tables*. Boston: Houghton Mifflin, 1910.
A–21	*Ibid.*
A–23	*PWSA*, p. 471.

Table A–1 Normal curve areas

Area under the standard normal curve from 0 to z, shown shaded, is $A(z)$.

Examples. If Z is the standard normal random variable and $z = 1.54$, then

$$A(z) = P(0 < Z < z) = 0.4382,$$
$$P(Z > z) = 0.0618,$$
$$P(Z < z) = 0.9382,$$
$$P(|Z| < z) = 0.8764.$$

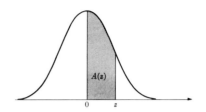

z	.00	.01	.02	.03	.04	.05	.06	.07	.08	.09
0.0	.0000	.0040	.0080	.0120	.0160	.0199	.0239	.0279	.0319	.0359
0.1	.0398	.0438	.0478	.0517	.0557	.0596	.0636	.0675	.0714	.0753
0.2	.0793	.0832	.0871	.0910	.0948	.0987	.1026	.1064	.1103	.1141
0.3	.1179	.1217	.1255	.1293	.1331	.1368	.1406	.1443	.1480	.1517
0.4	.1554	.1591	.1628	.1664	.1700	.1736	.1772	.1808	.1844	.1879
0.5	.1915	.1950	.1985	.2019	.2054	.2088	.2123	.2157	.2190	.2224
0.6	.2257	.2291	.2324	.2357	.2389	.2422	.2454	.2486	.2517	.2549
0.7	.2580	.2611	.2642	.2673	.2704	.2734	.2764	.2794	.2823	.2852
0.8	.2881	.2910	.2939	.2967	.2995	.3023	.3051	.3078	.3106	.3133
0.9	.3159	.3186	.3212	.3238	.3264	.3289	.3315	.3340	.3365	.3389
1.0	.3413	.3438	.3461	.3485	.3508	.3531	.3554	.3577	.3599	.3621
1.1	.3643	.3665	.3686	.3708	.3729	.3749	.3770	.3790	.3810	.3830
1.2	.3849	.3869	.3888	.3907	.3925	.3944	.3962	.3980	.3997	.4015
1.3	.4032	.4049	.4066	.4082	.4099	.4115	.4131	.4147	.4162	.4177
1.4	.4192	.4207	.4222	.4236	.4251	.4265	.4279	.4292	.4306	.4319
1.5	.4332	.4345	.4357	.4370	.4382	.4394	.4406	.4418	.4429	.4441
1.6	.4452	.4463	.4474	.4484	.4495	.4505	.4515	.4525	.4535	.4545
1.7	.4554	.4564	.4573	.4582	.4591	.4599	.4608	.4616	.4625	.4633
1.8	.4641	.4649	.4656	.4664	.4671	.4678	.4686	.4693	.4699	.4706
1.9	.4713	.4719	.4726	.4732	.4738	.4744	.4750	.4756	.4761	.4767
2.0	.4772	.4778	.4783	.4788	.4793	.4798	.4803	.4808	.4812	.4817
2.1	.4821	.4826	.4830	.4834	.4838	.4842	.4846	.4850	.4854	.4857
2.2	.4861	.4864	.4868	.4871	.4875	.4878	.4881	.4884	.4887	.4890
2.3	.4893	.4896	.4898	.4901	.4904	.4906	.4909	.4911	.4913	.4916
2.4	.4918	.4920	.4922	.4925	.4927	.4929	.4931	.4932	.4934	.4936
2.5	.4938	.4940	.4941	.4943	.4945	.4946	.4948	.4949	.4951	.4952
2.6	.4953	.4955	.4956	.4957	.4959	.4960	.4961	.4962	.4963	.4964
2.7	.4965	.4966	.4967	.4968	.4969	.4970	.4971	.4972	.4973	.4974
2.8	.4974	.4975	.4976	.4977	.4977	.4978	.4979	.4979	.4980	.4981
2.9	.4981	.4982	.4982	.4983	.4984	.4984	.4985	.4985	.4986	.4986
3.0	.4987	.4987	.4987	.4988	.4988	.4989	.4989	.4989	.4990	.4990

Table A–2 Inverse of the cumulative normal distribution

Cell entry is the value of z such that the area to the left of z, under the standard normal curve, is p, where the tenths digit of p is given in the left-hand column and the hundredths digit is given along the top.

Examples

1. $p = 0.16,$ $z = -0.994$
2. $p = 0.63,$ $z = +0.332$

p	.00	.01	.02	.03	.04	.05	.06	.07	.08	.09
.0	$-\infty$	−2.326	−2.054	−1.881	−1.751	−1.645	−1.555	−1.476	−1.405	−1.341
.1	−1.282	−1.227	−1.175	−1.126	−1.080	−1.036	−0.994	−0.954	−0.915	−0.878
.2	−0.842	−0.806	−0.772	−0.739	−0.706	−0.674	−0.643	−0.613	−0.583	−0.553
.3	−0.524	−0.496	−0.468	−0.440	−0.412	−0.385	−0.358	−0.332	−0.305	−0.279
.4	−0.253	−0.228	−0.202	−0.176	−0.151	−0.126	−0.100	−0.075	−0.050	−0.025
.5	0.000	0.025	0.050	0.075	0.100	0.126	0.151	0.176	0.202	0.228
.6	0.253	0.279	0.305	0.332	0.358	0.385	0.412	0.440	0.468	0.496
.7	0.524	0.553	0.583	0.613	0.643	0.674	0.706	0.739	0.772	0.806
.8	0.842	0.878	0.915	0.954	0.994	1.036	1.080	1.126	1.175	1.227
.9	1.282	1.341	1.405	1.476	1.555	1.645	1.751	1.881	2.054	2.326

Table A-3 Probability levels for Student t-distributions

Degrees of freedom	One-sample size†	Two-sided probability level				
		.50	.80	.90	.95	.98
		One-sided probability level				
		.75	.90	.95	.975	.99
1	2	1.00	3.08	6.31	12.71	31.82
2	3	.82	1.89	2.92	4.30	6.96
3	4	.76	1.64	2.35	3.18	4.54
4	5	.74	1.53	2.13	2.78	3.75
5	6	.73	1.48	2.02	2.57	3.36
6	7	.72	1.44	1.94	2.45	3.14
7	8	.71	1.41	1.89	2.36	3.00
8	9	.71	1.40	1.86	2.31	2.90
9	10	.70	1.38	1.83	2.26	2.82
10	11	.70	1.37	1.81	2.23	2.76
15	16	.69	1.34	1.75	2.13	2.60
30	31	.68	1.31	1.70	2.04	2.46
50	51	.68	1.30	1.68	2.01	2.40
100	101	.68	1.29	1.66	1.98	2.36
1000	1001	.67	1.28	1.65	1.96	2.33
∞‡	∞‡	.67	1.28	1.64	1.96	2.33

† For setting confidence limits on the mean of a single sample.
‡ Standard normal distribution.

Table A–4 Values of chi-square for various degrees of freedom

Examples. (a) The probability that a χ^2 random variable with 15 degrees of freedom exceeds 22.31 is $P(\chi^2_{15} \geq 22.31) = 0.10$. (b) The table shows that $P(\chi^2_{10} \geq 15)$ lies between 0.10 and 0.20. Interpolation gives $P(\chi^2_{10} \geq 15) = 0.14$, approximately.

Degrees of freedom	0.99	0.95	0.90	Probability levels 0.80	0.70	0.60	0.50
1	0.00	0.00	0.02	0.06	0.15	0.27	0.45
2	0.02	0.10	0.21	0.45	0.71	1.02	1.39
3	0.11	0.35	0.58	1.01	1.42	1.87	2.37
4	0.30	0.71	1.06	1.65	2.19	2.75	3.36
5	0.55	1.15	1.61	2.34	3.00	3.66	4.35
6	0.87	1.64	2.20	3.07	3.83	4.57	5.35
7	1.24	2.17	2.83	3.82	4.67	5.49	6.35
8	1.65	2.73	3.49	4.59	5.53	6.42	7.34
9	2.09	3.33	4.17	5.38	6.39	7.36	8.34
10	2.56	3.94	4.87	6.18	7.27	8.30	9.34
11	3.05	4.57	5.58	6.99	8.15	9.24	10.34
12	3.57	5.23	6.30	7.81	9.03	10.18	11.34
13	4.11	5.89	7.04	8.63	9.93	11.13	12.34
14	4.66	6.57	7.79	9.47	10.82	12.08	13.34
15	5.23	7.26	8.55	10.31	11.72	13.03	14.34
16	5.81	7.96	9.31	11.15	12.62	13.98	15.34
17	6.41	8.67	10.09	12.00	13.53	14.94	16.34
18	7.01	9.39	10.86	12.86	14.44	15.89	17.34
19	7.63	10.12	11.65	13.72	15.35	16.85	18.34
20	8.26	10.85	12.44	14.58	16.27	17.81	19.34
21	8.90	11.59	13.24	15.44	17.18	18.77	20.34
22	9.54	12.34	14.04	16.31	18.10	19.73	21.34
23	10.20	13.09	14.85	17.19	19.02	20.69	22.34
24	10.86	13.85	15.66	18.06	19.94	21.65	23.34
25	11.52	14.61	16.47	18.94	20.87	22.62	24.34
26	12.20	15.38	17.29	19.82	21.79	23.58	25.34
27	12.88	16.15	18.11	20.70	22.72	24.54	26.34
28	13.56	16.93	18.94	21.59	23.65	25.51	27.34
29	14.26	17.71	19.77	22.48	24.58	26.48	28.34
30	14.95	18.49	20.60	23.36	25.51	27.44	29.34
50	29.71	34.76	37.69	41.45	44.31	46.86	49.33
100	70.06	77.93	82.36	87.95	92.13	95.81	99.33
200	156.4	168.3	174.8	183.0	189.0	194.3	199.3
500	429.4	449.1	459.9	473.2	482.9	491.4	499.3
800	709.9	735.4	749.2	766.2	778.6	789.2	799.3
1000	898.9	927.6	943.1	962.2	976.1	988.1	999.3

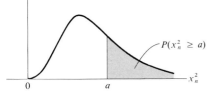

$P(x_n^2 \geq a)$

x_n^2

Degrees of freedom	Probability levels						
	0.40	0.30	0.20	0.10	0.05	0.01	0.001
1	0.71	1.07	1.64	2.71	3.84	6.63	10.83
2	1.83	2.41	3.22	4.61	5.99	9.21	13.82
3	2.95	3.66	4.64	6.25	7.81	11.34	16.27
4	4.04	4.88	5.99	7.78	9.49	13.28	18.47
5	5.13	6.06	7.29	9.24	11.07	15.09	20.52
6	6.21	7.23	8.56	10.64	12.59	16.81	22.46
7	7.28	8.38	9.80	12.02	14.07	18.48	24.32
8	8.35	9.52	11.03	13.36	15.51	20.09	26.12
9	9.41	10.66	12.24	14.68	16.92	21.67	27.88
10	10.47	11.78	13.44	15.99	18.31	23.21	29.59
11	11.53	12.90	14.63	17.28	19.68	24.72	31.26
12	12.58	14.01	15.81	18.55	21.03	26.22	32.91
13	13.64	15.12	16.98	19.81	22.36	27.69	34.53
14	14.69	16.22	18.15	21.06	23.68	29.14	36.12
15	15.73	17.32	19.31	22.31	25.00	30.58	37.70
16	16.78	18.42	20.47	23.54	26.30	32.00	39.25
17	17.82	19.51	21.61	24.77	27.59	33.41	40.79
18	18.87	20.60	22.76	25.99	28.87	34.81	42.31
19	19.91	21.69	23.90	27.20	30.14	36.19	43.82
20	20.95	22.77	25.04	28.41	31.41	37.57	45.31
21	21.99	23.86	26.17	29.62	32.67	38.93	46.80
22	23.03	24.94	27.30	30.81	33.92	40.29	48.27
23	24.07	26.02	28.43	32.01	35.17	41.64	49.73
24	25.11	27.10	29.55	33.20	36.42	42.98	51.18
25	26.14	28.17	30.68	34.38	37.65	44.31	52.62
26	27.18	29.25	31.79	35.56	38.89	45.64	54.05
27	28.21	30.32	32.91	36.74	40.11	46.96	55.48
28	29.25	31.39	34.03	37.92	41.34	48.28	56.89
29	30.28	32.46	35.14	39.09	42.56	49.59	58.30
30	31.32	33.53	36.25	40.26	43.77	50.89	59.70
50	51.89	54.72	58.16	63.17	67.50	76.15	86.66
100	102.9	106.9	111.7	118.5	124.3	135.8	149.4
200	204.4	210.0	216.6	226.0	234.0	249.4	267.5
500	507.4	516.1	526.4	540.9	553.1	576.5	603.4
800	809.5	820.5	833.5	851.7	866.9	896.0	929.3
1000	1011.	1023.	1037.	1058.	1075.	1107.	1144.

Table A–5 Three-place tables of the cumulative binomial distribution

$$P(X \geq r) = \sum_{x=r}^{n} b(x; n, p)$$

$$= \sum_{x=r}^{n} \binom{n}{x} p^x (1 - p)^{n-x}.$$

$P(X \geq r)$ is the probability of r or more successes in n independent binomial trials with probability p of success on a single trial.

Values of the function are given for $0 \leq r \leq n \leq 20$, for $p = 0.01$, 0.05, 0.10 (0.10) 0.90, 0.95, 0.99.

Each three-digit entry should be read with a decimal preceding it. For entries $1-$, the probability is larger thàn 0.9995 but less than 1. For entries $0+$, the probability is less than 0.0005, but greater than 0.

Examples

a) For $n = 16$, $p = 0.3$, $P(X \geq 4) = 0.754$.

b) For $n = 16$, $p = 0.3$, $P(X = 4) = P(X \geq 4) - P(X \geq 5)$
$$= 0.754 - 0.550 = 0.204.$$

$$\text{Cumulative terms, } \sum_{x=r}^{n} b(x;n,p)$$

n	r	.01	.05	.10	.20	.30	.40	p .50	.60	.70	.80	.90	.95	.99	r
2	0	1	1	1	1	1	1	1	1	1	1	1	1	1	0
	1	020	098	190	360	510	640	750	840	910	960	990	998	1−	1
	2	0+	002	010	040	090	160	250	360	490	640	810	902	980	2
3	0	1	1	1	1	1	1	1	1	1	1	1	1	1	0
	1	030	143	271	488	657	784	875	936	973	992	999	1−	1−	1
	2	0+	007	028	104	216	352	500	648	784	896	972	993	1−	2
	3	0+	0+	001	008	027	064	125	216	343	512	729	857	970	3
4	0	1	1	1	1	1	1	1	1	1	1	1	1	1	0
	1	039	185	344	590	760	870	938	974	992	998	1−	1−	1−	1
	2	001	014	052	181	348	525	688	821	916	973	996	1−	1−	2
	3	0+	0+	004	027	084	179	312	475	652	819	948	986	999	3
	4	0+	0+	0+	002	008	026	062	130	240	410	656	815	961	4
5	0	1	1	1	1	1	1	1	1	1	1	1	1	1	0
	1	049	226	410	672	832	922	969	990	998	1−	1−	1−	1−	1
	2	001	023	081	263	472	663	812	913	969	993	1−	1−	1−	2
	3	0+	001	009	058	163	317	500	683	837	942	991	999	1−	3
	4	0+	0+	0+	007	031	087	188	337	528	737	919	977	999	4
	5	0+	0+	0+	0+	002	010	031	078	168	328	590	774	951	5
6	0	1	1	1	1	1	1	1	1	1	1	1	1	1	0
	1	059	265	469	738	882	953	984	996	999	1−	1−	1−	1−	1
	2	001	033	114	345	580	767	891	959	989	998	1−	1−	1−	2
	3	0+	002	016	099	256	456	656	821	930	983	999	1−	1−	3
	4	0+	0+	001	017	070	179	344	544	744	901	984	998	1−	4
	5	0+	0+	0+	002	011	041	109	233	420	655	886	967	999	5
	6	0+	0+	0+	0+	001	004	016	047	118	262	531	735	941	6
7	0	1	1	1	1	1	1	1	1	1	1	1	1	1	0
	1	068	302	522	790	918	972	992	998	1−	1−	1−	1−	1−	1
	2	002	044	150	423	671	841	938	981	996	1−	1−	1−	1−	2
	3	0+	004	026	148	353	580	773	904	971	995	1−	1−	1−	3
	4	0+	0+	003	033	126	290	500	710	874	967	997	1−	1−	4
	5	0+	0+	0+	005	029	096	227	420	647	852	974	996	1−	5
	6	0+	0+	0+	0+	004	019	062	159	329	577	850	956	998	6
	7	0+	0+	0+	0+	0+	002	008	028	082	210	478	698	932	7
8	0	1	1	1	1	1	1	1	1	1	1	1	1	1	0
	1	077	337	570	832	942	983	996	999	1−	1−	1−	1−	1−	1
	2	003	057	187	497	745	894	965	991	999	1−	1−	1−	1−	2
	3	0+	006	038	203	448	685	855	950	989	999	1−	1−	1−	3
	4	0+	0+	005	056	194	406	637	826	942	990	1−	1−	1−	4
	5	0+	0+	0+	010	058	174	363	594	806	944	995	1−	1−	5
	6	0+	0+	0+	001	011	050	145	315	552	797	962	994	1−	6
	7	0+	0+	0+	0+	001	009	035	106	255	503	813	943	997	7
	8	0+	0+	0+	0+	0+	001	004	017	058	168	430	663	923	8

Cumulative terms, $\displaystyle\sum_{x=r}^{n} b(x; n, p)$

n	r	.01	.05	.10	.20	.30	.40	p .50	.60	.70	.80	.90	.95	.99	r
9	0	1	1	1	1	1	1	1	1	1	1	1	1	1	0
	1	086	370	613	866	960	990	998	1—	1—	1—	1—	1—	1—	1
	2	003	071	225	564	804	929	980	996	1—	1—	1—	1—	1—	2
	3	0+	008	053	262	537	768	910	975	996	1—	1—	1—	1—	3
	4	0+	001	008	086	270	517	746	901	975	997	1—	1—	1—	4
	5	0+	0+	001	020	099	267	500	733	901	980	999	1—	1—	5
	6	0+	0+	0+	003	025	099	254	483	730	914	992	999	1—	6
	7	0+	0+	0+	0+	004	025	090	232	463	738	947	992	1—	7
	8	0+	0+	0+	0+	0+	004	020	071	196	436	775	929	997	8
	9	0+	0+	0+	0+	0+	0+	002	010	040	134	387	630	914	9
10	0	1	1	1	1	1	1	1	1	1	1	1	1	1	0
	1	096	401	651	893	972	994	999	1—	1—	1—	1—	1—	1—	1
	2	004	086	264	624	851	954	989	998	1—	1—	1—	1—	1—	2
	3	0+	012	070	322	617	833	945	988	998	1—	1—	1—	1—	3
	4	0+	001	013	121	350	618	828	945	989	999	1—	1—	1—	4
	5	0+	0+	002	033	150	367	623	834	953	994	1—	1—	1—	5
	6	0+	0+	0+	006	047	166	377	633	850	967	998	1—	1—	6
	7	0+	0+	0+	001	011	055	172	382	650	879	987	999	1—	7
	8	0+	0+	0+	0+	002	012	055	167	383	678	930	988	1—	8
	9	0+	0+	0+	0+	0+	002	011	046	149	376	736	914	996	9
	10	0+	0+	0+	0+	0+	0+	001	006	028	107	349	599	904	10
11	0	1	1	1	1	1	1	1	1	1	1	1	1	1	0
	1	105	431	686	914	980	996	1—	1—	1—	1—	1—	1—	1—	1
	2	005	102	303	678	887	970	994	999	1—	1—	1—	1—	1—	2
	3	0+	015	090	383	687	881	967	994	999	1—	1—	1—	1—	3
	4	0+	002	019	161	430	704	887	971	996	1—	1—	1—	1—	4
	5	0+	0+	003	050	210	467	726	901	978	998	1—	1—	1—	5
	6	0+	0+	0+	012	078	247	500	753	922	988	1—	1—	1—	6
	7	0+	0+	0+	002	022	099	274	533	790	950	997	1—	1—	7
	8	0+	0+	0+	0+	004	029	113	296	570	839	981	998	1—	8
	9	0+	0+	0+	0+	001	006	033	119	313	617	910	985	1—	9
	10	0+	0+	0+	0+	0+	001	006	030	113	322	697	898	995	10
	11	0+	0+	0+	0+	0+	0+	0+	004	020	086	314	569	895	11
12	0	1	1	1	1	1	1	1	1	1	1	1	1	1	0
	1	114	460	718	931	986	998	1—	1—	1—	1—	1—	1—	1—	1
	2	006	118	341	725	915	980	997	1—	1—	1—	1—	1—	1—	2
	3	0+	020	111	442	747	917	981	997	1—	1—	1—	1—	1—	3
	4	0+	002	026	205	507	775	927	985	998	1—	1—	1—	1—	4
	5	0+	0+	004	073	276	562	806	943	991	999	1—	1—	1—	5
	6	0+	0+	001	019	118	335	613	842	961	996	1—	1—	1—	6
	7	0+	0+	0+	004	039	158	387	665	882	981	999	1—	1—	7
	8	0+	0+	0+	001	009	057	194	438	724	927	996	1—	1—	8
	9	0+	0+	0+	0+	002	015	073	225	493	795	974	998	1—	9

Cumulative terms, $\sum_{x=r}^{n} b(x; n, p)$

n	r	.01	.05	.10	.20	.30	.40	p .50	.60	.70	.80	.90	.95	.99	r
12	10	0+	0+	0+	0+	0+	003	019	083	253	558	889	980	1−	10
	11	0+	0+	0+	0+	0+	0+	003	020	085	275	659	882	994	11
	12	0+	0+	0+	0+	0+	0+	0+	002	014	069	282	540	886	12
13	0	1	1	1	1	1	1	1	1	1	1	1	1	1	0
	1	122	487	746	945	990	999	1−	1−	1−	1−	1−	1−	1−	1
	2	007	135	379	766	936	987	998	1−	1−	1−	1−	1−	1−	2
	3	0+	025	134	498	798	942	989	999	1−	1−	1−	1−	1−	3
	4	0+	003	034	253	579	831	954	992	999	1−	1−	1−	1−	4
	5	0+	0+	006	099	346	647	867	968	996	1−	1−	1−	1−	5
	6	0+	0+	001	030	165	426	709	902	982	999	1−	1−	1−	6
	7	0+	0+	0+	007	062	229	500	771	938	993	1−	1−	1−	7
	8	0+	0+	0+	001	018	098	291	574	835	970	999	1−	1−	8
	9	0+	0+	0+	0+	004	032	133	353	654	901	994	1−	1−	9
	10	0+	0+	0+	0+	001	008	046	169	421	747	966	997	1−	10
	11	0+	0+	0+	0+	0+	001	011	058	202	502	866	975	1−	11
	12	0+	0+	0+	0+	0+	0+	002	013	064	234	621	865	993	12
	13	0+	0+	0+	0+	0+	0+	0+	001	010	055	254	513	878	13
14	0	1	1	1	1	1	1	1	1	1	1	1	1	1	0
	1	131	512	771	956	993	999	1−	1−	1−	1−	1−	1−	1−	1
	2	008	153	415	802	953	992	999	1−	1−	1−	1−	1−	1−	2
	3	0+	030	158	552	839	960	994	999	1−	1−	1−	1−	1−	3
	4	0+	004	044	302	645	876	971	996	1−	1−	1−	1−	1−	4
	5	0+	0+	009	130	416	721	910	982	998	1−	1−	1−	1−	5
	6	0+	0+	001	044	219	514	788	942	992	1−	1−	1−	1−	6
	7	0+	0+	0+	012	093	308	605	850	969	998	1−	1−	1−	7
	8	0+	0+	0+	002	031	150	395	692	907	988	1−	1−	1−	8
	9	0+	0+	0+	0+	008	058	212	486	781	956	999	1−	1−	9
	10	0+	0+	0+	0+	002	018	090	279	584	870	991	1−	1−	10
	11	0+	0+	0+	0+	0+	004	029	124	355	698	956	996	1−	11
	12	0+	0+	0+	0+	0+	001	006	040	161	448	842	970	1−	12
	13	0+	0+	0+	0+	0+	0+	001	008	047	198	585	847	992	13
	14	0+	0+	0+	0+	0+	0+	0+	001	007	044	229	488	869	14
15	0	1	1	1	1	1	1	1	1	1	1	1	1	1	0
	1	140	537	794	965	995	1−	1−	1−	1−	1−	1−	1−	1−	1
	2	010	171	451	833	965	995	1−	1−	1−	1−	1−	1−	1−	2
	3	0+	036	184	602	873	973	996	1−	1−	1−	1−	1−	1−	3
	4	0+	005	056	352	703	909	982	998	1−	1−	1−	1−	1−	4
	5	0+	001	013	164	485	783	941	991	999	1−	1−	1−	1−	5
	6	0+	0+	002	061	278	597	849	966	996	1−	1−	1−	1−	6
	7	0+	0+	0+	018	131	390	696	905	985	999	1−	1−	1−	7
	8	0+	0+	0+	004	050	213	500	787	950	996	1−	1−	1−	8
	9	0+	0+	0+	001	015	095	304	610	869	982	1−	1−	1−	9

$$\text{Cumulative terms, } \sum_{x=r}^{n} b(x; n, p)$$

n	r	.01	.05	.10	.20	.30	.40	p .50	.60	.70	.80	.90	.95	.99	r
15	10	0+	0+	0+	0+	004	034	151	403	722	939	998	1−	1−	10
	11	0+	0+	0+	0+	001	009	059	217	515	836	987	999	1−	11
	12	0+	0+	0+	0+	0+	002	018	091	297	648	944	995	1−	12
	13	0+	0+	0+	0+	0+	0+	004	027	127	398	816	964	1−	13
	14	0+	0+	0+	0+	0+	0+	0+	005	035	167	549	829	990	14
	15	0+	0+	0+	0+	0+	0+	0+	0+	005	035	206	463	860	15
16	0	1	1	1	1	1	1	1	1	1	1	1	1	1	0
	1	149	560	815	972	997	1−	1−	1−	1−	1−	1−	1−	1−	1
	2	011	189	485	859	974	997	1−	1−	1−	1−	1−	1−	1−	2
	3	001	043	211	648	901	982	998	1−	1−	1−	1−	1−	1−	3
	4	0+	007	068	402	754	935	989	999	1−	1−	1−	1−	1−	4
	5	0+	001	017	202	550	833	962	995	1−	1−	1−	1−	1−	5
	6	0+	0+	003	082	340	671	895	981	998	1−	1−	1−	1−	6
	7	0+	0+	001	027	175	473	773	942	993	1−	1−	1−	1−	7
	8	0+	0+	0+	007	074	284	598	858	974	999	1−	1−	1−	8
	9	0+	0+	0+	001	026	142	402	716	926	993	1−	1−	1−	9
	10	0+	0+	0+	0+	007	058	227	527	825	973	999	1−	1−	10
	11	0+	0+	0+	0+	002	019	105	329	660	918	997	1−	1−	11
	12	0+	0+	0+	0+	0+	005	038	167	450	798	983	999	1−	12
	13	0+	0+	0+	0+	0+	001	011	065	246	598	932	993	1−	13
	14	0+	0+	0+	0+	0+	0+	002	018	099	352	789	957	999	14
	15	0+	0+	0+	0+	0+	0+	0+	003	026	141	515	811	989	15
	16	0+	0+	0+	0+	0+	0+	0+	0+	003	028	185	440	851	16
17	0	1	1	1	1	1	1	1	1	1	1	1	1	1	0
	1	157	582	833	977	998	1−	1−	1−	1−	1−	1−	1−	1−	1
	2	012	208	518	882	981	998	1−	1−	1−	1−	1−	1−	1−	2
	3	001	050	238	690	923	988	999	1−	1−	1−	1−	1−	1−	3
	4	0+	009	083	451	798	954	994	1−	1−	1−	1−	1−	1−	4
	5	0+	001	022	242	611	874	975	997	1−	1−	1−	1−	1−	5
	6	0+	0+	005	106	403	736	928	989	999	1−	1−	1−	1−	6
	7	0+	0+	001	038	225	552	834	965	997	1−	1−	1−	1−	7
	8	0+	0+	0+	011	105	359	685	908	987	1−	1−	1−	1−	8
	9	0+	0+	0+	003	040	199	500	801	960	997	1−	1−	1−	9
	10	0+	0+	0+	0+	013	092	315	641	895	989	1−	1−	1−	10
	11	0+	0+	0+	0+	003	035	166	448	775	962	999	1−	1−	11
	12	0+	0+	0+	0+	001	011	072	264	597	894	995	1−	1−	12
	13	0+	0+	0+	0+	0+	003	025	126	389	758	978	999	1−	13
	14	0+	0+	0+	0+	0+	0+	006	046	202	549	917	991	1−	14
	15	0+	0+	0+	0+	0+	0+	001	012	077	310	762	950	999	15
	16	0+	0+	0+	0+	0+	0+	0+	002	019	118	482	792	988	16
	17	0+	0+	0+	0+	0+	0+	0+	0+	002	023	167	418	843	17

Cumulative terms, $\sum_{x=r}^{n} b(x; n, p)$

n	r	.01	.05	.10	.20	.30	.40	.50	.60	.70	.80	.90	.95	.99	r
18	0	1	1	1	1	1	1	1	1	1	1	1	1	1	0
	1	165	603	850	982	998	1−	1−	1−	1−	1−	1−	1−	1−	1
	2	014	226	550	901	986	999	1−	1−	1−	1−	1−	1−	1−	2
	3	001	058	266	729	940	992	999	1−	1−	1−	1−	1−	1−	3
	4	0+	011	098	499	835	967	996	1−	1−	1−	1−	1−	1−	4
	5	0+	002	028	284	667	906	985	999	1−	1−	1−	1−	1−	5
	6	0+	0+	006	133	466	791	952	994	1−	1−	1−	1−	1−	6
	7	0+	0+	001	051	278	626	881	980	999	1−	1−	1−	1−	7
	8	0+	0+	0+	016	141	437	760	942	994	1−	1−	1−	1−	8
	9	0+	0+	0+	004	060	263	593	865	979	999	1−	1−	1−	9
	10	0+	0+	0+	001	021	135	407	737	940	996	1−	1−	1−	10
	11	0+	0+	0+	0+	006	058	240	563	859	984	1−	1−	1−	11
	12	0+	0+	0+	0+	001	020	119	374	722	949	999	1−	1−	12
	13	0+	0+	0+	0+	0+	006	048	209	534	867	994	1−	1−	13
	14	0+	0+	0+	0+	0+	001	015	094	333	716	972	998	1−	14
	15	0+	0+	0+	0+	0+	0+	004	033	165	501	902	989	1−	15
	16	0+	0+	0+	0+	0+	0+	001	008	060	271	734	942	999	16
	17	0+	0+	0+	0+	0+	0+	0+	001	014	099	450	774	986	17
	18	0+	0+	0+	0+	0+	0+	0+	0+	002	018	150	397	835	18
19	0	1	1	1	1	1	1	1	1	1	1	1	1	1	0
	1	174	623	865	986	999	1−	1−	1−	1−	1−	1−	1−	1−	1
	2	015	245	580	917	990	999	1−	1−	1−	1−	1−	1−	1−	2
	3	001	067	295	763	954	995	1−	1−	1−	1−	1−	1−	1−	3
	4	0+	013	115	545	867	977	998	1−	1−	1−	1−	1−	1−	4
	5	0+	002	035	327	718	930	990	999	1−	1−	1−	1−	1−	5
	6	0+	0+	009	163	526	837	968	997	1−	1−	1−	1−	1−	6
	7	0+	0+	002	068	334	692	916	988	999	1−	1−	1−	1−	7
	8	0+	0+	0+	023	182	512	820	965	997	1−	1−	1−	1−	8
	9	0+	0+	0+	007	084	333	676	912	989	1−	1−	1−	1−	9
	10	0+	0+	0+	002	033	186	500	814	967	998	1−	1−	1−	10
	11	0+	0+	0+	0+	011	088	324	667	916	993	1−	1−	1−	11
	12	0+	0+	0+	0+	003	035	180	488	818	977	1−	1−	1−	12
	13	0+	0+	0+	0+	001	012	084	308	666	932	998	1−	1−	13
	14	0+	0+	0+	0+	0+	003	032	163	474	837	991	1−	1−	14
	15	0+	0+	0+	0+	0+	001	010	070	282	673	965	998	1−	15
	16	0+	0+	0+	0+	0+	0+	002	023	133	455	885	987	1−	16
	17	0+	0+	0+	0+	0+	0+	0+	005	046	237	705	933	999	17
	18	0+	0+	0+	0+	0+	0+	0+	001	010	083	420	755	985	18
	19	0+	0+	0+	0+	0+	0+	0+	0+	001	014	135	377	826	19
20	0	1	1	1	1	1	1	1	1	1	1	1	1	1	0
	1	182	642	878	988	999	1−	1−	1−	1−	1−	1−	1−	1−	1
	2	017	264	608	931	992	999	1−	1−	1−	1−	1−	1−	1−	2
	3	001	075	323	794	965	996	1−	1−	1−	1−	1−	1−	1−	3
	4	0+	016	133	589	893	984	999	1−	1−	1−	1−	1−	1−	4

Cumulative terms, $\sum_{x=r}^{n} b(x; n, p)$

n	r	.01	.05	.10	.20	.30	.40	p .50	.60	.70	.80	.90	.95	.99	r
20	5	0+	003	043	370	762	949	994	1—	1—	1—	1—	1—	1—	5
	6	0+	0+	011	196	584	874	979	998	1—	1—	1—	1—	1—	6
	7	0+	0+	002	087	392	750	942	994	1—	1—	1—	1—	1—	7
	8	0+	0+	0+	032	228	584	868	979	999	1—	1—	1—	1—	8
	9	0+	0+	0+	010	113	404	748	943	995	1—	1—	1—	1—	9
	10	0+	0+	0+	003	048	245	588	872	983	999	1—	1—	1—	10
	11	0+	0+	0+	001	017	128	412	755	952	997	1—	1—	1—	11
	12	0+	0+	0+	0+	005	057	252	596	887	990	1—	1—	1—	12
	13	0+	0+	0+	0+	001	021	132	416	772	968	1—	1—	1—	13
	14	0+	0+	0+	0+	0+	006	058	250	608	913	998	1—	1—	14
	15	0+	0+	0+	0+	0+	002	021	126	416	804	989	1—	1—	15
	16	0+	0+	0+	0+	0+	0+	006	051	238	630	957	997	1—	16
	17	0+	0+	0+	0+	0+	0+	001	016	107	411	867	984	1—	17
	18	0+	0+	0+	0+	0+	0+	0+	004	035	206	677	925	999	18
	19	0+	0+	0+	0+	0+	0+	0+	001	008	069	392	736	983	19
	20	0+	0+	0+	0+	0+	0+	0+	0+	001	012	122	358	818	20

Table A-6 Three-place tables of the cumulative binomial distribution for $p = \frac{1}{2}$:

$$P(X \le l) = \sum_{x=0}^{l} \binom{n}{x}\left(\frac{1}{2}\right)^{n}.$$

Note. (1) The probabilities in Table A-6 are cumulated in the direction opposite to that used in Table A-5. (2) Each three-digit entry should be read with a decimal point preceding it. (3) Since the distribution is symmetrical about $n/2$, we need tabulate values of $P(X \le l)$ only for $l < n/2$. If $l \ge n/2$, we get probabilities by using the following symmetric properties:

$$P(X = l) = P(X = n - l) \quad \text{or} \quad P(X \le l) = P(X \ge n - l).$$

Examples. If $n = 10$ and $p = \frac{1}{2}$, then

$$P(X \le 7) = 1 - P(X \ge 8) = 1 - P(X \le 2) = 1 - 0.055 = 0.945,$$

$$P(X = 6) = P(X = 4) = P(X \le 4) - P(X \le 3) = 0.377 - 0.172 = 0.205;$$

one-tail: $P(X \ge 8) = P(X \le 2) = 0.055,$

two-tail: $P(X \le 2 \text{ or } X \ge 8) = 2(0.055) = 0.110.$

Binomial: $p = \frac{1}{2}$

One-tail *P*-values ($\times 1000$) for relatively extreme values in the sign test or binomial. (Double for two-tail values; subtract from 1000 to obtain values for other tail.)

Number in less frequent class

n	0	1	2	3	4
1	500				
2	250				
3	125	500			
4	062	312			
5	031	188	500		
6	016	109	344		
7	008	062	227	500	
8	004	035	145	363	
9	002	020	090	254	500
10	001	011	055	172	377

n	1	2	3	4	5	6	7	8	9
11	006	033	113	274	500				
12	003	019	073	194	387				
13	002	011	046	133	291	500			
14	001	006	029	090	212	395			
15		004	018	059	151	304	500		
16		002	011	038	105	227	402		
17		001	006	025	072	166	315	500	
18		001	004	015	048	119	240	407	
19			002	010	032	084	180	324	500
20			001	006	021	058	132	252	412

n	3	4	5	6	7	8	9	10	11	12	13	14
21	001	004	013	039	095	192	332	500				
22		002	008	026	067	143	262	416				
23		001	005	017	047	105	202	339	500			
24		001	003	011	032	076	154	271	419			
25			002	007	022	054	115	212	345	500		
26			001	005	014	038	084	163	279	423		
27			001	003	010	026	061	124	221	351	500	
28				002	006	018	044	092	172	286	425	
29				001	004	012	031	068	132	229	356	500
30				001	003	008	021	049	100	181	292	428

326   Cumulative binomial for $p = \frac{1}{2}$

Binomial: $p = \frac{1}{2}$

Number in less frequent class

n	7	8	9	10	11	12	13	14	15	16	17
31	002	005	015	035	075	141	237	360	500		
32	001	004	010	025	055	108	189	298	430		
33	001	002	007	018	040	081	148	243	364	500	
34		001	005	012	029	061	115	196	304	432	
35		001	003	008	020	045	088	155	250	368	500

n	8	9	10	11	12	13	14	15	16	17	18	19
36	001	002	006	014	033	066	121	203	309	434		
37		001	004	010	024	049	094	162	256	371	500	
38		001	003	007	017	036	072	128	209	314	436	
39		001	002	005	012	027	054	100	168	261	375	500
40			001	003	008	019	040	077	134	215	318	437

n	10	11	12	13	14	15	16	17	18	19	20	21	22
41	001	002	006	014	030	059	106	174	266	378	500		
42		001	004	010	022	044	082	140	220	322	439		
43		001	003	007	016	033	063	111	180	271	380	500	
44		001	002	005	011	024	048	087	146	226	326	440	
45			001	003	008	018	036	068	116	186	276	383	500

n	12	13	14	15	16	17	18	19	20	21	22	23	24
46	001	002	006	013	027	052	092	151	231	329	441		
47	001	002	004	009	020	039	072	121	191	280	385	500	
48		001	003	007	015	030	056	097	156	235	333	443	
49		001	002	005	011	022	043	076	126	196	284	388	500
50			001	003	008	016	032	059	101	161	240	336	444

n	14	15	16	17	18	19	20	21	22	23	24	25	26	27
51	001	002	005	012	024	046	080	131	201	288	390	500		
52	001	002	004	009	018	035	063	106	166	244	339	445		
53		001	003	006	014	027	049	084	136	205	292	392	500	
54		001	002	005	010	020	038	067	110	170	248	342	446	
55		001	001	003	007	015	029	052	089	140	209	295	394	500

Binomial: $p = \frac{1}{2}$

Number in less frequent class

n	16	17	18	19	20	21	22	23	24	25	26	27	28	29	
56	001	002	005	011	022	041	070	114	175	252	344	447			
57	001	002	004	008	017	031	056	092	145	214	298	396	500		
58		001	003	006	012	024	043	074	119	179	256	347	448		
59		001	002	004	009	018	034	059	096	149	217	301	397	500	
60		001	001	003	007	014	026	046	078	123	183	259	349	449	

	18	19	20	21	22	23	24	25	26	27	28	29	30	31	32
61	001	002	005	010	020	036	062	100	153	221	304	399	500		
62	001	002	004	008	015	028	049	081	126	187	263	352	450		
63		001	003	006	011	021	038	065	104	157	225	307	401	500	
64		001	002	004	008	016	030	052	084	130	191	266	354	450	
65		001	001	003	006	012	023	041	068	107	161	229	310	402	500

	20	21	22	23	24	25	26	27	28	29	30	31	32	33	34
66	001	002	005	009	018	032	054	088	134	195	269	356	451		
67	001	002	003	007	014	025	043	071	111	164	232	313	404	500	
68		001	002	005	010	019	034	057	091	137	198	272	358	452	
69		001	002	004	008	015	027	046	074	114	168	235	315	405	500
70		001	001	003	006	011	021	036	060	094	141	201	275	360	452

	22	23	24	25	26	27	28	29	30	31	32	33	34	35	
71	001	002	004	008	016	028	048	077	118	171	238	318	406	500	
72	001	001	003	006	012	022	038	062	097	144	205	278	362	453	

	23	24	25	26	27	28	29	30	31	32	33	34	35	36	37
73	001	002	005	009	017	030	050	080	121	175	241	320	408	500	
74	001	002	004	007	013	024	040	065	100	148	208	281	364	454	
75	001	001	003	005	010	018	032	053	083	124	178	244	322	409	500

Table A–7 Four-place tables of the multinomial distribution:

$$P(x_1, x_2, x_3) = \frac{n!}{x_1!\, x_2!\, x_3!}\, p_1^{x_1} p_2^{x_2} p_3^{x_3}.$$

Values of the probability are given for $n = 2, 3, 4, 5$, and 10; for selected values of p_i ($\sum p_i = 1$); and for all values of x_1, x_2, x_3 such that $x_1 + x_2 + x_3 = n$.

Each four-digit entry should be read with a decimal point preceding it. For entries $0+$, the probability is less than 0.00005 but greater than 0.

Examples. For $p_1 = 0.1$, $p_2 = 0.2$, $p_3 = 0.7$;

if $n = 2$, $P(0, 0, 2) = 0.4900$;

if $n = 4$, $P(2, 1, 1) = 0.0168$;

if $n = 5$, $P(3, 0, 2) = 0.0049$;

if $n = 10$, $P(4, 5, 1) = 0.0000$.

$$P(x_1, x_2, x_3) = \frac{n!}{x_1!\, x_2!\, x_3!}\, p_1^{x_1} p_2^{x_2} p_3^{x_3}$$

				p_1:0.1	0.1	0.1	0.1	0.2	0.2	0.2	0.3	$\frac{1}{3}$	$\frac{1}{4}$	$\frac{1}{6}$
				p_2:0.1	0.2	0.3	0.4	0.2	0.3	0.4	0.3	$\frac{1}{3}$	$\frac{1}{4}$	$\frac{2}{6}$
n	x_1	x_2	x_3	p_3:0.8	0.7	0.6	0.5	0.6	0.5	0.4	0.4	$\frac{1}{3}$	$\frac{2}{4}$	$\frac{3}{6}$
2	2	0	0	0100	0100	0100	0100	0400	0400	0400	0900	1111	0625	0278
	1	1	0	0200	0400	0600	0800	0800	1200	1600	1800	2222	1250	1111
	1	0	1	1600	1400	1200	1000	2400	2000	1600	2400	2222	2500	1667
	0	2	0	0100	0400	0900	1600	0400	0900	1600	0900	1111	0625	1111
	0	1	1	1600	2800	3600	4000	2400	3000	3200	2400	2222	2500	3333
	0	0	2	6400	4900	3600	2500	3600	2500	1600	1600	1111	2500	2500
3	3	0	0	0010	0010	0010	0010	0080	0080	0080	0270	0370	0156	0046
	2	1	0	0030	0060	0090	0120	0240	0360	0480	0810	1111	0469	0278
	2	0	1	0240	0210	0180	0150	0720	0600	0480	1080	1111	0938	0417
	1	2	0	0030	0120	0270	0480	0240	0540	0960	0810	1111	0469	0556
	1	1	1	0480	0840	1080	1200	1440	1800	1920	2160	2222	1875	1667
	1	0	2	1920	1470	1080	0750	2160	1500	0960	1440	1111	1875	1250
	0	3	0	0010	0080	0270	0640	0080	0270	0640	0270	0370	0156	0370
	0	2	1	0240	0840	1620	2400	0720	1350	1920	1080	1111	0938	1667
	0	1	2	1920	2940	3240	3000	2160	2250	1920	1440	1111	1875	2500
	0	0	3	5120	3430	2160	1250	2160	1250	0640	0640	0370	1250	1250

$$P(x_1, x_2, x_3) = \frac{n!}{x_1! \, x_2! \, x_3!} \, p_1^{x_1} p_2^{x_2} p_3^{x_3}$$

n	x_1	x_2	x_3	p_1:0.1 p_2:0.1 p_3:0.8	0.1 0.2 0.7	0.1 0.3 0.6	0.1 0.4 0.5	0.2 0.2 0.6	0.2 0.3 0.5	0.2 0.4 0.4	0.3 0.3 0.4	$\frac{1}{3}$ $\frac{1}{3}$ $\frac{1}{3}$	$\frac{1}{4}$ $\frac{1}{4}$ $\frac{2}{4}$	$\frac{1}{6}$ $\frac{2}{6}$ $\frac{3}{6}$
4	4	0	0	0001	0001	0001	0001	0016	0016	0016	0081	0123	0039	0008
	3	1	0	0004	0008	0012	0016	0064	0096	0128	0324	0494	0156	0062
	3	0	1	0032	0028	0024	0020	0192	0160	0128	0432	0494	0312	0093
	2	2	0	0006	0024	0054	0096	0096	0216	0384	0486	0741	0234	0185
	2	1	1	0096	0168	0216	0240	0576	0720	0768	1296	1481	0938	0556
	2	0	2	0384	0294	0216	0150	0864	0600	0384	0864	0741	0938	0417
	1	3	0	0004	0032	0108	0256	0064	0216	0512	0324	0494	0156	0247
	1	2	1	0096	0336	0648	0960	0576	1080	1536	1296	1481	0938	1111
	1	1	2	0768	1176	1296	1200	1728	1800	1536	1728	1481	1875	1667
	1	0	3	2048	1372	0864	0500	1728	1000	0512	0768	0494	1250	0833
	0	4	0	0001	0016	0081	0256	0016	0081	0256	0081	0123	0039	0123
	0	3	1	0032	0224	0648	1280	0192	0540	1024	0432	0494	0312	0741
	0	2	2	0384	1176	1944	2400	0864	1350	1536	0864	0741	0938	1667
	0	1	3	2048	2744	2592	2000	1728	1500	1024	0768	0494	1250	1667
	0	0	4	4096	2401	1296	0625	1296	0625	0256	0256	0123	0625	0625

$$P(x_1, x_2, x_3) = \frac{n!}{x_1! \, x_2! \, x_3!} \, p_1^{x_1} p_2^{x_2} p_3^{x_3}$$

n	x_1	x_2	x_3	p_1:0.1 p_2:0.1 p_3:0.8	0.1 0.2 0.7	0.1 0.3 0.6	0.1 0.4 0.5	0.2 0.2 0.6	0.2 0.3 0.5	0.2 0.4 0.4	0.3 0.3 0.4	$\frac{1}{3}$ $\frac{1}{3}$ $\frac{1}{3}$	$\frac{1}{4}$ $\frac{1}{4}$ $\frac{2}{4}$	$\frac{1}{6}$ $\frac{2}{6}$ $\frac{3}{6}$
5	5	0	0	0+	0+	0+	0+	0003	0003	0003	0024	0041	0010	0001
	4	1	0	0001	0001	0002	0002	0016	0024	0032	0122	0206	0049	0013
	4	0	1	0004	0004	0003	0002	0048	0040	0032	0162	0206	0098	0019
	3	2	0	0001	0004	0009	0016	0032	0072	0128	0243	0412	0098	0051
	3	1	1	0016	0028	0036	0040	0192	0240	0256	0648	0823	0391	0154
	3	0	2	0064	0049	0036	0025	0288	0200	0128	0432	0412	0391	0116
	2	3	0	0001	0008	0027	0064	0032	0108	0256	0243	0412	0098	0103
	2	2	1	0024	0084	0162	0240	0288	0540	0768	0972	1235	0586	0463
	2	1	2	0192	0294	0324	0300	0864	0900	0768	1296	1235	1172	0694
	2	0	3	0512	0343	0216	0125	0864	0500	0256	0576	0412	0781	0347
	1	4	0	0001	0008	0040	0128	0016	0081	0256	0122	0206	0049	0103
	1	3	1	0016	0112	0324	0640	0192	0540	1024	0648	0823	0391	0617
	1	2	2	0192	0588	0972	1200	0864	1350	1536	1296	1235	1172	1389
	1	1	3	1024	1372	1296	1000	1728	1500	1024	1152	0823	1562	1389
	1	0	4	2048	1201	0648	0312	1296	0625	0256	0384	0206	0781	0521
	0	5	0	0+	0003	0024	0102	0003	0024	0102	0024	0041	0010	0041
	0	4	1	0004	0056	0243	0640	0048	0202	0512	0162	0206	0098	0309
	0	3	2	0064	0392	0972	1600	0288	0675	1024	0432	0412	0391	0926
	0	2	3	0512	1372	1944	2000	0864	1125	1024	0576	0412	0781	1389
	0	1	4	2048	2401	1944	1250	1296	0938	0512	0384	0206	0781	1042
	0	0	5	3277	1681	0778	0312	0778	0312	0102	0102	0041	0312	0312

$$P(x_1, x_2, x_3) = \frac{n!}{x_1!\,x_2!\,x_3!}\, p_1^{x_1} p_2^{x_2} p_3^{x_3}$$

				p_1: 0.1	0.1	0.1	0.1	0.2	0.2	0.2	0.3	$\frac{1}{3}$	$\frac{1}{4}$	$\frac{1}{6}$
				p_2: 0.1	0.2	0.3	0.4	0.2	0.3	0.4	0.3	$\frac{1}{3}$	$\frac{1}{4}$	$\frac{2}{6}$
n	x_1	x_2	x_3	p_3: 0.8	0.7	0.6	0.5	0.6	0.5	0.4	0.4	$\frac{1}{3}$	$\frac{2}{4}$	$\frac{3}{6}$
10	10	0	0	0+	0+	0+	0+	0+	0+	0+	0+	0+	0+	0+
	9	1	0	0+	0+	0+	0+	0+	0+	0+	0001	0002	0+	0+
	9	0	1	0+	0+	0+	0+	0+	0+	0+	0001	0002	0+	0+
	8	2	0	0+	0+	0+	0+	0+	0+	0+	0003	0008	0+	0+
	8	1	1	0+	0+	0+	0+	0+	0+	0+	0007	0015	0002	0+
	8	0	2	0+	0+	0+	0+	0+	0+	0+	0005	0008	0002	0+
	7	3	0	0+	0+	0+	0+	0+	0+	0001	0007	0020	0001	0+
	7	2	1	0+	0+	0+	0+	0001	0002	0003	0028	0061	0007	0001
	7	1	2	0+	0+	0+	0+	0003	0003	0003	0038	0061	0014	0001
	7	0	3	0+	0+	0+	0+	0003	0002	0001	0017	0020	0009	0001
	6	4	0	0+	0+	0+	0+	0+	0001	0003	0012	0036	0002	0001
	6	3	1	0+	0+	0+	0+	0003	0007	0014	0066	0142	0016	0003
	6	2	2	0+	0+	0+	0001	0012	0018	0021	0132	0213	0048	0008
	6	1	3	0+	0001	0001	0+	0023	0020	0014	0118	0142	0064	0008
	6	0	4	0001	0001	0+	0+	0017	0008	0003	0039	0036	0032	0003
	5	5	0	0+	0+	0+	0+	0+	0002	0008	0015	0043	0002	0001
	5	4	1	0+	0+	0001	0002	0004	0016	0041	0099	0213	0024	0010
	5	3	2	0+	0001	0002	0004	0023	0054	0083	0265	0427	0096	0030
	5	2	3	0001	0003	0005	0005	0070	0091	0083	0353	0427	0192	0045
	5	1	4	0005	0006	0005	0003	0105	0076	0041	0235	0213	0192	0034
	5	0	5	0008	0004	0002	0001	0063	0025	0008	0063	0043	0077	0010
	4	6	0	0+	0+	0+	0001	0+	0002	0014	0012	0036	0002	0002
10	4	5	1	0+	0+	0002	0006	0004	0024	0083	0099	0213	0024	0020
	4	4	2	0+	0002	0009	0020	0029	0102	0206	0331	0533	0120	0075
	4	3	3	0002	0012	0024	0034	0116	0227	0275	0588	0711	0320	0150
	4	2	4	0013	0030	0037	0032	0261	0284	0206	0588	0533	0481	0169
	4	1	5	0041	0042	0029	0016	0314	0189	0083	0314	0213	0385	0101
	4	0	6	0055	0025	0010	0003	0157	0052	0014	0070	0036	0128	0025
	3	7	0	0+	0+	0+	0002	0+	0002	0016	0007	0020	0001	0003
	3	6	1	0+	0+	0004	0017	0003	0024	0110	0066	0142	0016	0027
	3	5	2	0+	0004	0022	0065	0023	0122	0330	0265	0427	0096	0120
	3	4	3	0002	0023	0073	0134	0116	0340	0551	0588	0711	0320	0300
	3	3	4	0017	0081	0147	0168	0348	0567	0551	0784	0711	0641	0450
	3	2	5	0083	0169	0176	0126	0627	0567	0330	0627	0427	0769	0405
	3	1	6	0220	0198	0118	0052	0627	0315	0110	0279	0142	0513	0203
	3	0	7	0252	0099	0034	0009	0269	0075	0016	0053	0020	0146	0043
	2	8	0	0+	0+	0+	0003	0+	0001	0012	0003	0008	0+	0002
	2	7	1	0+	0+	0005	0029	0001	0016	0094	0028	0061	0007	0023
	2	6	2	0+	0004	0033	0129	0012	0092	0330	0132	0213	0048	0120
	2	5	3	0001	0028	0132	0323	0070	0306	0661	0353	0427	0192	0360

$$P(x_1, x_2, x_3) = \frac{n!}{x_1! \, x_2! \, x_3!} \, p_1^{x_1} p_2^{x_2} p_3^{x_3}$$

n	x_1	x_2	x_3	p_1:0.1 p_2:0.1 p_3:0.8	0.1 0.2 0.7	0.1 0.3 0.6	0.1 0.4 0.5	0.2 0.2 0.6	0.2 0.3 0.5	0.2 0.4 0.4	0.3 0.3 0.4	$\frac{1}{3}$ $\frac{1}{3}$ $\frac{1}{3}$	$\frac{1}{4}$ $\frac{1}{4}$ $\frac{2}{4}$	$\frac{1}{6}$ $\frac{2}{6}$ $\frac{3}{6}$
10	2	4	4	0013	0121	0331	0504	0261	0638	0826	0588	0533	0481	0675
	2	3	5	0083	0339	0529	0504	0627	0850	0661	0627	0427	0769	0810
	2	2	6	0330	0593	0529	0315	0941	0709	0330	0418	0213	0769	0608
	2	1	7	0755	0593	0302	0112	0806	0338	0094	0159	0061	0439	0260
10	2	0	8	0755	0259	0076	0018	0302	0070	0012	0027	0008	0110	0049
	1	9	0	0+	0+	0+	0003	0+	0+	0005	0001	0002	0+	0001
	1	8	1	0+	0+	0004	0029	0+	0006	0047	0007	0015	0002	0011
	1	7	2	0+	0002	0028	0147	0003	0039	0189	0038	0061	0014	0069
	1	6	3	0+	0018	0132	0430	0023	0153	0440	0118	0142	0064	0240
	1	5	4	0005	0097	0397	0806	0105	0383	0661	0235	0213	0192	0540
	1	4	5	0041	0339	0794	1008	0314	0638	0661	0314	0213	0385	0810
	1	3	6	0220	0791	1058	0840	0627	0709	0440	0279	0142	0513	0810
	1	2	7	0755	1186	0907	0450	0806	0506	0189	0159	0061	0439	0521
	1	1	8	1510	1038	0453	0141	0605	0211	0047	0053	0015	0220	0195
	1	0	9	1342	0404	0101	0020	0202	0039	0005	0008	0002	0049	0033
	0	10	0	0+	0+	0+	0001	0+	0+	0001	0+	0+	0+	0+
	0	9	1	0+	0+	0001	0013	0+	0001	0010	0001	0002	0+	0003
	0	8	2	0+	0001	0011	0074	0+	0007	0047	0005	0008	0002	0017
	0	7	3	0+	0005	0057	0246	0003	0033	0126	0017	0020	0009	0069
	0	6	4	0001	0032	0198	0538	0017	0096	0220	0039	0036	0032	0180
	0	5	5	0008	0136	0476	0806	0063	0191	0264	0063	0043	0077	0324
	0	4	6	0055	0395	0794	0840	0157	0266	0220	0070	0036	0128	0405
	0	3	7	0252	0791	0907	0600	0269	0253	0126	0053	0020	0146	0347
	0	2	8	0755	1038	0680	0281	0302	0158	0047	0027	0008	0110	0195
	0	1	9	1342	0807	0302	0078	0202	0059	0010	0008	0002	0049	0065
	0	0	10	1074	0282	0060	0010	0060	0010	0001	0001	0+	0010	0010

Table A–8   Three-place tables for the Poisson distribution

For each value of m, the first column gives values of the individual terms

$$P(X = x) = \frac{e^{-m}m^x}{x!},$$

and the second column gives the cumulative

$$P(X \geq r) = \sum_{x=r}^{\infty} \frac{e^{-m}m^x}{x!}.$$

Values of probabilities are given for x (or r) = 0, 1, 2, ..., and for

$$m = 0.1(0.1)3.5(0.5)7(1)14.$$

Each three-digit entry should be read with a decimal point preceding it. If no entry appears in a column for a particular value of x, then the value of the corresponding individual term or cumulative is less than 0.0005. For entries $1-$, the probability is larger than 0.9995 but less than 1.

Examples. For $m = 8$,

$P(X = 7) = 0.140,$

$P(X = 19) = 0.000,$ or less than 0.0005,

$P(X \geq 0) = 1$ (exactly),

$P(X \geq 1) = 1-,$ or more than 0.9995,

$P(X \geq 12) = 0.112.$

	$m = 0.1$		$m = 0.2$		$m = 0.3$		$m = 0.4$		$m = 0.5$		$m = 0.6$		$m = 0.7$		
x	Ind.	Cum.	Ind.	Cum.	Ind.	Cum.	Ind.	Cum.	Ind.	Cum.	Ind.	Cum.	Ind.	Cum.	x
0	905	1.	819	1.	741	1.	670	1.	607	1.	549	1.	497	1.	0
1	090	095	164	181	222	259	268	330	303	393	329	451	348	503	1
2	005	005	016	018	033	037	054	062	076	090	099	122	122	156	2
3			001	001	003	004	007	008	013	014	020	023	028	034	3
4							001	001	002	002	003	003	005	006	4
5													001	001	5

	$m = 0.8$		$m = 0.9$		$m = 1.0$		$m = 1.1$		$m = 1.2$		$m = 1.3$		$m = 1.4$		
0	449	1.	407	1.	368	1.	333	1.	301	1.	273	1.	247	1.	0
1	359	551	366	593	368	632	366	667	361	699	354	727	345	753	1
2	144	191	165	228	184	264	201	301	217	337	230	373	242	408	2
3	038	047	049	063	061	080	074	100	087	121	100	143	113	167	3
4	008	009	011	013	015	019	020	026	026	034	032	043	039	054	4
5	001	001	002	002	003	004	004	005	006	008	008	011	011	014	5
6					001	001	001	001	001	002	002	002	003	003	6
7													001	001	7

	m = 1.5		m = 1.6		m = 1.7		m = 1.8		m = 1.9		m = 2.0		m = 2.1		
x	Ind.	Cum.	Ind.	Cum.	Ind.	Cum.	Ind.	Cum.	Ind.	Cum.	Ind.	Cum.	Ind.	Cum.	x
0	223	1.	202	1.	183	1.	165	1.	150	1.	135	1.	122	1.	0
1	335	777	323	798	311	817	298	835	284	850	271	865	257	878	1
2	251	442	258	475	264	507	268	537	270	566	271	594	270	620	2
3	126	191	138	217	150	243	161	269	171	296	180	323	189	350	3
4	047	066	055	079	064	093	072	109	081	125	090	143	099	161	4
5	014	019	018	024	022	030	026	036	031	044	036	053	042	062	5
6	004	004	005	006	006	008	008	010	010	013	012	017	015	020	6
7	001	001	001	001	001	002	002	003	003	003	003	005	004	006	7
8									001	001	001	001	001	001	8

	m = 2.2		m = 2.3		m = 2.4		m = 2.5		m = 2.6		m = 2.7		m = 2.8		
0	111	1.	100	1.	091	1.	082	1.	074	1.	067	1.	061	1.	0
1	244	889	231	900	218	909	205	918	193	926	181	933	170	939	1
2	268	645	265	669	261	692	257	713	251	733	245	751	238	769	2
3	197	377	203	404	209	430	214	456	218	482	220	506	222	531	3
4	108	181	117	201	125	221	134	242	141	264	149	286	156	308	4
5	048	072	054	084	060	096	067	109	074	123	080	137	087	152	5
6	017	025	021	030	024	036	028	042	032	049	036	057	041	065	6
7	005	007	007	009	008	012	010	014	012	017	014	021	016	024	7
8	002	002	002	003	002	003	003	004	004	005	005	007	006	008	8
9				001	001	001	001	001	001	001	001	002	002	002	9
10													001		001	10

	m = 2.9		m = 3.0		m = 3.1		m = 3.2		m = 3.3		m = 3.4		m = 3.5		
0	055	1.	050	1.	045	1.	041	1.	037	1.	033	1.	030	1.	0
1	160	945	149	950	140	955	130	959	122	963	113	967	106	970	1
2	231	785	224	801	216	815	209	829	201	841	193	853	185	864	2
3	224	554	224	577	224	599	223	620	221	641	219	660	216	679	3
4	162	330	168	353	173	375	178	397	182	420	186	442	189	463	4
5	094	168	101	185	107	202	114	219	120	237	126	256	132	275	5
6	045	074	050	084	056	094	061	105	066	117	072	129	077	142	6
7	019	029	022	034	025	039	028	045	031	051	035	058	039	065	7
8	007	010	008	012	010	014	011	017	013	020	015	023	017	027	8
9	002	003	003	004	003	005	004	006	005	007	006	008	007	010	9
10	001	001	001	001	001	001	001	002	002	002	002	003	002	003	10
11											001	001	001	001	11

x	m = 4.0 Ind.	Cum.	m = 4.5 Ind.	Cum.	m = 5.0 Ind.	Cum.	m = 5.5 Ind.	Cum.	m = 6.0 Ind.	Cum.	m = 6.5 Ind.	Cum.	m = 7.0 Ind.	Cum.	x
0	018	1.	011	1.	007	1.	004	1.	002	1.	002	1.	001	1.	0
1	073	982	050	989	034	993	022	996	015	998	010	998	006	999	1
2	147	908	112	939	084	960	062	973	045	983	032	989	022	993	2
3	195	762	169	826	140	875	113	912	089	938	069	957	052	970	3
4	195	567	190	658	175	735	156	798	134	849	112	888	091	918	4
5	156	371	171	468	175	560	171	642	161	715	145	776	128	827	5
6	104	215	128	297	146	384	157	471	161	554	157	631	149	699	6
7	060	111	082	169	104	238	123	314	138	394	146	473	149	550	7
8	030	051	046	087	065	133	085	191	103	256	119	327	130	401	8
9	013	021	023	040	036	068	052	106	069	153	086	208	101	271	9
10	005	008	010	017	018	032	029	054	041	084	056	123	071	170	10
11	002	003	004	007	008	014	014	025	023	043	033	067	045	099	11
12	001	001	002	002	003	005	007	011	011	020	018	034	026	053	12
13			001	001	001	002	003	004	005	009	009	016	014	027	13
14						001	001	002	002	004	004	007	007	013	14
15								001	001	001	002	003	003	006	15
16										001	001	001	001	002	16
17													001	001	17

x	m = 8.0 Ind.	Cum.	m = 9.0 Ind.	Cum.	m = 10.0 Ind.	Cum.	m = 11.0 Ind.	Cum.	m = 12.0 Ind.	Cum.	m = 13.0 Ind.	Cum.	m = 14.0 Ind.	Cum.	x
0		1.		1.		1.		1.		1.		1.		1.	0
1	003	1 −	001	1 −		1 −		1 −		1 −		1 −		1 −	1
2	011	997	005	999	002	1 −	001	1 −		1 −		1 −		1 −	2
3	029	986	015	994	008	997	004	999	002	999	001	1 −		1 −	3
4	057	958	034	979	019	990	010	995	005	998	003	999	001	1 −	4
5	092	900	061	945	038	971	022	985	013	992	007	996	004	998	5
6	122	809	091	884	063	933	041	962	025	980	015	989	009	994	6
7	140	687	117	793	090	870	065	921	044	954	028	974	017	986	7
8	140	547	132	676	113	780	089	857	066	910	046	946	030	968	8
9	124	407	132	544	125	667	109	768	087	845	066	900	047	938	9
10	099	283	119	413	125	542	119	659	105	758	086	834	066	891	10
11	072	184	097	294	114	417	119	540	114	653	101	748	084	824	11
12	048	112	073	197	095	303	109	421	114	538	110	647	098	740	12
13	030	064	050	124	073	208	093	311	106	424	110	537	106	642	13
14	017	034	032	074	052	136	073	219	090	318	102	427	106	536	14
15	009	017	019	041	035	083	053	146	072	228	088	325	099	430	15
16	005	008	011	022	022	049	037	093	054	156	072	236	087	331	16
17	002	004	006	011	013	027	024	056	038	101	055	165	071	244	17
18	001	002	003	005	007	014	015	032	026	063	040	110	055	173	18
19		001	001	002	004	007	008	018	016	037	027	070	041	117	19
20			001	001	002	003	005	009	010	021	018	043	029	077	20
21					001	002	002	005	006	012	011	025	019	048	21
22						001	001	002	003	006	006	014	012	029	22
23							001	001	002	003	004	008	007	017	23
24									001	001	002	004	004	009	24
25										001	001	002	002	005	25
26											001	001	001	003	26
27													001	001	27
28														001	28

Table A–9 One-sided cumulative probabilities for the Mann-Whitney statistic, *T*

Sample *A* has *m* observations, sample *B* has *n* observations, and *T* denotes the rank sum of *B*.

    The four-digit entries in the table are $10^4 P(T \leq l)$ and $10^4 P(T \geq u)$, where (l, u) is the number pair shown immediately to the left of the probability.

    For $n = 1$ and $n = 2$, see formulas (1a), (1b), (2a), and (2b) in Section 3–4.

Example. For $m = 7$ and $n = 3$, find the value of *T* that gives a one-sided significance level nearest to, but above, 0.05.

Solution. In the column labeled 0.05 for $m = 7$ and $n = 3$, we find $P(T \leq 9) = P(T \geq 24) = 0.0583$. Hence, for one-sided testing on the high side, 24 is the required value of *T*. If both $T \leq 9$ and $T \geq 24$ are used, the test is two-sided with level $2(0.0583) = 0.1166$.

$n = 3$		Probability levels nearest to					
m	0.10		0.05		0.025		0.005
3	(7, 14)	1000	(6, 15)	0500			
4	(7, 17)	0571	(6, 18)	0286	–		
	(8, 16)	1143	(7, 17)	0571	(6, 18)	0286	
5	(8, 19)	0714	(7, 20)	0357	(6, 21)	0179	
	(9, 18)	1250	(8, 19)	0714	(7, 20)	0357	
6	(9, 21)	0833	(8, 22)	0476	(7, 23)	0238	–
	(10, 20)	1310	(9, 21)	0833	(8, 22)	0476	(6, 24) 0119
7	(10, 23)	0917	(8, 25)	0333	(7, 26)	0167	–
	(11, 22)	1333	(9, 24)	0583	(8, 25)	0333	(6, 27) 0083
8	(11, 25)	0970	(9, 27)	0424	(8, 28)	0242	–
	(12, 24)	1394	(10, 26)	0667	(9, 27)	0424	(6, 30) 0061

$n = 4$							
m							
4	(12, 24)	0571	(11, 25)	0286	(10, 26)	0143	–
	(13, 23)	1000	(12, 24)	0571	(11, 25)	0286	(10, 26) 0143
	(14, 22)	1714					
5	(14, 26)	0952	(12, 28)	0317	(11, 29)	0159	–
	(15, 25)	1429	(13, 27)	0556	(12, 28)	0317	(10, 30) 0079
6	(15, 29)	0857	(13, 31)	0333	(12, 32)	0190	(10, 34) 0048
	(16, 28)	1286	(14, 30)	0571	(13, 31)	0333	(11, 33) 0095
7	(16, 32)	0818	(14, 34)	0364	(13, 35)	0212	(10, 38) 0030
	(17, 31)	1152	(15, 33)	0545	(14, 34)	0364	(11, 37) 0061
8	(17, 35)	0768	(15, 37)	0364	(14, 38)	0242	(11, 41) 0040
	(18, 34)	1071	(16, 36)	0545	(15, 37)	0364	(12, 40) 0081

338   Probabilities for Mann-Whitney, _T_

$n = 5$							
			Probability levels nearest to				
m	0.10		0.05		0.025		0.005
5	(20, 35) 0754		(19, 36) 0476		(17, 38) 0159		(15, 40) 0040
	(21, 34) 1111		(20, 35) 0754		(18, 37) 0278		(16, 39) 0079
6	(22, 38) 0887		(20, 40) 0411		(18, 42) 0152		(16, 44) 0043
	(23, 37) 1234		(21, 39) 0628		(19, 41) 0260		(17, 43) 0087
7	(23, 42) 0745		(21, 44) 0366		(20, 45) 0240		(16, 49) 0025
	(24, 41) 1010		(22, 43) 0530		(21, 44) 0366		(17, 48) 0051
8	(25, 45) 0855		(23, 47) 0466		(21, 49) 0225		(17, 53) 0031
	(26, 44) 1111		(24, 46) 0637		(22, 48) 0326		(18, 52) 0054

$n = 6$							
m							
6	(30, 48) 0898		(28, 50) 0465		(26, 52) 0206		(23, 55) 0043
	(31, 47) 1201		(29, 49) 0660		(27, 51) 0325		(24, 54) 0076
7	(32, 52) 0903		(29, 55) 0367		(27, 57) 0175		(24, 60) 0041
	(33, 51) 1171		(30, 54) 0507		(28, 56) 0256		(25, 59) 0070
8	(34, 56) 0906		(31, 59) 0406		(29, 61) 0213		(25, 65) 0040
	(35, 55) 1142		(32, 58) 0539		(30, 60) 0296		(26, 64) 0063

$n = 7$							
m							
7	(41, 64) 0825		(39, 66) 0487		(36, 69) 0189		(32, 73) 0035
	(42, 63) 1043		(40, 65) 0641		(37, 68) 0265		(33, 72) 0055
8	(44, 68) 0946		(41, 71) 0469		(38, 74) 0200		(34, 78) 0047
	(45, 67) 1159		(42, 70) 0603		(39, 73) 0270		(35, 77) 0070

$n = 8$							
m							
8	(55, 81) 0974		(51, 85) 0415		(49, 87) 0249		(43, 93) 0035
	(56, 80) 1172		(52, 84) 0524		(50, 86) 0325		(44, 92) 0052

Table A–10 Mean and standard deviation for the Mann-Whitney statistic, T

$$\mu_T = n(m + n + 1)/2$$

$$\sigma_T = \sqrt{mn(m + n + 1)/12}$$

Samples A and B have m and n observations, respectively, and $m \geq n$.

Example. If $m = 5$ and $n = 4$, the table shows $\mu_T = 20.0$ and $\sigma_T = 4.08$.

m n	μ_T	σ_T	m n	μ_T	σ_T	m n	μ_T	σ_T	m n	μ_T	σ_T
1 1	1.5	0.50	8 1	5.0	2.58	11 5	42.5	8.83	14 1	8.0	4.32
			2	11.0	3.83	6	54.0	9.95	2	17.0	6.30
2 1	2.0	0.82	3	18.0	4.90	7	66.5	11.04	3	27.0	7.94
2	5.0	1.29	4	26.0	5.89	8	80.0	12.11	4	38.0	9.42
			5	35.0	6.83	9	94.5	13.16	5	50.0	10.80
3 1	2.5	1.12	6	45.0	7.75	10	110.0	14.20	6	63.0	12.12
2	6.0	1.73	7	56.0	8.64	11	126.5	15.23	7	77.0	13.40
3	10.5	2.29	8	68.0	9.52				8	92.0	14.65
						12 1	7.0	3.74	9	108.0	15.87
4 1	3.0	1.41	9 1	5.5	2.87	2	15.0	5.48	10	125.0	17.08
2	7.0	2.16	2	12.0	4.24	3	24.0	6.93	11	143.0	18.27
3	12.0	2.83	3	19.5	5.41	4	34.0	8.25	12	162.0	19.44
4	18.0	3.46	4	28.0	6.48	5	45.0	9.49	13	182.0	20.61
			5	37.5	7.50	6	57.0	10.68	14	203.0	21.76
5 1	3.5	1.71	6	48.0	8.49	7	70.0	11.83			
2	8.0	2.58	7	59.5	9.45	8	84.0	12.96	15 1	8.5	4.61
3	13.5	3.35	8	72.0	10.39	9	99.0	14.07	2	18.0	6.71
4	20.0	4.08	9	85.5	11.32	10	115.0	15.17	3	28.5	8.44
5	27.5	4.79				11	132.0	16.25	4	40.0	10.00
			10 1	6.0	3.16	12	150.0	17.32	5	52.5	11.46
6 1	4.0	2.00	2	13.0	4.65				6	66.0	12.85
2	9.0	3.00	3	21.0	5.92	13 1	7.5	4.03	7	80.5	14.19
3	15.0	3.87	4	30.0	7.07	2	16.0	5.89	8	96.0	15.49
4	22.0	4.69	5	40.0	8.16	3	25.5	7.43	9	112.5	16.77
5	30.0	5.48	6	51.0	9.22	4	36.0	8.83	10	130.0	18.03
6	39.0	6.24	7	63.0	10.25	5	47.5	10.14	11	148.5	19.27
			8	76.0	11.25	6	60.0	11.40	12	168.0	20.49
7 1	4.5	2.29	9	90.0	12.25	7	73.5	12.62	13	188.5	21.71
2	10.0	3.42	10	105.0	13.23	8	88.0	13.81	14	210.0	22.91
3	16.5	4.39				9	103.5	14.97	15	232.5	24.11
4	24.0	5.29	11 1	6.5	3.45	10	120.0	16.12			
5	32.5	6.16	2	14.0	5.07	11	137.5	17.26			
6	42.0	7.00	3	22.5	6.42	12	156.0	18.38			
7	52.5	7.83	4	32.0	7.66	13	175.5	19.50			

Table A–11 Probabilities for the Wilcoxon signed rank statistics, W and W_S

W = sum of all signed ranks.

W_S = smaller of the sum of the positive signed ranks and the sum of the negative signed ranks, with the signs disregarded.

For any line, $2P(W \geq b) = P(W_S \leq a)$.

    Probabilities in the table are those nearest the levels 0.01, 0.025, 0.05, and 0.10.

Examples

One-sided test. For $n = 6$ and $w = 13$, $P(W \geq 13) = 0.109$.

Two-sided test. For $n = 6$ and $w_S = 4$, $P(W_S \leq 4) = 0.219$.

One-sided test. For $n = 9$ and $w = 25$, $P(W \geq 25)$ is between 0.049 and 0.102. Interpolation gives $P(W \geq 25) = 0.084$.

n	a	$P(W_S \leq a)$ 2-sided for smallest absolute value	b	$P(W \geq b)$ 1-sided for total sum of signed ranks	n	a	$P(W_S \leq a)$ 2-sided for smallest absolute value	b	$P(W \geq b)$ 1-sided for total sum of signed ranks
1	0	1.	1	0.500	12	10	0.021	58	0.010
2	0	0.500	3	0.250		14	0.052	50	0.026
						17	0.092	44	0.046
3	0	0.250	6	0.125		22	0.204	34	0.102
4	0	0.125	10	0.062	13	13	0.021	65	0.011
	1	0.250	8	0.125		17	0.048	57	0.024
5	0	0.062	15	0.031		21	0.094	49	0.047
	1	0.125	13	0.062		26	0.191	39	0.095
	2	0.188	11	0.094	14	16	0.020	73	0.010
6	0	0.031	21	0.016		21	0.049	63	0.025
	1	0.062	19	0.031		26	0.104	53	0.052
	2	0.094	17	0.047		31	0.194	43	0.097
	4	0.219	13	0.109	15	20	0.022	80	0.011
7	0	0.016	28	0.008		25	0.048	70	0.024
	2	0.047	24	0.023		30	0.095	60	0.047
	4	0.109	20	0.055		37	0.208	46	0.104
	6	0.219	16	0.109	16	24	0.021	88	0.011
8	2	0.023	32	0.012		30	0.051	76	0.025
	4	0.055	28	0.027		36	0.105	64	0.052
	6	0.109	24	0.055		42	0.193	52	0.096
	8	0.195	20	0.098	17	28	0.020	97	0.010
9	3	0.020	39	0.010		35	0.051	83	0.025
	6	0.055	33	0.027		41	0.098	71	0.049
	8	0.098	29	0.049		49	0.207	55	0.103
	11	0.203	23	0.102	18	33	0.021	105	0.010
10	5	0.020	45	0.010		40	0.048	91	0.024
	8	0.049	39	0.024		47	0.099	77	0.049
	11	0.105	33	0.053		55	0.196	61	0.098
	14	0.193	27	0.097	19	38	0.020	114	0.010
11	7	0.019	52	0.009		46	0.049	98	0.025
	11	0.054	44	0.027		54	0.104	82	0.052
	14	0.102	38	0.051		62	0.196	66	0.098
	18	0.206	30	0.013	20	43	0.019	124	0.010
						52	0.048	106	0.024
						60	0.097	90	0.049
						70	0.202	70	0.101

Table A–12 Standard deviation of the Wilcoxon signed rank statistic, W

$W = $ sum of signed ranks.

$$\sigma_W = \sqrt{n(n + 1)(2n + 1)/6} \qquad (\mu_W = 0)$$

Example. For 8 matched pairs, $\sigma_W = 14.28$.

n	σ_W	n	σ_W	n	σ_W	n	σ_W
1	1.00	14	31.86	27	83.25	40	148.80
2	2.24	15	35.21	28	87.83	41	154.34
3	3.74	16	38.68	29	92.49	42	159.95
4	5.48	17	42.25	30	97.24	43	165.63
5	7.42	18	45.92	31	102.06	44	171.38
6	9.54	19	49.70	32	106.96	45	177.19
7	11.83	20	53.57	33	111.93	46	183.06
8	14.28	21	57.54	34	116.98	47	189.00
9	16.88	22	61.60	35	122.11	48	195.00
10	19.62	23	65.76	36	127.30	49	201.06
11	22.49	24	70.00	37	132.57	50	207.18
12	25.50	25	74.33	38	137.91		
13	28.62	26	78.75	39	143.32		

Table A–13 Probabilities for sums of squared differences of ranks, D, and values of rank correlation coefficients, R_S

The random variable D has values $\sum d^2$, and the random variable R_S has values r_S.
   The table lists probabilities falling nearest the one-sided 0.025, 0.05, and 0.10 levels.

Examples. If $n = 6$ and $\sum d^2 = 8$, we have $r_S = 0.771$, $P(D \leq 8) = P(R_S \geq 0.771) = 0.051$.
   If $n = 7$ and $r_S = -0.571$, $P(R_S \leq -0.571) = 0.100$.

	r_S positive				r_S negative		
n	$\sum d^2$	r_S	$P(D \leq \sum d^2)$ $= P(R_S \geq r_S)$		$\sum d^2$	r_S	$P(D \geq \sum d^2)$ $= P(R_S \leq r_S)$
2	0	1	0.500		2	-1	0.500
3	0	1	0.167		8	-1	0.167
4	0	1	0.042		18	-0.8	0.167
	2	0.8	0.167		20	-1	0.042
5	0	1	0.008		34	-0.7	0.117
	2	0.9	0.042		38	-0.9	0.042
	6	0.7	0.117		40	-1	0.008
6	6	0.829	0.029		58	-0.657	0.088
	8	0.771	0.051		62	-0.771	0.051
	12	0.657	0.088		64	-0.829	0.029
7	12	0.786	0.024		88	-0.571	0.100
	18	0.679	0.055		94	-0.679	0.055
	24	0.571	0.100		100	-0.786	0.024
8	22	0.738	0.023		128	-0.524	0.098
	30	0.643	0.048		138	-0.643	0.048
	40	0.524	0.098		146	-0.738	0.023
9	38	0.683	0.025		178	-0.483	0.097
	48	0.600	0.048		192	-0.600	0.048
	62	0.483	0.097		202	-0.683	0.025
10	58	0.648	0.024		238	-0.442	0.102
	72	0.564	0.048		258	-0.564	0.048
	92	0.442	0.102		272	-0.648	0.024
11	86	0.609	0.026		312	-0.418	0.102
	104	0.527	0.050		336	-0.527	0.050
	128	0.418	0.102		354	-0.609	0.026

Table A–14 Probabilities for the Kruskal-Wallis statistic, *H*: three or four categories

$$H = \frac{12}{N(N+1)} \sum_{i=1}^{c} \frac{R_i^2}{n_i} - 3(N+1)$$

P is the probability that *H* has a value as large as, or larger than, the corresponding tabulated value.

Values of $\sum (R_i^2/n_i)$ and *h* are given for probabilities nearest 0.99, 0.95, 0.90, 0.75, 0.50, 0.25, 0.10, 0.05, and 0.01.

Example. If $N = 12$, $n_1 = 5$, $n_2 = 4$, $n_3 = 3$, and $\sum (R_i^2/n_i) = 580.2$, we have $h = 5.63$ and $P(H \geq 5.63) = 0.050$.

Values intermediate to those tabulated are handled by interpolation.

Kruskal-Wallis statistic: 3 categories

N	n_1	n_2	n_3	$\sum R_i^2/n_i$	h	P	N	n_1	n_2	n_3	$\sum R_i^2/n_i$	h	P
4	2	1	1	25.5	0.30	1.	6	3	2	1	73.8	0.10	1.
				28.0	1.80	0.833					74.3	0.24	0.933
				29.5	2.70	0.500					75.0	0.43	0.900
5	3	1	1	46.3	0.53	1.					77.8	1.24	0.700
				47.0	0.80	0.800					81.0	2.14	0.533
				50.3	2.13	0.700					84.3	3.10	0.267
				53.0	3.20	0.300					88.5	4.29	0.100
5	2	2	1	45.0	0	1.	6	2	2	2	73.5	0	1.
				46.0	0.40	0.933					74.5	0.29	0.933
				46.5	0.60	0.867					76.5	0.86	0.800
				48.5	1.40	0.733					80.5	2.00	0.533
				50.0	2.00	0.600					86.5	3.71	0.200
				51.0	2.40	0.467					89.5	4.57	0.067
				52.5	3.00	0.333	7	5	1	1	113.2	0.26	1.
				54.0	3.60	0.200					114.0	0.43	0.905
6	4	1	1	74.0	0.14	1.					117.2	1.11	0.762
				76.2	0.79	0.933					122.8	2.31	0.524
				77.0	1.00	0.800					125.2	2.83	0.333
				82.2	2.50	0.467					130.0	3.86	0.143
				86.0	3.57	0.200							

N	n_1	n_2	n_3	$\sum R_i^2/n_i$	h	P
7	4	2	1	112.0	0	1.
				113.2	0.27	0.933
				113.5	0.32	0.895
				117.0	1.07	0.743
				121.5	2.04	0.495
				125.5	2.89	0.267
				130.8	4.02	0.114
				134.5	4.82	0.057
7	3	3	1	112.7	0.14	0.986
				113.3	0.29	0.957
				114.7	0.57	0.871
				117.3	1.14	0.743
				121.3	2.00	0.514
				126.7	3.14	0.243
				133.3	4.57	0.100
				136.0	5.14	0.043
7	3	2	2	112.0	0	1.
				112.8	0.18	0.971
				113.0	0.21	0.895
				115.3	0.71	0.743
				119.5	1.61	0.524
				128.0	3.43	0.248
				132.8	4.46	0.105
				134.0	4.71	0.048
				137.0	5.36	0.029
8	6	1	1	162.5	0.08	1.
				164.7	0.44	0.964
				165.2	0.53	0.893
				170.0	1.33	0.714
				174.5	2.08	0.500
				180.7	3.11	0.250
				186.5	4.08	0.107
8	5	2	1	162.3	0.05	1.
				163.2	0.20	0.940
				164.7	0.45	0.905
				168.0	1.00	0.750
				173.5	1.92	0.488
				178.8	2.80	0.286
				187.2	4.20	0.095
				192.0	5.00	0.048
				193.5	5.25	0.036

N	n_1	n_2	n_3	$\sum R_i^2/n_i$	h	P
8	4	3	1	162.3	0.06	1.
				163.2	0.21	0.950
				164.6	0.43	0.900
				167.2	0.88	0.743
				173.0	1.83	0.514
				178.6	2.76	0.229
				186.3	4.06	0.093
				193.2	5.21	0.050
				197.0	5.83	0.021
8	4	2	2	162.0	0	1.
				162.8	0.12	0.971
				164.0	0.33	0.890
				166.0	0.67	0.757
				172.8	1.79	0.514
				180.8	3.12	0.248
				188.8	4.46	0.100
				192.8	5.12	0.052
				198.0	6.00	0.014
8	3	3	2	162.2	0.03	1.
				163.3	0.22	0.946
				163.5	0.25	0.896
				166.2	0.69	0.757
				172.8	1.81	0.511
				180.8	3.14	0.243
				189.3	4.56	0.100
				192.8	5.14	0.061
				199.5	6.25	0.011
9	7	1	1	226.1	0.15	1.
				227.0	0.27	0.944
				229.6	0.61	0.917
				233.0	1.07	0.750
				241.6	2.21	0.500
				245.6	2.74	0.306
				257.0	4.27	0.083
9	6	2	1	225.7	0.09	0.984
				226.2	0.16	0.960
				227.7	0.36	0.889
				231.2	0.82	0.730
				238.5	1.80	0.500
				247.2	2.96	0.230
				256.5	4.20	0.095
				261.2	4.82	0.048
				267.0	5.60	0.024

N	n_1	n_2	n_3	$\sum R_i^2/n_i$	h	P	N	n_1	n_2	n_3	$\sum R_i^2/n_i$	h	P
9	5	3	1	225.5	0.07	0.992	9	3	3	3	225.7	0.09	0.993
				226.3	0.18	0.952					227.0	0.27	0.929
				227.5	0.34	0.889					227.7	0.36	0.879
				231.1	0.82	0.750					231.0	0.80	0.721
				238.3	1.78	0.488					237.7	1.69	0.511
				246.3	2.84	0.258					249.0	3.20	0.254
				255.1	4.02	0.095					259.7	4.62	0.100
				261.5	4.87	0.052					267.0	5.60	0.050
				273.0	6.40	0.012					273.7	6.49	0.011
9	5	2	2	225.7	0.09	0.984	10	8	1	1	303.0	0.05	1.
				226.0	0.13	0.937					305.5	0.33	0.933
				226.8	0.24	0.913					307.0	0.49	0.889
				230.8	0.77	0.759					313.0	1.15	0.733
				237.7	1.69	0.495					321.1	2.03	0.489
				248.5	3.13	0.254					329.5	2.95	0.244
				257.8	4.37	0.090					343.0	4.42	0.067
				262.8	5.04	0.056	10	7	2	1	302.8	0.03	1.
				274.0	6.53	0.008					305.1	0.28	0.939
9	4	4	1	225.5	0.07	0.987					305.3	0.30	0.900
				226.2	0.17	0.968					310.8	0.90	0.744
				227.2	0.30	0.911					319.0	1.80	0.494
				231.5	0.87	0.759					328.5	2.84	0.244
				238.2	1.77	0.498					341.0	4.20	0.100
				245.2	2.70	0.260					345.6	4.71	0.050
				255.5	4.07	0.102					356.5	5.89	0.017
				262.2	4.97	0.048	10	6	3	1	303.0	0.05	0.983
				275.0	6.67	0.010					304.2	0.18	0.952
9	4	3	2	225.6	0.08	0.987					305.5	0.33	0.921
				225.8	0.11	0.944					309.7	0.78	0.752
				227.1	0.28	0.902					318.8	1.78	0.502
				230.2	0.70	0.756					329.5	2.95	0.255
				237.1	1.61	0.502					338.3	3.91	0.095
				248.2	3.10	0.251					347.0	4.85	0.050
				258.3	4.44	0.102					362.8	6.58	0.012
				265.5	5.40	0.051							
				272.2	6.30	0.011							

N	n_1	n_2	n_3	$\sum R_i^2/n_i$	h	P
10	6	2	2	303.2	0.07	0.984
				303.5	0.11	0.946
				305.2	0.29	0.911
				309.2	0.73	0.743
				318.5	1.75	0.500
				330.2	3.02	0.244
				343.2	4.44	0.108
				348.5	5.02	0.051
				362.5	6.55	0.011
10	5	4	1	303.0	0.06	0.983
				304.2	0.19	0.952
				305.2	0.29	0.906
				310.0	0.82	0.762
				318.8	1.78	0.498
				329.0	2.90	0.251
				338.8	3.96	0.102
				347.0	4.86	0.056
				365.2	6.84	0.011
10	5	3	2	303.1	0.07	0.981
				303.7	0.13	0.951
				305.0	0.28	0.901
				309.0	0.71	0.743
				317.3	1.61	0.502
				329.8	2.98	0.252
				343.7	4.49	0.101
				350.6	5.25	0.049
				365.0	6.82	0.010
10	4	4	2	303.0	0.05	0.988
				304.2	0.19	0.940
				305.0	0.27	0.893
				309.2	0.74	0.757
				317.5	1.64	0.510
				330.5	3.05	0.239
				344.2	4.55	0.098
				350.5	5.24	0.052
				365.5	6.87	0.011
10	4	3	3	302.9	0.05	0.984
				304.0	0.16	0.941
				305.6	0.34	0.895
				308.9	0.70	0.764
				317.3	1.62	0.497
				329.6	2.95	0.253
				345.6	4.70	0.101
				355.0	5.73	0.050
				364.3	6.75	0.010
11	9	1	1	397.1	0.10	1.
				400.4	0.40	0.945
				401.1	0.46	0.909
				407.8	1.07	0.745
				416.4	1.86	0.509
				426.4	2.77	0.255
				437.1	3.74	0.127
				446.0	4.55	0.055
11	8	2	1	396.6	0.06	0.990
				397.5	0.14	0.958
				400.1	0.38	0.901
				406.1	0.92	0.739
				415.6	1.78	0.501
				426.6	2.78	0.251
				440.1	4.01	0.093
				446.6	4.60	0 053
				463.5	6.14	0.012
11	7	3	1	396.5	0.04	0.994
				397.9	0.17	0.955
				400.3	0.39	0.903
				405.9	0.90	0.765
				415.1	1.74	0.503
				427.9	2.90	0.230
				438.3	3.84	0.105
				450.5	4.95	0.047
				469.1	6.65	0.011

N	n_1	n_2	n_3	$\sum R_i^2/n_i$	*h*	*P*	*N*	n_1	n_2	n_3	$\sum R_i^2/n_i$	*h*	*P*
11	7	2	2	396.6	0.06	0.990	11	5	4	2	396.4	0.04	0.992
				397.0	0.09	0.958					397.6	0.14	0.952
				398.6	0.24	0.895					398.8	0.25	0.902
				403.3	0.66	0.746					402.8	0.62	0.749
				413.6	1.60	0.509					413.5	1.59	0.499
				429.8	3.07	0.251					428.0	2.91	0.249
				445.8	4.53	0.099					445.7	4.52	0.101
				452.6	5.14	0.044					454.0	5.27	0.051
				468.6	6.60	0.011					474.3	7.12	0.010
11	6	4	1	396.4	0.04	0.993	11	5	3	3	396.5	0.05	0.994
				397.7	0.15	0.955					397.9	0.17	0.948
				399.9	0.36	0.895					398.7	0.24	0.902
				405.9	0.90	0.757					403.5	0.68	0.765
				415.2	1.75	0.499					412.8	1.53	0.505
				426.7	2.79	0.254					428.7	2.97	0.242
				440.4	4.04	0.094					445.9	4.53	0.097
				450.4	4.95	0.047					456.7	5.52	0.051
				473.9	7.08	0.010					472.8	6.98	0.011
11	6	3	2	396.5	0.05	0.992	11	4	4	3	396.5	0.05	0.993
				397.5	0.14	0.951					397.6	0.14	0.959
				398.7	0.24	0.900					399.6	0.33	0.890
				403.5	0.68	0.747					403.3	0.67	0.742
				414.0	1.64	0.496					413.0	1.55	0.503
				428.0	2.91	0.253					428.2	2.93	0.250
				446.0	4.55	0.101					446.0	4.55	0.099
				453.5	5.23	0.052					457.3	5.58	0.051
				472.7	6.97	0.009					474.6	7.14	0.010
11	5	5	1	396.4	0.04	0.994	12	10	1	1	509.6	0.20	0.985
				398.0	0.18	0.944					509.9	0.22	0.955
				400.4	0.40	0.885					513.9	0.53	0.909
				406.0	0.91	0.752					521.1	1.08	0.742
				415.2	1.75	0.493					531.5	1.88	0.500
				428.0	2.91	0.242					544.4	2.88	0.273
				440.4	4.04	0.105					557.6	3.89	0.106
				450.0	4.91	0.053					567.5	4.65	0.046
				476.4	7.31	0.009							

N	n_1	n_2	n_3	$\sum R_i^2/n_i$	h	P	N	n_1	n_2	n_3	$\sum R_i^2/n_i$	h	P
12	9	2	1	507.8	0.06	0.985	12	7	3	2	507.6	0.05	0.987
				509.5	0.19	0.958					508.3	0.10	0.954
				511.3	0.33	0.894					510.0	0.23	0.902
				517.8	0.83	0.748					516.0	0.69	0.752
				531.1	1.85	0.491					527.3	1.56	0.494
				545.5	2.96	0.248					546.0	3.00	0.251
				557.8	3.91	0.100					566.0	4.54	0.101
				569.9	4.84	0.048					576.5	5.34	0.050
				589.5	6.35	0.009					595.9	6.84	0.010
12	8	3	1	507.5	0.04	0.989	12	6	5	1	507.5	0.04	0.990
				509.8	0.22	0.946					509.5	0.19	0.950
				511.1	0.32	0.902					511.2	0.32	0.900
				517.5	0.80	0.746					518.0	0.84	0.752
				529.8	1.76	0.508					529.9	1.76	0.505
				544.1	2.86	0.256					543.3	2.79	0.263
				559.1	4.01	0.099					558.0	3.92	0.104
				570.5	4.88	0.048					569.9	4.84	0.051
				595.5	6.80	0.009					598.0	7.00	0.010
12	8	2	2	507.6	0.05	0.990	12	6	4	2	507.4	0.03	0.993
				508.5	0.12	0.952					508.7	0.13	0.951
				509.6	0.20	0.911					510.0	0.23	0.909
				516.6	0.74	0.757					516.4	0.72	0.743
				529.0	1.69	0.494					528.4	1.65	0.502
				546.6	3.05	0.251					546.0	3.00	0.252
				565.0	4.46	0.101					565.4	4.49	0.100
				572.6	5.05	0.053					575.4	5.26	0.050
				591.6	6.51	0.010					602.4	7.34	0.010
12	7	4	1	507.5	0.04	0.989	12	6	3	3	507.3	0.03	0.990
				509.5	0.20	0.947					508.5	0.12	0.962
				511.2	0.33	0.897					510.7	0.28	0.888
				517.6	0.81	0.754					516.0	0.69	0.761
				529.8	1.76	0.505					526.7	1.51	0.503
				543.5	2.81	0.260					545.8	2.99	0.249
				559.2	4.02	0.103					566.7	4.59	0.098
				571.3	4.95	0.052					580.0	5.62	0.050
				597.8	6.99	0.010					600.5	7.19	0.010

N	n_1	n_2	n_3	$\sum R_i^2/n_i$	h	P		N	n_1	n_2	n_3	$\sum R_i^2/n_i$	h	P
12	5	5	2	507.6	0.05	0.988		13	9	3	1	638.1	0.07	0.987
				509.2	0.17	0.947						639.4	0.16	0.950
				510.3	0.25	0.896						642.3	0.35	0.894
				516.0	0.69	0.749						649.0	0.79	0.754
				528.4	1.65	0.496						664.8	1.83	0.496
				546.3	3.02	0.243						680.3	2.86	0.247
				565.6	4.51	0.100						698.8	4.07	0.099
				575.2	5.25	0.051						710.8	4.86	0.052
				601.5	7.27	0.010						737.0	6.59	0.010
12	5	4	3	507.4	0.03	0.990		13	9	2	2	637.6	0.04	0.993
				508.5	0.12	0.953						639.4	0.16	0.944
				510.3	0.26	0.900						640.5	0.23	0.905
				515.4	0.64	0.754						646.8	0.64	0.755
				526.6	1.51	0.495						661.1	1.59	0.497
				545.3	2.95	0.251						684.9	3.16	0.248
				566.1	4.55	0.099						704.3	4.44	0.101
				580.2	5.63	0.050						714.0	5.08	0.051
				603.8	7.44	0.010						738.4	6.69	0.011
12	4	4	4	507.5	0.04	0.994		13	8	4	1	638.1	0.07	0.987
				509.0	0.15	0.941						639.1	0.14	0.951
				510.5	0.27	0.913						642.0	0.33	0.901
				516.5	0.73	0.746						649.5	0.82	0.748
				526.5	1.50	0.510						664.2	1.80	0.502
				545.0	2.92	0.252						679.4	2.79	0.255
				567.5	4.65	0.097						698.2	4.04	0.099
				581.0	5.69	0.049						712.2	4.96	0.050
				605.0	7.54	0.011						742.5	6.96	0.010
13	11	1	1	638.1	0.07	1.		13	8	3	2	637.5	0.03	0.995
				641.4	0.29	0.962						639.0	0.13	0.947
				642.8	0.38	0.910						641.0	0.26	0.896
				654.5	1.15	0.744						647.0	0.66	0.752
				666.5	1.94	0.500						660.5	1.55	0.500
				682.8	3.02	0.231						682.8	3.02	0.253
				698.1	4.03	0.090						704.5	4.45	0.101
				709.0	4.75	0.038						717.6	5.32	0.050
13	10	2	1	637.6	0.04	0.993						743.0	6.99	0.010
				639.4	0.16	0.946								
				641.6	0.30	0.904								
				649.4	0.82	0.751								
				664.9	1.84	0.513								
				680.5	2.87	0.247								
				697.6	4.00	0.105								
				710.4	4.84	0.049								
				734.5	6.43	0.009								

N	n_1	n_2	n_3	$\sum R_i^2/n_i$	h	P	N	n_1	n_2	n_3	$\sum R_i^2/n_i$	h	P
13	7	5	1	638.1	0.08	0.988	13	6	5	2	637.7	0.04	0.986
				638.9	0.13	0.954					639.4	0.16	0.942
				641.8	0.32	0.904					640.3	0.22	0.903
				649.3	0.81	0.749					646.7	0.64	0.751
				664.2	1.79	0.500					660.5	1.55	0.499
				679.3	2.79	0.251					682.7	3.01	0.250
				698.2	4.04	0.098					704.9	4.47	0.100
				712.8	5.00	0.051					717.7	5.32	0.051
				742.9	6.99	0.010					747.7	7.30	0.010
13	7	4	2	637.6	0.04	0.986	13	6	4	3	637.5	0.03	0.988
				638.6	0.11	0.949					638.8	0.12	0.950
				640.5	0.23	0.899					640.8	0.25	0.907
				646.6	0.63	0.753					646.9	0.65	0.748
				660.5	1.55	0.502					660.2	1.53	0.499
				682.3	2.99	0.251					680.9	2.90	0.251
				706.0	4.55	0.100					706.8	4.60	0.100
				718.3	5.36	0.051					722.0	5.60	0.050
				748.0	7.32	0.010					750.2	7.47	0.010
13	7	3	3	637.5	0.03	0.996	13	5	5	3	637.5	0.04	0.989
				638.9	0.13	0.948					638.7	0.11	0.951
				641.3	0.28	0.895					640.6	0.24	0.902
				648.0	0.72	0.747					647.0	0.66	0.751
				660.0	1.51	0.498					659.9	1.51	0.497
				681.6	2.94	0.249					681.5	2.94	0.246
				705.6	4.52	0.101					705.9	4.55	0.100
				722.0	5.60	0.050					722.3	5.63	0.051
				746.6	7.23	0.010					751.4	7.54	0.010
13	6	6	1	638.2	0.08	0.987	13	5	4	4	637.4	0.03	0.996
				640.0	0.20	0.946					638.8	0.12	0.952
				642.2	0.34	0.897					640.4	0.23	0.903
				649.0	0.79	0.744					646.7	0.64	0.745
				664.2	1.79	0.504					660.2	1.53	0.501
				679.0	2.77	0.256					681.2	2.92	0.249
				697.7	4.00	0.098					707.0	4.62	0.100
				710.7	4.86	0.051					722.2	5.62	0.050
				744.2	7.07	0.010					754.7	7.76	0.009

Kruskal-Wallis statistic: 4 categories

N	n_1	n_2	n_3	n_4	$\sum R_i^2/n_i$	h	P	N	n_1	n_2	n_3	n_4	$\sum R_i^2/n_i$	h	P
5	2	1	1	1	47.0	0.80	1.	8	5	1	1	1	165.2	0.53	1.
					50.5	2.20	0.900						166.8	0.80	0.964
					53.0	3.20	0.700						170.8	1.47	0.893
					54.5	3.80	0.400						174.8	2.13	0.750
6	3	1	1	1	77.0	1.00	1.						180.8	3.13	0.536
					78.3	1.38	0.900						186.8	4.13	0.286
					82.3	2.52	0.800						194.0	5.33	0.071
					86.3	3.67	0.500	8	4	2	1	1	163.2	0.21	0.993
					89.0	4.43	0.200						166.0	0.67	0.945
6	2	2	1	1	74.0	0.14	1.						169.2	1.21	0.900
					76.5	0.86	0.956						173.5	1.92	0.748
					78.0	1.2Ɔ	0.911						179.2	2.88	0.502
					82.0	2.43	0.711						187.2	4.21	0.243
					86.0	3.57	0.467						193.5	5.25	0.090
					88.5	4.29	0.267						197.0	5.83	0.043
					90.0	4.71	0.133						198.5	6.08	0.029
7	4	1	1	1	114.0	0.43	1.	8	3	3	1	1	164.7	0.44	0.989
					117.2	1.12	0.971						166.7	0.78	0.946
					119.0	1.50	0.914						168.0	1.00	0.896
					121.2	1.98	0.771						173.3	1.89	0.771
					126.0	3.00	0.543						180.7	3.11	0.489
					131.2	4.12	0.286						186.7	4.11	0.289
					135.0	4.93	0.114						194.0	5.33	0.096
7	3	2	1	1	113.5	0.32	0.981						197.3	5.89	0.064
					114.8	0.61	0.962						200.0	6.33	0.021
					117.5	1.18	0.886	8	3	2	2	1	163.3	0.22	0.988
					121.5	2.04	0.743						165.0	0.50	0.955
					126.8	3.18	0.476						166.8	0.81	0.900
					131.5	4.18	0.210						172.8	1.81	0.760
					136.0	5.14	0.086						179.3	2.89	0.500
					137.5	5.46	0.057						187.5	4.25	0.255
7	2	2	2	1	113.0	0.21	0.990						194.3	5.39	0.110
					114.5	0.54	0.952						196.8	5.81	0.057
					116.0	0.86	0.895						201.0	6.50	0.014
					121.0	1.93	0.752								
					126.5	3.11	0.476								
					133.0	4.50	0.257								
					135.5	5.04	0.124								
					138.5	5.68	0.038								

N	n_1	n_2	n_3	n_4	$\sum R_i^2/n_i$	h	P		N	n_1	n_2	n_3	n_4	$\sum R_i^2/n_i$	h	P
8	2	2	2	2	163.0	0.17	0.990		9	4	2	2	1	226.0	0.13	0.990
					165.0	0.50	0.952							228.5	0.47	0.952
					167.0	0.83	0.886							231.2	0.83	0.898
					171.0	1.50	0.771							237.8	1.70	0.755
					180.0	3.00	0.495							246.0	2.80	0.511
					189.0	4.50	0.267							256.5	4.20	0.252
					195.0	5.50	0.114							266.5	5.53	0.098
					199.0	6.17	0.038							270.0	6.00	0.057
					202.0	6.67	0.010							277.5	7.00	0.010
9	6	1	1	1	230.2	0.69	0.988		9	3	3	2	1	226.3	0.18	0.992
					231.7	0.89	0.940							229.7	0.62	0.950
					234.2	1.22	0.905							231.2	0.82	0.901
					241.7	2.22	0.738							237.7	1.69	0.753
					247.5	3.00	0.488							246.3	2.84	0.500
					255.5	4.07	0.250							256.3	4.18	0.250
					261.7	4.89	0.119							267.2	5.62	0.101
					267.5	5.67	0.048							271.2	6.16	0.056
9	5	2	1	1	226.5	0.20	0.992							277.8	7.04	0.011
					230.2	0.69	0.943		9	3	2	2	2	226.0	0.13	0.987
					232.2	0.96	0.903							228.0	0.40	0.948
					239.3	1.91	0.746							230.5	0.73	0.900
					246.2	2.83	0.499							236.3	1.51	0.749
					255.3	4.04	0.258							246.0	2.80	0.506
					263.3	5.11	0.106							257.8	4.38	0.248
					268.2	5.76	0.056							267.3	5.64	0.100
					274.5	6.60	0.016							272.5	6.33	0.048
9	4	3	1	1	227.0	0.27	0.990							278.5	7.13	0.008
					230.2	0.70	0.950									
					232.2	0.97	0.907									
					238.3	1.78	0.744									
					246.3	2.84	0.506									
					255.6	4.08	0.250									
					263.0	5.07	0.095									
					271.3	6.18	0.049									
					278.0	7.07	0.010									

Table A–15 Probabilities for Friedman's chi-square statistic, χ_r^2

$$\chi_r^2 = \frac{12}{mI(I+1)} \sum R_i^2 - 3m(I+1),$$

where R_i is the sum of the m ranks assigned to item i, $i = 1, 2, \ldots, I$. In the table, χ^2 denotes a value of the random variable χ_r^2. Values of $P(\chi_r^2 \geq \chi^2)$ are given for levels 0.01, 0.05, 0.10, 0.25, 0.50, 0.75, and 0.95, and for $I = 3$ and $I = 4$.

Example. If $I = 4$, $m = 5$, and $\chi^2 = 4.20$, we have $P(\chi_r^2 \geq 4.20) = 0.266$. For values of χ^2 between those tabulated, use interpolation.

$I = 3$

m	χ^2	P	m	χ^2	P	m	χ^2	P	m	χ^2	P
2	0.	1.	7	0.29	0.964	11	0.18	0.976	15	0.13	0.982
	1.00	0.833		0.86	0.768		0.73	0.732		0.93	0.711
	3.00	0.500		2.00	0.486		1.64	0.470		1.60	0.513
	4.00	0.167		3.43	0.237		2.91	0.256		2.80	0.267
				4.57	0.112		4.91	0.100		4.93	0.096
3	0.67	0.944		6.00	0.051		6.54	0.043		6.40	0.047
	2.00	0.528		8.86	0.008		8.91	0.011		8.93	0.010
	4.67	0.194									
	6.00	0.028	8	0.25	0.967	12	0.17	0.978			
				0.75	0.794		0.67	0.751			
4	0.50	0.931		1.75	0.531		1.50	0.500			
	1.50	0.653		3.25	0.236		3.17	0.249			
	2.00	0.431		4.75	0.120		4.67	0.108			
	3.50	0.273		6.25	0.047		6.17	0.050			
	4.50	0.125		9.00	0.010		8.67	0.011			
	6.50	0.042									
	8.00	0.005	9	0.22	0.971	13	0.15	0.980			
				0.67	0.814		0.62	0.767			
5	0.40	0.954		1.56	0.569		1.38	0.527			
	1.20	0.691		2.89	0.278		2.92	0.278			
	1.60	0.522		4.67	0.107		4.77	0.098			
	3.60	0.182		6.22	0.048		6.00	0.050			
	5.20	0.093		8.67	0.010		8.67	0.012			
	6.40	0.039									
	8.40	0.008	10	0.20	0.974	14	0.14	0.981			
				0.80	0.710		0.57	0.781			
6	0.33	0.956		1.80	0.436		1.71	0.489			
	1.00	0.740		3.20	0.222		3.00	0.242			
	2.33	0.430		5.00	0.092		5.14	0.089			
	3.00	0.252		6.20	0.046		6.14	0.049			
	5.33	0.072		8.60	0.012		9.00	0.010			
	6.33	0.052									
	9.00	0.008									

$I = 4$

m	χ^2	P	m	χ^2	P
2	0.60	0.958	8	0.45	0.957
	1.80	0.792		1.35	0.754
	3.00	0.542		2.55	0.500
	4.80	0.208		4.20	0.247
	6.00	0.042		6.30	0.098
				7.65	0.049
3	0.60	0.958		10.35	0.010
	1.80	0.727			
	2.60	0.524			
	4.20	0.293			
	6.60	0.075			
	7.00	0.054			
	8.20	0.017			
4	0.60	0.930			
	1.50	0.753			
	2.70	0.513			
	4.50	0.237			
	6.00	0.106			
	7.50	0.054			
	9.30	0.011			
5	0.60	0.944			
	1.32	0.769			
	2.52	0.520			
	4.20	0.266			
	6.12	0.102			
	7.80	0.049			
	9.96	0.009			
6	0.40	0.952			
	1.40	0.779			
	2.60	0.517			
	4.20	0.259			
	6.20	0.109			
	7.60	0.043			
	10.20	0.010			
7	0.43	0.964			
	1.46	0.754			
	2.49	0.533			
	4.37	0.239			
	6.26	0.101			
	7.63	0.051			
	10.37	0.009			

Table A–16 Expected values of order statistics for samples of size n from the standard normal distribution

[Number of standard deviations above (+) or below (−) the mean]

Numbers in left margin are the numbers of the order statistics taken from the right— greatest to least. These numbers are values of $n - i + 1$; the $n - i + 1$st order statistic from the right is the ith from the left, and vice versa.

Example. Find the mean difference between the 18th and 7th order statistic in a sample of size 20 from the standard normal.

Solution. Under $n = 20$, the number listed for 7 is 0.45, which gives -0.45 for the 7th smallest; $20 - 18 + 1 = 3$, and so $+1.13$ is the mean for the 18th order statistic. The difference is $1.13 - (-0.45) = 1.58$ standard deviations.

$n - i + 1$ \\ n	1	2	3	4	5	6	7	8	9	10
1	0	0.56	0.85	1.03	1.16	1.27	1.35	1.42	1.49	1.54
2		−0.56	0.00	0.30	0.50	0.64	0.76	0.85	0.93	1.00
3			−0.85	−0.30	0.00	0.20	0.35	0.47	0.57	0.66
4				−1.03	−0.50	−0.20	0.00	0.15	0.27	0.38
5					−1.16	−0.64	−0.35	−0.15	0.00	0.12
6						−1.27	−0.76	−0.47	−0.27	−0.12

$n - i + 1$ \\ n	11	12	13	14	15	16	17	18	19	20
1	1.59	1.63	1.67	1.70	1.74	1.77	1.79	1.82	1.84	1.87
2	1.06	1.12	1.16	1.21	1.25	1.28	1.32	1.35	1.38	1.41
3	0.73	0.79	0.85	0.90	0.95	0.99	1.03	1.07	1.10	1.13
4	0.46	0.54	0.60	0.66	0.71	0.76	0.81	0.85	0.89	0.92
5	0.22	0.31	0.39	0.46	0.52	0.57	0.62	0.66	0.71	0.75
6	0.00	0.10	0.19	0.27	0.34	0.40	0.45	0.50	0.55	0.59
7	−0.22	−0.10	0.00	0.09	0.17	0.23	0.30	0.35	0.40	0.45
8	−0.46	−0.31	−0.19	−0.09	0.00	0.08	0.15	0.21	0.26	0.31
9	−0.73	−0.54	−0.39	−0.27	−0.17	−0.08	0.00	0.07	0.13	0.19
10	−1.06	−0.79	−0.60	−0.46	−0.34	−0.23	−0.15	−0.07	0.00	0.06
11	−1.59	−1.12	−0.85	−0.66	−0.52	−0.40	−0.30	−0.21	−0.13	−0.06

$n-i+1$ \ n	21	22	23	24	25	26	27	28	29	30
1	1.89	1.91	1.93	1.95	1.97	1.98	2.00	2.01	2.03	2.04
2	1.43	1.46	1.48	1.50	1.52	1.54	1.56	1.58	1.60	1.62
3	1.16	1.19	1.21	1.24	1.26	1.29	1.31	1.33	1.35	1.36
4	0.95	0.98	1.01	1.04	1.07	1.09	1.11	1.14	1.16	1.18
5	0.78	0.82	0.85	0.88	0.91	0.93	0.96	0.98	1.00	1.03
6	0.63	0.67	0.70	0.73	0.76	0.79	0.82	0.85	0.87	0.89
7	0.49	0.53	0.57	0.60	0.64	0.67	0.70	0.73	0.75	0.78
8	0.36	0.41	0.45	0.48	0.52	0.55	0.58	0.61	0.64	0.67
9	0.24	0.29	0.33	0.37	0.41	0.44	0.48	0.51	0.54	0.57
10	0.12	0.17	0.22	0.26	0.30	0.34	0.38	0.41	0.44	0.47
11	0.00	0.06	0.11	0.16	0.20	0.24	0.28	0.32	0.35	0.38
12	−0.12	−0.06	0.00	0.05	0.10	0.14	0.19	0.22	0.26	0.29
13	−0.24	−0.17	−0.11	−0.05	0.00	0.05	0.09	0.13	0.17	0.21
14	−0.36	−0.29	−0.22	−0.16	−0.10	−0.05	0.00	0.04	0.09	0.12
15	−0.49	−0.41	−0.33	−0.26	−0.20	−0.14	−0.09	−0.04	0.00	0.04
16	−0.63	−0.53	−0.45	−0.37	−0.30	−0.24	−0.19	−0.13	−0.09	−0.04

$n-i+1$ \ n	31	32	33	34	35	36	37	38	39	40
1	2.06	2.07	2.08	2.09	2.11	2.12	2.13	2.14	2.15	2.16
2	1.63	1.65	1.66	1.68	1.69	1.70	1.72	1.73	1.74	1.75
3	1.38	1.40	1.42	1.43	1.45	1.46	1.48	1.49	1.50	1.52
4	1.20	1.22	1.23	1.25	1.27	1.28	1.30	1.32	1.33	1.34
5	1.05	1.07	1.09	1.11	1.12	1.14	1.16	1.17	1.19	1.20
6	0.92	0.94	0.96	0.98	1.00	1.02	1.03	1.05	1.07	1.08
7	0.80	0.82	0.85	0.87	0.89	0.91	0.92	0.94	0.96	0.98
8	0.69	0.72	0.74	0.76	0.79˙	0.81	0.83	0.85	0.86	0.88
9	0.60	0.62	0.65	0.67	0.69	0.71	0.73	0.75	0.77	0.79
10	0.50	0.53	0.56	0.58	0.60	0.63	0.65	0.67	0.69	0.71
11	0.41	0.44	0.47	0.50	0.52	0.54	0.57	0.59	0.61	0.63
12	0.33	0.36	0.39	0.41	0.44	0.47	0.49	0.51	0.54	0.56
13	0.24	0.28	0.31	0.34	0.36	0.39	0.42	0.44	0.46	0.49
14	0.16	0.20	0.23	0.26	0.29	0.32	0.34	0.37	0.39	0.42
15	0.08	0.12	0.15	0.18	0.22	0.24	0.27	0.30	0.33	0.35
16	0.00	0.04	0.08	0.11	0.14	0.17	0.20	0.23	0.26	0.28
17	−0.08	−0.04	0.00	0.04	0.07	0.10	0.14	0.16	0.19	0.22
18	−0.16	−0.12	−0.08	−0.04	0.00	0.03	0.07	0.10	0.13	0.16
19	−0.24	−0.20	−0.15	−0.11	−0.07	−0.03	0.00	0.03	0.06	0.09
20	−0.33	−0.28	−0.23	−0.18	−0.14	−0.10	−0.07	−0.03	0.00	0.03
21	−0.41	−0.36	−0.31	−0.26	−0.22	−0.17	−0.14	−0.10	−0.06	−0.03

358 **Expected values of order statistics**

$n-i+1$ \ n	41	42	43	44	45	46	47	48	49	50
1	2.17	2.18	2.19	2.20	2.21	2.22	2.22	2.23	2.24	2.25
2	1.76	1.78	1.79	1.80	1.81	1.82	1.83	1.84	1.85	1.85
3	1.53	1.54	1.55	1.57	1.58	1.59	1.60	1.61	1.62	1.63
4	1.36	1.37	1.38	1.40	1.41	1.42	1.43	1.44	1.45	1.46
5	1.22	1.23	1.25	1.26	1.27	1.28	1.30	1.31	1.32	1.33
6	1.10	1.11	1.13	1.14	1.16	1.17	1.18	1.19	1.21	1.22
7	0.99	1.01	1.02	1.04	1.05	1.07	1.08	1.09	1.11	1.12
8	0.90	0.91	0.93	0.95	0.96	0.98	0.99	1.00	1.02	1.03
9	0.81	0.83	0.84	0.86	0.88	0.89	0.91	0.92	0.94	0.95
10	0.73	0.75	0.76	0.78	0.80	0.81	0.83	0.84	0.86	0.87
11	0.65	0.67	0.69	0.71	0.72	0.74	0.76	0.77	0.79	0.80
12	0.58	0.60	0.62	0.64	0.65	0.67	0.69	0.70	0.72	0.74
13	0.51	0.53'	0.55	0.57	0.59	0.60	0.62	0.64	0.66	0.67
14	0.44	0.46	0.48	0.50	0.52	0.54	0.56	0.58	0.59	0.61
15	0.37	0.40	0.42	0.44	0.46	0.48	0.50	0.52	0.53	0.55
16	0.31	0.33	0.36	0.38	0.40	0.42	0.44	0.46	0.48	0.49
17	0.25	0.27	0.29	0.32	0.34	0.36	0.38	0.40	0.42	0.44
18	0.18	0.21	0.23	0.26	0.28	0.30	0.32	0.34	0.36	0.38
19	0.12	0.15	0.17	0.20	0.22	0.25	0.27	0.29	0.31	0.33
20	0.06	0.09	0.12	0.14	0.17	0.19	0.21	0.24	0.26	0.28
21	0.00	0.03	0.06	0.09	0.11	0.14	0.16	0.18	0.21	0.23
22	−0.06	−0.03	0.00	0.03	0.06	0.08	0.11	0.13	0.15	0.18
23	−0.12	−0.09	−0.06	−0.03	0.00	0.03	0.05	0.08	0.10	0.13
24	−0.18	−0.15	−0.12	−0.09	−0.06	−0.03	0.00	0.03	0.05	0.07
25	−0.25	−0.21	−0.17	−0.14	−0.11	−0.08	−0.05	−0.03	0.00	0.02
26	−0.31	−0.27	−0.23	−0.20	−0.17	−0.14	−0.11	−0.08	−0.05	−0.02

$n-i+1$ \ n	60	70	80	90	100
1	2.32	2.38	2.43	2.47	2.51
2	1.94	2.00	2.06	2.11	2.15
3	1.72	1.79	1.85	1.90	1.95
4	1.56	1.63	1.70	1.75	1.80
5	1.43	1.51	1.58	1.64	1.69
6	1.32	1.41	1.48	1.54	1.59
7	1.23	1.32	1.39	1.45	1.51
8	1.14	1.24	1.31	1.38	1.43
9	1.07	1.16	1.24	1.31	1.37
10	1.00	1.10	1.18	1.25	1.31

n $n-i+1$	60	70	80	90	100
11	0.93	1.03	1.12	1.19	1.25
12	0.87	0.97	1.06	1.13	1.20
13	0.81	0.92	1.01	1.08	1.15
14	0.75	0.86	0.96	1.03	1.10
15	0.70	0.81	0.91	0.99	1.06
16	0.65	0.76	0.86	0.94	1.01
17	0.60	0.72	0.82	0.90	0.97
18	0.55	0.67	0.77	0.86	0.93
19	0.50	0.63	0.73	0.82	0.89
20	0.45	0.59	0.69	0.78	0.86
21	0.41	0.54	0.65	0.74	0.82
22	0.36	0.50	0.61	0.71	0.79
23	0.32	0.46	0.58	0.67	0.75
24	0.27	0.42	0.54	0.64	0.72
25	0.23	0.38	0.51	0.60	0.69
26	0.19	0.35	0.47	0.57	0.66
27	0.15	0.31	0.44	0.54	0.63
28	0.10	0.27	0.40	0.51	0.60
29	0.06	0.23	0.37	0.48	0.57
30	0.02	0.20	0.33	0.44	0.54
31	−0.02	0.16	0.30	0.41	0.51
32	−0.06	0.13	0.27	0.38	0.48
33	−0.10	0.09	0.24	0.35	0.45
34	−0.15	0.05	0.20	0.33	0.43
35	−0.19	0.02	0.17	0.30	0.40
36	−0.23	−0.02	0.14	0.27	0.37
37	−0.27	−0.05	0.11	0.24	0.34
38	−0.32	−0.09	0.08	0.21	0.32
39	−0.36	−0.13	0.05	0.18	0.29
40	−0.41	−0.16	0.02	0.15	0.27
41	−0.45	−0.20	−0.02	0.13	0.24
42	−0.50	−0.23	−0.05	0.10	0.21
43	−0.55	−0.27	−0.08	0.07	0.19
44	−0.60	−0.31	−0.11	0.04	0.16
45	−0.65	−0.35	−0.14	0.01	0.14
46	−0.70	−0.38	−0.17	−0.01	0.11
47	−0.75	−0.42	−0.20	−0.04	0.09
48	−0.81	−0.46	−0.24	−0.07	0.06
49	−0.87	−0.50	−0.27	−0.10	0.04
50	−0.93	−0.54	−0.30	−0.13	0.01
51	−1.00	−0.59	−0.33	−0.15	−0.01

Table A–17 Covariances ($\times 10^3$) between the ith and jth order statistics in a sample of size n drawn from the standard normal distribution (variances in boldface)

$n = 2$

i \\ j	1	2
1	**682**	318
2	318	**682**

$n = 3$

i \\ j	1	2	3
1	**559**	276	165
2	276	**449**	276
3	165	276	**559**

$n = 4$

i \\ j	1	2	3	4
1	**492**	246	158	105
2	246	**360**	236	158
3	158	236	**360**	246
4	105	158	246	**492**

$n = 5$

i \\ j	1	2	3	4	5
1	**448**	224	148	106	074
2	224	**312**	208	150	106
3	148	208	**287**	208	148
4	106	150	208	**312**	224
5	074	106	148	224	**448**

$n = 6$

i \\ j	1	2	3	4	5	6
1	**416**	209	139	102	077	056
2	209	**280**	189	140	106	077
3	139	189	**246**	183	140	102
4	102	140	183	**246**	189	139
5	077	106	140	189	**280**	209
6	056	077	102	139	209	**416**

$n = 7$

$i \backslash j$	1	2	3	4	5	6	7
1	**392**	196	132	098	077	060	045
2	196	**257**	174	131	102	080	060
3	132	174	**220**	166	130	102	077
4	098	131	166	**210**	166	131	098
5	077	102	130	166	**220**	174	132
6	060	080	102	131	174	**257**	196
7	045	060	077	098	132	196	**392**

$n = 8$

$i \backslash j$	1	2	3	4	5	6	7	8
1	**373**	186	126	095	075	060	048	037
2	186	**239**	163	123	098	079	063	048
3	126	163	**201**	152	121	098	079	060
4	095	123	152	**187**	149	121	098	075
5	075	098	121	149	**187**	152	123	095
6	060	079	098	121	152	**201**	163	126
7	048	063	079	098	123	163	**239**	186
8	037	048	060	075	095	126	186	**373**

$n = 9$

$i \backslash j$	1	2	3	4	5	6	7	8	9
1	**357**	178	121	091	073	059	049	040	031
2	178	**226**	154	117	093	077	063	052	040
3	121	154	**186**	142	114	093	077	063	049
4	091	117	142	**171**	137	113	093	077	059
5	073	093	114	137	**166**	137	114	093	073
6	059	077	093	113	137	**171**	142	117	091
7	049	063	077	093	114	142	**186**	154	121
8	040	052	063	077	093	117	154	**226**	178
9	031	040	049	059	073	091	121	178	**357**

362   **Covariances between order statistics**

$$n = 10$$

$\diagdown$_i	1	2	3	4	5	6	7	8	9	10
1	**344**	171	116	088	071	058	049	041	034	027
2	171	**215**	147	112	090	074	062	052	043	034
3	116	147	**175**	134	108	089	075	063	052	041
4	088	112	134	**158**	128	106	089	075	062	049
5	071	090	108	128	**151**	126	106	089	074	058
6	058	074	089	106	126	**151**	128	108	090	071
7	049	062	075	089	106	128	**158**	134	112	088
8	041	052	063	075	089	108	134	**175**	147	116
9	034	043	052	062	074	090	112	147	**215**	171
10	027	034	041	049	058	071	088	116	171	**344**

Table A–18 Divisors d_n for the range compared with $\sqrt{n}$ to estimate standard deviation; and a divisor for range to make a substitute s for a substitute t-test.

Sample size	d_n	$\sqrt{n}$	Divisor to get substitute s	Sample size	d_n
2	1.13	1.41	1.41	110	5.08
3	1.69	1.73	1.91	120	5.14
4	2.06	2.00	2.24	130	5.20
5	2.33	2.24	2.48	140	5.25
6	2.53	2.45	2.67	150	5.30
7	2.70	2.65	2.83	160	5.34
8	2.85	2.83	2.96	170	5.38
9	2.97	3.00	3.08	180	5.42
10	3.08	3.16	3.18	190	5.46
11	3.17	3.32	3.27	200	5.49
12	3.26	3.46	3.35	250	5.64
13	3.34	3.61	3.42	300	5.76
14	3.41	3.74	3.49	350	5.85
15	3.47	3.87	3.55	400	5.94
16	3.53	4.00	3.61	450	6.01
17	3.59	4.12	3.66	500	6.07
18	3.64	4.24	3.71	600	6.18
19	3.69	4.36	3.76	700	6.28
20	3.73	4.47	3.81	800	6.35
30	4.09		4.17	900	6.42
40	4.32		4.40	1000	6.48
50	4.50		4.57		
60	4.64		4.71		
70	4.75		4.84		
80	4.85		4.94		
90	4.94		5.02		
100	5.02		5.08		

Table A–19 Ordinates of the normal distribution ($\times 10^4$)

$$f(z) = \frac{1}{\sqrt{2\pi}} e^{-z^2/2}$$

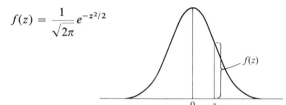

Example.

$z = 0.43, f(z) = 0.3637$

$z = -0.43, f(z) = 0.3637$

z	0.00	0.01	0.02	0.03	0.04	0.05	0.06	0.07	0.08	0.09
0.0	3989	3989	3989	3988	3986	3984	3982	3980	3977	3973
0.1	3970	3965	3961	3956	3951	3945	3939	3932	3925	3918
0.2	3910	3902	3894	3885	3876	3867	3857	3847	3836	3825
0.3	3814	3802	3790	3778	3765	3752	3739	3725	3712	3697
0.4	3683	3668	3653	3637	3621	3605	3589	3572	3555	3538
0.5	3521	3503	3485	3467	3448	3429	3410	3391	3372	3352
0.6	3332	3312	3292	3271	3251	3230	3209	3187	3166	3144
0.7	3123	3101	3079	3056	3034	3011	2989	2966	2943	2920
0.8	2897	2874	2850	2827	2803	2780	2756	2732	2709	2685
0.9	2661	2637	2613	2589	2565	2541	2516	2492	2468	2444
1.0	2420	2396	2371	2347	2323	2299	2275	2251	2227	2203
1.1	2179	2155	2131	2107	2083	2059	2036	2012	1989	1965
1.2	1942	1919	1895	1872	1849	1826	1804	1781	1758	1736
1.3	1714	1691	1669	1647	1626	1604	1582	1561	1539	1518
1.4	1497	1476	1456	1435	1415	1394	1374	1354	1334	1315
1.5	1295	1276	1257	1238	1219	1200	1182	1163	1145	1127
1.6	1109	1092	1074	1057	1040	1023	1006	0989	0973	0957
1.7	0940	0925	0909	0893	0878	0863	0848	0833	0818	0804
1.8	0790	0775	0761	0748	0734	0721	0707	0694	0681	0669
1.9	0656	0644	0632	0620	0608	0596	0584	0573	0562	0551
2.0	0540	0529	0519	0508	0498	0488	0478	0468	0459	0449
2.1	0440	0431	0422	0413	0404	0396	0387	0379	0371	0363
2.2	0355	0347	0339	0332	0325	0317	0310	0303	0297	0290
2.3	0283	0277	0270	0264	0258	0252	0246	0241	0235	0229
2.4	0224	0219	0213	0208	0203	0198	0194	0189	0184	0180
2.5	0175	0171	0167	0163	0158	0154	0151	0147	0143	0139
2.6	0136	0132	0129	0126	0122	0119	0116	0113	0110	0107
2.7	0104	0101	0099	0096	0093	0091	0088	0086	0084	0081
2.8	0079	0077	0075	0073	0071	0069	0067	0065	0063	0061
2.9	0060	0058	0056	0055	0053	0051	0050	0048	0047	0046
3.0	0044	0043	0042	0040	0039	0038	0037	0036	0035	0034

Table A–20 Squares. Moving the decimal point *one* place in n is equivalent to moving it *two* places in n^2.

Examples. $3.2^2 = 10.24$, $32^2 = 1024$, $0.32^2 = 0.1024$

n	0	1	2	3	4	5	6	7	8	9	10	|Tenths 1	2	3	4	5
1.0	1.000	1.020	1.040	1.061	1.082	1.103	1.124	1.145	1.166	1.188	1.210	2	4	6	8	10
1.1	1.210	1.232	1.254	1.277	1.300	1.323	1.346	1.369	1.392	1.416	1.440	2	5	7	9	11
1.2	1.440	1.464	1.488	1.513	1.538	1.563	1.588	1.613	1.638	1.664	1.690	2	5	7	10	12
1.3	1.690	1.716	1.742	1.769	1.796	1.823	1.850	1.877	1.904	1.932	1.960	3	5	8	11	13
1.4	1.960	1.988	2.016	2.045	2.074	2.103	2.132	2.161	2.190	2.220	2.250	3	6	9	12	14
1.5	2.250	2.280	2.310	2.341	2.372	2.403	2.434	2.465	2.496	2.528	2.560	3	6	9	12	15
1.6	2.560	2.592	2.624	2.657	2.690	2.723	2.756	2.789	2.822	2.856	2.890	3	7	10	13	16
1.7	2.890	2.924	2.958	2.993	3.028	3.063	3.098	3.133	3.168	3.204	3.240	3	7	10	14	17
1.8	3.240	3.276	3.312	3.349	3.386	3.423	3.460	3.497	3.534	3.572	3.610	4	7	11	15	18
1.9	3.61	3.65	3.69	3.72	3.76	3.80	3.84	3.88	3.92	3.96	4.00	0	1	1	2	2
2.0	4.00	4.04	4.08	4.12	4.16	4.20	4.24	4.28	4.33	4.37	4.41	0	1	1	2	2
2.1	4.41	4.45	4.49	4.54	4.58	4.62	4.67	4.71	4.75	4.80	4.84	0	1	1	2	2
2.2	4.84	4.88	4.93	4.97	5.02	5.06	5.11	5.15	5.20	5.24	5.29	0	1	1	2	2
2.3	5.29	5.34	5.38	5.43	5.48	5.52	5.57	5.62	5.66	5.71	5.76	0	1	1	2	2
2.4	5.76	5.81	5.86	5.90	5.95	6.00	6.05	6.10	6.15	6.20	6.25	0	1	1	2	2
2.5	6.25	6.30	6.35	6.40	6.45	6.50	6.55	6.60	6.66	6.71	6.76	1	1	2	2	3
2.6	6.76	6.81	6.86	6.92	6.97	7.02	7.08	7.13	7.18	7.24	7.29	1	1	2	2	3
2.7	7.29	7.34	7.40	7.45	7.51	7.56	7.62	7.67	7.73	7.78	7.84	1	1	2	2	3
2.8	7.84	7.90	7.95	8.01	8.07	8.12	8.18	8.24	8.29	8.35	8.41	1	1	2	2	3
2.9	8.41	8.47	8.53	8.58	8.64	8.70	8.76	8.82	8.88	8.94	9.00	1	1	2	2	3
3.0	9.00	9.06	9.12	9.18	9.24	9.30	9.36	9.42	9.49	9.55	9.61	1	1	2	2	3
3.1	9.61	9.67	9.73	9.80	9.86	9.92	9.99	10.05	10.11	10.18	10.24	1	1	2	3	3
3.2	10.24	10.30	10.37	10.43	10.50	10.56	10.63	10.69	10.76	10.82	10.89	1	1	2	3	3
3.3	10.89	10.96	11.02	11.09	11.16	11.22	11.29	11.36	11.42	11.49	11.56	1	1	2	3	3
3.4	11.56	11.63	11.70	11.76	11.83	11.90	11.97	12.04	12.11	12.18	12.25	1	1	2	3	3
3.5	12.25	12.32	12.39	12.46	12.53	12.60	12.67	12.74	12.82	12.89	12.96	1	1	2	3	4
3.6	12.96	13.03	13.10	13.18	13.25	13.32	13.40	13.47	13.54	13.62	13.69	1	1	2	3	4
3.7	13.69	13.76	13.84	13.91	13.99	14.06	14.14	14.21	14.29	14.36	14.44	1	1	2	3	4
3.8	14.44	14.52	14.59	14.67	14.75	14.82	14.90	14.98	15.05	15.13	15.21	1	2	2	3	4
3.9	15.21	15.29	15.37	15.44	15.52	15.60	15.68	15.76	15.84	15.92	16.00	1	2	2	3	4
4.0	16.00	16.08	16.16	16.24	16.32	16.40	16.48	16.56	16.65	16.73	16.81	1	2	2	3	4
4.1	16.81	16.89	16.97	17.06	17.14	17.22	17.31	17.39	17.47	17.56	17.64	1	2	2	3	4
4.2	17.64	17.72	17.81	17.89	17.98	18.06	18.15	18.23	18.32	18.40	18.49	1	2	3	3	4
4.3	18.49	18.58	18.66	18.75	18.84	18.92	19.01	19.10	19.18	19.27	19.36	1	2	3	3	4
4.4	19.36	19.45	19.54	19.62	19.71	19.80	19.89	19.98	20.07	20.16	20.25	1	2	3	4	4
4.5	20.25	20.34	20.43	20.52	20.61	20.70	20.79	20.88	20.98	21.07	21.16	1	2	3	4	5
4.6	21.16	21.25	21.34	21.44	21.53	21.62	21.72	21.81	21.90	22.00	22.09	1	2	3	4	5
4.7	22.09	22.18	22.28	22.37	22.47	22.56	22.66	22.75	22.85	22.94	23.04	1	2	3	4	5
4.8	23.04	23.14	23.23	23.33	23.43	23.52	23.62	23.72	23.81	23.91	24.01	1	2	3	4	5
4.9	24.01	24.11	24.21	24.30	24.40	24.50	24.60	24.70	24.80	24.90	25.00	1	2	3	4	5
5.0	25.00	25.10	25.20	25.30	25.40	25.50	25.60	25.70	25.81	25.91	26.01	1	2	3	4	5
5.1	26.01	26.11	26.21	26.32	26.42	26.52	26.63	26.73	26.83	26.94	27.04	1	2	3	4	5
5.2	27.04	27.14	27.25	27.35	27.46	27.56	27.67	27.77	27.88	27.98	28.09	1	2	3	4	5
5.3	28.09	28.20	28.30	28.41	28.52	28.62	28.73	28.84	28.94	29.05	29.16	1	2	3	4	5
5.4	29.16	29.27	29.38	29.48	29.59	29.70	29.81	29.92	30.03	30.14	30.25	1	2	3	4	5
5.5	30.25	30.36	30.47	30.58	30.69	30.80	30.91	31.02	31.14	31.25	31.36	1	2	3	4	6
5.6	31.36	31.47	31.58	31.70	31.81	31.92	32.04	32.15	32.26	32.38	32.49	1	2	3	5	6
5.7	32.49	32.60	32.72	32.83	32.95	33.06	33.18	33.29	33.41	33.52	33.64	1	2	3	5	6
5.8	33.64	33.76	33.87	33.99	34.11	34.22	34.34	34.46	34.57	34.69	34.81	1	2	4	5	6
5.9	34.81	34.93	35.05	35.16	35.28	35.40	35.52	35.64	35.76	35.88	36.00	1	2	4	5	6
6.	36.0	37.2	38.4	39.7	41.0	42.3	43.6	44.9	46.2	47.6	49.0	1	3	4	5	6
7.	49.0	50.4	51.8	53.3	54.8	56.3	57.8	59.3	60.8	62.4	64.0	1	3	4	6	7
8.	64.0	65.6	67.2	68.9	70.6	72.3	74.0	75.7	77.4	79.2	81.0	2	3	5	7	8
9.	81.0	82.8	84.6	86.5	88.4	90.3	92.2	94.1	96.0	98.0	100.0	2	4	6	8	9

Table A–21 Square roots. Moving the decimal point *two* places in *n* is equivalent to moving it *one place* in $\sqrt{n}$.

Examples. $\sqrt{2.56} = 1.600, \sqrt{256} = 16.00, \sqrt{0.0256} = 0.16$

n	0	1	2	3	4	5	6	7	8	9	10	Tenths of the tabular difference 1	2	3	4	5
.1	.316	.332	.346	.361	.374	.387						1	3	4	6	7
							.387	.400	.412	.424	.436	.447	1	2	4	5	6
.2	.447	.458	.469	.480	.490	.500	.510	.520	.529	.539	.548	1	2	3	4	5
.3	.548	.557	.566	.574	.583	.592	.600	.608	.616	.624	.632	1	2	3	3	4
.4	.632	.640	.648	.656	.663	.671	.678	.686	.693	.700	.707	1	1	2	3	4
.5	.707	.714	.721	.728	.735	.742	.748	.755	.762	.768	.775	1	1	2	3	3
.6	.775	.781	.787	.794	.800	.806	.812	.819	.825	.831	.837	1	1	2	2	3
.7	.837	.843	.849	.854	.860	.866	.872	.877	.883	.889	.894	1	1	2	2	3
.8	.894	.900	.906	.911	.917	.922	.927	.933	.938	.943	0.949	1	1	2	2	3
.9	0.949	0.954	0.959	0.964	0.970	0.975	0.980	0.985	0.990	0.995	1.000	1	1	2	2	3
1.0	1.000	1.005	1.010	1.015	1.020	1.025	1.030	1.034	1.039	1.044	1.049	0	1	1	2	2
1.1	1.049	1.054	1.058	1.063	1.068	1.072	1.077	1.082	1.086	1.091	1.095	0	1	1	2	2
1.2	1.095	1.100	1.105	1.109	1.114	1.118	1.122	1.127	1.131	1.136	1.140	0	1	1	2	2
1.3	1.140	1.145	1.149	1.153	1.158	1.162	1.166	1.170	1.175	1.179	1.183	0	1	1	2	2
1.4	1.183	1.187	1.192	1.196	1.200	1.204	1.208	1.212	1.217	1.221	1.225	0	1	1	2	2
1.5	1.225	1.229	1.233	1.237	1.241	1.245	1.249	1.253	1.257	1.261	1.265	0	1	1	2	2
1.6	1.265	1.269	1.273	1.277	1.281	1.285	1.288	1.292	1.296	1.300	1.304	0	1	1	2	2
1.7	1.304	1.308	1.311	1.315	1.319	1.323	1.327	1.330	1.334	1.338	1.342	0	1	1	2	2
1.8	1.342	1.345	1.349	1.353	1.356	1.360	1.364	1.367	1.371	1.375	1.378	0	1	1	1	2
1.9	1.378	1.382	1.386	1.389	1.393	1.396	1.400	1.404	1.407	1.411	1.414	0	1	1	1	2
2.0	1.414	1.418	1.421	1.425	1.428	1.432	1.435	1.439	1.442	1.446	1.449	0	1	1	1	2
2.1	1.449	1.453	1.456	1.459	1.463	1.466	1.470	1.473	1.476	1.480	1.483	0	1	1	1	2
2.2	1.483	1.487	1.490	1.493	1.497	1.500	1.503	1.507	1.510	1.513	1.517	0	1	1	1	2
2.3	1.517	1.520	1.523	1.526	1.530	1.533	1.536	1.539	1.543	1.546	1.549	0	1	1	1	2
2.4	1.549	1.552	1.556	1.559	1.562	1.565	1.568	1.572	1.575	1.578	1.581	0	1	1	1	2
2.5	1.581	1.584	1.587	1.591	1.594	1.597	1.600	1.603	1.606	1.609	1.612	0	1	1	1	2
2.6	1.612	1.616	1.619	1.622	1.625	1.628	1.631	1.634	1.637	1.640	1.643	0	1	1	1	2
2.7	1.643	1.646	1.649	1.652	1.655	1.658	1.661	1.664	1.667	1.670	1.673	0	1	1	1	2
2.8	1.673	1.676	1.679	1.682	1.685	1.688	1.691	1.694	1.697	1.700	1.703	0	1	1	1	1
2.9	1.703	1.706	1.709	1.712	1.715	1.718	1.720	1.723	1.726	1.729	1.732	0	1	1	1	1
3.0	1.732	1.735	1.738	1.741	1.744	1.746	1.749	1.752	1.755	1.758	1.761	0	1	1	1	1
3.1	1.761	1.764	1.766	1.769	1.772	1.775	1.778	1.780	1.783	1.786	1.789	0	1	1	1	1
3.2	1.789	1.792	1.794	1.797	1.800	1.803	1.806	1.808	1.811	1.814	1.817	0	1	1	1	1
3.3	1.817	1.819	1.822	1.825	1.828	1.830	1.833	1.836	1.838	1.841	1.844	0	1	1	1	1
3.4	1.844	1.847	1.849	1.852	1.855	1.857	1.860	1.863	1.865	1.868	1.871	0	1	1	1	1
3.5	1.871	1.873	1.876	1.879	1.881	1.884	1.887	1.889	1.892	1.895	1.897	0	1	1	1	1
3.6	1.897	1.900	1.903	1.905	1.908	1.910	1.913	1.916	1.918	1.921	1.924	0	1	1	1	1
3.7	1.924	1.926	1.929	1.931	1.934	1.936	1.939	1.942	1.944	1.947	1.949	0	1	1	1	1
3.8	1.949	1.952	1.954	1.957	1.960	1.962	1.965	1.967	1.970	1.972	1.975	0	1	1	1	1
3.9	1.975	1.977	1.980	1.982	1.985	1.987	1.990	1.992	1.995	1.997	2.000	0	1	1	1	1
4.0	2.000	2.002	2.005	2.007	2.010	2.012	2.015	2.017	2.020	2.022	2.025	0	0	1	1	1
4.1	2.025	2.027	2.030	2.032	2.035	2.037	2.040	2.042	2.045	2.047	2.049	0	0	1	1	1
4.2	2.049	2.052	2.054	2.057	2.059	2.062	2.064	2.066	2.069	2.071	2.074	0	0	1	1	1
4.3	2.074	2.076	2.078	2.081	2.083	2.086	2.088	2.090	2.093	2.095	2.098	0	0	1	1	1
4.4	2.098	2.100	2.102	2.105	2.107	2.110	2.112	2.114	2.117	2.119	2.121	0	0	1	1	1
4.5	2.121	2.124	2.126	2.128	2.131	2.133	2.135	2.138	2.140	2.142	2.145	0	0	1	1	1
4.6	2.145	2.147	2.149	2.152	2.154	2.156	2.159	2.161	2.163	2.166	2.168	0	0	1	1	1
4.7	2.168	2.170	2.173	2.175	2.177	2.179	2.182	2.184	2.186	2.189	2.191	0	0	1	1	1
4.8	2.191	2.193	2.195	2.198	2.200	2.202	2.205	2.207	2.209	2.211	2.214	0	0	1	1	1
4.9	2.214	2.216	2.218	2.220	2.223	2.225	2.227	2.229	2.232	2.234	2.236	0	0	1	1	1
5.	2.236	2.258	2.280	2.302	2.324	2.345	2.366	2.387	2.408	2.429	2.449	2	4	6	9	11
6.	2.449	2.470	2.490	2.510	2.530	2.550	2.569	2.588	2.608	2.627	2.646	2	4	6	8	10
7.	2.646	2.665	2.683	2.702	2.720	2.739	2.757	2.775	2.793	2.811	2.828	2	4	5	7	9
8.	2.828	2.846	2.864	2.881	2.898	2.915	2.933	2.950	2.966	2.983	3.000	2	3	5	7	9
9.	3.000	3.017	3.033	3.050	3.066	3.082	3.098	3.114	3.130	3.146	3.162	2	3	5	6	8

Table A–22   1000 random digits†

Line	Column number				
number	00–09	10–19	20–29	30–39	40–49
00	15544 80712	97742 21500	97081 42451	50623 56071	28882 28739
01	01011 21285	04729 39986	73150 31548	30168 76189	56996 19210
02	47435 53308	40718 29050	74858 64517	93573 51058	68501 42723
03	91312 75137	86274 59834	69844 19853	06917 17413	44474 86530
04	12775 08768	80791 16298	22934 09630	98862 39746	64623 32768
05	31466 43761	94872 92230	52367 13205	38634 55882	77518 36252
06	09300 43847	40881 51243	97810 18903	53914 31688	06220 40422
07	73582 13810	57784 72454	68997 72229	30340 08844	53924 89630
08	11092 81392	58189 22697	41063 09451	09789 00637	06450 85990
09	93322 98567	00116 35605	66790 52965	62877 21740	56476 49296
10	80134 12484	67089 08674	70753 90959	45842 59844	45214 36505
11	97888 31797	95037 84400	76041 96668	75920 68482	56855 97417
12	92612 27082	59459 69380	98654 20407	88151 56263	27126 63797
13	72744 45586	43279 44218	83638 05422	00995 70217	78925 39097
14	96256 70653	45285 26293	78305 80252	03625 40159	68760 84716
15	07851 47452	66742 83331	54701 06573	98169 37499	67756 68301
16	25594 41552	96475 56151	02089 33748	65239 89956	89559 33687
17	65358 15155	59374 80940	03411 94656	69440 47156	77115 99463
18	09402 31008	53424 21928	02198 61201	02457 87214	59750 51330
19	97424 90765	01634 37328	41243 33564	17884 94747	93650 77668

† A good source of random digits and random normal deviates is *A Million Random Digits with 100,000 Normal Deviates*, The RAND Corporation. Glencoe, Ill.: Free Press of Glencoe, 1955.

Table A–23 200 random normal deviates, $\mu = 0$, $\sigma = 1$†

Line number	0	1	2	3	4	5	6	7	8	9
					Column number					
00	−0.56	+1.51	−0.35	−0.51	−0.27	+0.24	+0.13	+2.10	−0.28	−0.98
01	−0.55	+0.91	−0.55	+1.82	+0.03	+1.26	−1.27	−0.90	−0.77	+0.88
02	+0.74	−1.33	+0.19	−0.33	−0.86	−1.39	+0.07	+0.35	+0.46	−0.24
03	+2.56	−0.11	+1.37	−0.86	−0.40	+0.10	+0.13	−0.32	+2.00	+0.32
04	−1.29	−0.02	+0.17	+1.14	−2.00	−0.19	−0.38	+0.88	−0.34	−0.62
05	+1.25	+0.81	+0.16	−1.43	+1.76	+0.31	−0.38	−0.04	+1.61	−0.33
06	−0.76	+0.82	−1.20	−1.39	−1.23	+2.55	+0.08	+1.29	+2.06	+0.76
07	−1.15	+0.36	+0.57	−2.46	+0.22	+0.54	+1.33	+1.65	−0.35	+0.34
08	−1.77	+0.25	−1.16	−0.81	+1.13	+0.80	+0.61	+1.57	+0.31	+0.91
09	−0.06	+1.21	−0.34	−0.19	+1.03	+0.35	+1.40	−0.13	+0.47	−1.09
10	−1.23	+0.30	−1.13	+0.98	−0.29	+0.05	−0.85	−0.20	+2.00	+0.28
11	−0.14	−0.57	+0.46	−0.12	−1.84	+0.20	+0.41	−0.57	−1.52	+0.75
12	−0.02	−0.44	−0.15	+0.99	+0.29	+0.07	+1.36	+0.28	−0.02	−1.20
13	+1.45	+2.24	−1.01	−0.81	−0.07	−0.60	−0.20	+1.04	−0.57	+0.08
14	−0.64	+1.49	−0.37	−0.92	+0.58	+0.69	−0.19	+1.89	−0.06	−0.56
15	−0.41	+0.56	−0.22	−0.45	−0.09	+1.20	−0.81	−0.42	+0.70	−0.71
16	+0.03	−1.77	−0.52	−0.41	+1.87	−0.19	−1.66	−0.31	+1.27	+0.55
17	+1.41	−0.51	−0.21	−0.86	−0.01	+0.87	+0.58	+0.48	−0.02	+0.11
18	−0.54	−0.10	−0.52	−0.76	+1.53	−1.07	−0.67	+0.70	−1.17	+0.54
19	+1.23	+0.31	−1.97	+0.81	−0.78	+0.52	−2.33	+1.30	+0.51	+2.02

† A good source of random digits and random normal deviates is *A Million Random Digits with 100,000 Normal Deviates*, The RAND Corporation. Glencoe, Ill.: Free Press of Glencoe, 1955.

Glossary

≈   approximately equals

$|x|$ absolute value of x

$[x]$ greatest integer contained in x, where noted

$\binom{n}{r}$ binomial coefficient

d.f.   degrees of freedom

$|$ given

★   a long or difficult exercise, project, or section

Random variables, probabilities, and distributions

N number in population

n number in sample

$P(E)$ probability of event E

X (also Y, U, V, etc.) a random variable

x (also y, u, v, etc.) a value of the capitalized random variable

F (also G, H, etc.) cumulative distribution function (cdf)

f (also g, h, etc.) probability density function (pdf)

Z often standard normal random variable ($\mu = 0$, $\sigma = 1$)

z a value of Z

Sample and population means, variances, covariances, and correlations

$\overline{X}$ sample mean of X

$\overline{x}$ a value of $\overline{X}$

$E(X)$ expected value of X; $E(X) = \mu_X$

Var (X)	variance of X
μ	population mean, often with subscript
σ	population standard deviation, often with subscript
s	sample standard deviation, often with subscript
σ^2	population variance
s^2	sample variance
Cov (X, Y)	covariance of X and Y
ρ	population correlation coefficient
$\rho_{X,Y}$	correlation between X and Y
r	sample correlation coefficient
r_S	rank correlation coefficient

Special statistics and parameters

χ^2	chi-square statistic, often with d.f. as subscript
χ_r^2	Friedman index
H	Kruskal-Wallis statistic
$\hat{p}$	maximum likelihood estimate (MLE) of p
$\bar{p}$	observed fraction of successes
t	Student's statistic
T	often the Mann-Whitney statistic
W	often the Wilcoxon statistic
W_S	alternative Wilcoxon statistic
U_i	ith order statistic
Y_i	probability to the left of U_i
x_p	pth quantile, $P(X \leq x_p) = p$, for continuous distributions
$\tilde{x}$	sample median

References

A. Elementary introductions to probability and statistics

Freund, J. E., *Statistics*. Englewood Cliffs, N.J.: Prentice-Hall, 1970.

Goldberg, S., *Probability: An Introduction*. Englewood Cliffs, N.J.: Prentice-Hall, 1960.

Hodges, J. L., Jr., and E. L. Lehmann, *Basic Concepts of Probability and Statistics*. San Francisco: Holden-Day, 1970.

Mosteller, F., R. E. K. Rourke, and G. B. Thomas, Jr., *Probability with Statistical Applications*, 2nd ed. Reading, Mass.: Addison-Wesley, 1970.

Noether, G. E., *Introduction to Statistics*. Boston: Houghton-Mifflin, 1971.

Wonnacott, T. H., and R. J. Wonnacott, *Introductory Statistics*. New York: Wiley, 1969.

B. Statistical methods

Bradley, J. V., *Distribution-Free Statistical Tests*. Englewood Cliffs, N.J.: Prentice-Hall, 1968.

Conover, W. J., *Practical Nonparametric Statistics*. New York: Wiley, 1971.

David, H. A., *Order Statistics*. New York: Wiley, 1970.

Dixon, W. J., and F. J. Massey, Jr., *Introduction to Statistical Analysis*, 3rd ed. New York: McGraw-Hill, 1969.

Gibbons, J., *Nonparametric Statistical Inference*. New York: McGraw-Hill, 1971.

Hajek, J., *Nonparametric Statistics*. San Francisco: Holden-Day, 1969.

Kendall, M. G., *Rank Correlation Methods*, 2nd ed. New York: Hafner, 1955.

Kraft, C. H., and C. van Eeden, *A Nonparametric Introduction to Statistics*. New York: Macmillan, 1968.

Mosteller, F., and J. W. Tukey, "Data analysis, including statistics," Chapter 10 in G. Lindzey and E. Aronson, *The Handbook of Social Psychology*, 2nd ed., Vol. 2. Reading, Mass.: Addison-Wesley, 1968, pp. 80–203.

Noether, G. E., *Elements of Nonparametric Statistics.* New York: Wiley, 1967.

Savage, I. R., *Bibliography of Nonparametric Statistics.* Cambridge, Mass.: Harvard University Press, 1962.

Siegel, S., *Nonparametric Statistics for the Behavioral Sciences.* New York: McGraw-Hill, 1956.

Snedecor, G. W., and W. G. Cochran, *Statistical Methods*, 6th ed. Ames, Iowa: Iowa State College Press, 1967. (Applications to experiments in agriculture and biology.)

Short Answers to Selected Odd-Numbered Exercises

Section 1–3

1. 16.1%
3. Table 1–1 omits class intervals containing less than 0.05% of the cases.

Section 1–4

1. 0.3413   3. 0.3808   5. 0.3808   7. 0.6892
9. (a) 0.1151, (b) 0.0668, (c) 0.9544   11. $k \approx 0.524$; $k \approx 1.5$

Section 1–5

1. One-sided descriptive level 2/6   3. 5/84   5. (a) 3/10, (b) 4/10
7. 4/10   9. $m! \, n!/(m + n)!$

Section 1–6

1. (a) 0.032, (b) 0.037   3. 0.057, 0.114

Section 1–7

1. 0.0192, 0.0384   7. $P(|\bar{Y} - \bar{A}| \geq 0.041) \approx 0.20$

Section 2–2

1. 0.984, 0.026   3. $0.8/\sqrt{n}$ 7. 0.255, 0.345

Section 2–3

1. No   3. No

Section 2–4

1. 0.930   3. 0.253

5.

	Significance level	Power for $p = 0.8$
a)	0.042	0.804
b)	0.021	0.804
c)	0.058	0.913
d)	0.132	0.968

7. 0.048   9. Reduce rejection number

Section 2–5

1. (a) 0.055, (b) 0.678   3. 0.069, 0.678   5. (a) 0.35, (b) 0.55, (c) 0.86

Section 2–6

3. $P(t \geq 1.38) = 0.10$, 11 d.f.
$P(X \geq 8 | p = \frac{1}{2}) = 0.194$

Section 2–7

1. 0.0339, 0.1186   3. 0.0484   5. (a) 0.0348, (b) 0.000363, (c) 95.9
7. $-1/5$

Section 2–8

1. 0.1664   3. 2, 3, 5; 0.5599   5. (a) 0.3649, (b) 0.006, (c) 0.9998

Section 2–9

7. 0.032

Section 3–1

1. $N(N + 1)/2$ 3. Sum to $N(N + 1)/2$

Section 3–2

1. 0.60, 0.77
3. $P(t)$: $\frac{1}{6} \; \frac{1}{6} \; \frac{2}{6} \; \frac{1}{6} \; \frac{1}{6}$
     t 3 4 5 6 7
5. $\dfrac{(m + n)!}{m! \, n!}$
7. (a) 34, (b) 7, (c) 7/210 = 0.033

Section 3–3

1. 0.238, 0.238   3. (a) 0.012, (b) 0.206, (c) 0.464   5. $T \geq 44$
7. 0.970   9. 0.0238, close agreement   11. Accept

Section 3–4

1. 4/11, 4/11   3. 1/3

7. $P(T = t | t \text{ odd and } t \leq N + 1) = \dfrac{t - 1}{N(N - 1)}$

$P(T = t | t \text{ odd and } t \geq N + 1) = \dfrac{2N - t + 1}{N(N - 1)}$

$P(T = t | t \text{ even and } t \leq N + 1) = \dfrac{t - 2}{N(N - 1)}$

$P(T = t | t \text{ even and } t \geq N + 1) = \dfrac{2N - t}{N(N - 1)}$

Section 3–6

1. True   3. False   5. (a) 0.984, (b) 0.956, (c) 0.866, (d) 0.702
7. $a = 0$ 9. 0.944

Section 4–1

1. $\mu_T = 6, \sigma_T^2 = 3$ 3. $\mu_T = 5, \sigma_T^2 = 5/3$
5. 13.5, 11.25   7. $T = 118, \mu_T = 105, \sigma_T^2 = 175$

Section 4–2

1. 0.247, 0.258   3. 0.0512, 0.048
5. $P(T \geq 57) \approx P(Z \geq 2.80) = 0.003$ 7. $P(T \geq 49) \approx P(Z \geq 1.52) = 0.064$
9. $P(T \geq 34) \approx P(Z \geq 1.25) = 0.106$ close to 0.111

Section 4–3

1. $\mu_V = 0.918$ 5. $E(V) = \frac{1}{2}$ 7. $V = 0.76$

Section 4–4

1. $\mu_T = 6, \sigma_T^2 = 2.85$ 3. $\mu_T = 6, \sigma_T^2 = 2.25$ 5. 65
7. $10\frac{5}{112}$ 9. $P \approx 0.28$, accept

Section 5–1

1. $W = 7$ 3. $W = -7, W_S = 7$ 5. $W = 24, W_S = 27$

Section 5–2

1. $P(W = -6) = 0.125$ 5. $P(W \geq 19) = 2/64, P(W = 21) = 1/64$

Section 5–3

1. (a) 0.078, (b) 0.064   3. (a) 0.010, (b) 0.012   5. (a) accept, (b) reject
7. (a) No, (b) Yes

Section 5–4

1. $h = 0$ 3. $n = 4$: $h \approx 2.3$; $n = 10$: $h \approx 1.1$; $n = 25$: $h \approx 0.7$
5. $(140 - 125)/10.46 = 1.43$ 9. 0.51 (exact), 0.507 (simulated)

Section 6–1

1. (a) 18, (b) 0.1   3. Part (a) agrees; part (b) $n = 5$ gave 40, $n = 6$ gave 70.
5. (a) 8, (b) 20, (c) 40, (d) 70   7. 0.77   9. $r_S' = 100\, r_S$

Section 6–2

3. 0.92
5. Let F be finishing ranking and A, B, C rankings of the three contestants. $r_{FA} = 0.37$, $r_{FB} = 0.54$, $r_{FC} = 0.14$. C gives prize to B.
7. 0.75   9. $2n(n^2 - 1)/6$

Section 6–3

1. 1, 1/2, 1/2, $-1/2$, $-1/2$, -1
3. 2/6 counting 1 and -1 as perfect; 1/6 if only 1 is counted.
5. $E(r_S) = 0$; Var $(r_S) = 1/2$

Section 6–4

1. $P(r_S \geq 0.81) \approx P(Z \geq 2.15) = 0.016$; reject   3. 0.059; accept
5. 0.098   7. $P(r_S \geq 0.054) \approx 0.41$, and we could accept the null hypothesis.

Section 6–5

1. $p \approx 0.36$ gives for $n = 5$: power 0.13 and for $n = 10$: power 0.25.

Section 7–2

1. 0, 407.7; 1, 256.3; 2, 81.1; 3 or more, 19.9   3. $\bar{x} = 0.63$, $s^2 = 0.65$
5. $P(X = 0) \approx 0.010$
   $P(X \leq 3) \approx 0.315$
7. $s^2 = 1.21$, near $\bar{x} = 1.05$
9. (a) $P(X = 0) = 0.819$, (b) $P(X \geq 3) = 0.001$
11. $(1/m)P(X \geq x + 1|m)$, where P is a Poisson cumulative probability.
13. 2 or "2 to 1"; 0.2 or "1 to 5"   15. 0.05   17. (a) 0.189, (b) 0.235
19. $P(X \geq 21|m = 14) = 0.048$, and so 20 lines should be provided.
21. $P(X \geq 2|m = 1) = 0.264$

Section 7–3

1. (a) 0.126, 0.191; (b) 0.128, 0.184
3. (a) $P(0) = 0.607$, $P(1) = 0.303$, $P(X \geq 1) = 0.393$

5. $P(X = 0|m = 3) = 0.050; n = 97$
7. (a) 0.122, (b) 0.135; lack of independence

Section 7–4

1. 0.819   3. (a) 0.063, (b) 0.933

Section 7–5

1. 0.06, 0.3539; 0.05, 0.3   3. 0.026   5. 0.018

Section 8–1

1. Since the observation is over 6 standard deviations from the mean, it is far from the expected value.
3. 61/4   5. 5.0975 (or 5.10)
7. $\chi_{10}^2 = 9.05$, close to the theoretical mean of 10   9. $\chi_3^2 = 24$, partial answer

Section 8–2

1. (a) 0.99, (b) 0.30, (c) 0.70, (d) 0.166, (e) 0.834
3. $P(Z^2 > 2.71) = P(|Z| > 1.646) = 0.0998 \approx 0.10$ 5. 0.048   7. 0.247

Section 8–3

1. 0.80   3. 0.957   5. 0.865

Section 8–4

1. 0.90   3. 0.871
5. More extreme one-sided deviations occur about $\frac{1}{2}\%$ of the time.
7. $\mu_T = 150$, $\sigma_T^2 = 3000$

Section 8–5

1. $P(6.4 < \sigma^2 < 18.4) = 0.90$
3. $P(15.8 < \sigma^2 < 46.9) = 0.90$. But unreasonable that all died within 10 hours of true mean. Possibly data are wrong for s^2.
5. $1060 < \sigma < 2690$; accept

Section 9–1

1. $\chi_1^2 = 8$, $P(\chi_1^2 \geq 8) = 0.007$ 3. $\chi_5^2 = 8.46$, $P(\chi_5^2 \geq 8.46) = 0.14$

Section 9–4 (Partial answers)

1. $P(\chi_1^2 > 9) = 0.005$ 3. $P(Z \geq 2.75) = 0.003$ 5. 0.007
7. $\chi_1^2 = 1.05$; no   9. Collapses to $(x - np)^2/npq$.

Section 9–5

1. 6.04, decrease of 21%

Short answers to selected odd-numbered exercises

Section 10–1

1. $P(\chi_5^2 > 5) \approx 0.42$. Close fit.   3. 102,167

Section 10–2

1. $6.58 < 23$

3. $P(\chi_2^2 > 0.83) = 0.66$. Poorer fits occur about 2/3 of the time when the model is exactly true.

5. $P(\chi_3^2 > 3.51) = 0.32$ for the two sets fitted by a common p. The fit is close.

7. $P(\chi_7^2 > 17.5) = 0.019$. Not a close fit, but not bad either, considering the large sample size.

Section 10–3

1. (a) $267.0 \times (0.3413) = 91.1$ (b) Pooling gives 5, not pooling, 7.
   (c) Pooling gives $P(\chi_5^2 > 3.04) \approx 0.70$.

3. $12169(0.152) = 1849.7$ (rounding leads to inaccuracy)

Section 10–5

3.

Part of speech	Type of score		
	Low	Medium	High
Verb	3	6	11
Noun	4	5	2
Possessive pronoun	1	1	5
Preposition	8	7	2
Conjunction	4	5	1

5. Scores are similarly distributed.

Section 11–2

1. Hypothesis of independence is unreasonable because the value of chi-square is much too large.

3. $\chi_1^2 = 14.8$. Strong evidence for dependence.

5. $P(\chi_1^2 > 4.00) = 0.048$; seeding seems more likely to increase the frequency of hail than reduce it.

Section 11–3

1. $\chi^2 = 13.93$, still large   3. $P(\chi^2 \geq 3.08) = 0.084$, up from 0.048

Section 11–4

1. $\chi_4^2 = 1.85$. Agreement very close.

3. $P(\chi_8^2 > 12) = 0.84$. Good fit. Days and types of creels seem independent as far as catch is concerned.

5. We pooled scores $-1, -2, -3$ and got $\chi^2 = 1.01$, a close fit for 4 d.f., and conclude no evidence that taste preferences differed by sex.

7. $\chi_3^2 = 3.50$. Good fit.

9. $P(\chi_2^2 > 19.28) < 0.01$. Strong evidence of dependence.

Section 12–1

1. $D = 1/2$

Section 12–2

1. $D = 287/60$ 3. $H = 287/780, E(H) = 2$

5. $H = 3/35, E(H) = 2$ 7. $H = 9.97, E(H) = 4$

Section 12–3

1. $P(H \geq 9.2) \approx P(\chi_5^2 \geq 9.2) = 0.10$

5. $P(H \geq 9.97) \approx P(\chi_4^2 \geq 9.97) = 0.045$, strong evidence against equality of fuel consumption

Section 12–4

1. $R_1 = 9 + 5 + 1 = 15$
   $R_2 = 8 + 4 + 3 = 15$
   $R_3 = 7 + 2 + 6 = 15$

   $R_1 = 2 + 4 + 9 = 15$
   $R_2 = 1 + 6 + 8 = 15$
   $R_3 = 7 + 5 + 3 = 15$

   $R_1 = 7 + 2 + 6 = 15$
   $R_2 = 9 + 5 + 1 = 15$
   $R_3 = 8 + 4 + 3 = 15$

3. From Table A–14, $P(H \geq 6.75) < 0.011$
   From Table A–4, $P(H \geq 6.75) \approx P(\chi_2^2 \geq 6.75) = 0.041$

5. $H = 1.07, P(H \geq 1.07) = 0.66$

7. $P(H \geq 4.32) \approx P(\chi_5^2 \geq 4.32) = 0.50$

9. $P(H \geq 3.55) \approx P(\chi_2^2 > 3.55) = 0.18$.   Fourth Humanities department looks different.

Section 13–1

1. All entries in any column are either smaller than or larger than all the entries in each other column. So each row must have the same ranking.

Section 13–2

1. Average $r_S = 0.06$

3. Columns 1 and 2 give $r_S = 0.20$; columns 1 and 3 give $r_S = 1$; and columns 2 and 3 give 0.20; average $r_S = 7/15$.

5. As n increases, the average of the largest rank correlation coefficients tends to zero.

Section 13–3

1. $\chi_9^2 \approx 8.96$ 3. $T = 0$

Section 14–1

1. Average observed in Table 14–1 is 1.71, to be compared with the theoretical mean 1.87.
3. The symmetry of the normal and the complementarity of 10 and 11 for $n = 20$ imply that $E(U_{10}) + E(U_{11}) = 0$ for the standard normal. Table 14–1 gives $U_{11} = -0.02$, and $U_{10} = -0.17$, averaging to -0.086. This differs from 0 because of sampling error.

Section 14–2

1. $C = \displaystyle\sum_{k=2}^{8} \binom{10}{k} \left(\frac{1}{2}\right)^{10} = 0.978$; confidence statement: $7.1 < x_{0.5} < 11.3$.

3. Let $n = 2m$, m a positive integer; then the two middle measurements are u_m and u_{m+1}, and $L = m$. Then

$$C = P(L = m) = \binom{2m}{m}\left(\frac{1}{2}\right)^{2m}.$$

5. Using the normal approximation for $n = 64$, we find $i = 24$ and $P(L < 24) \approx 0.016$ instead of 0.025.

7. Choosing u_i and u_{20-j+1} with $i = 1$ and $20 - j + 1 = 7$ gives C about 0.90. The confidence statement for this experiment is: $-1.78 < x_{0.2} < -0.51$.

9. $C = \displaystyle\sum_{k=i}^{n-j} \binom{n}{k} p^k (1 - p)^{n-k}$.

11. Using the normal approximation u_{40} and u_{61} (or u_{41} and u_{60}) gives about 95% confidence. $u_{40} = -0.36$, $u_{61} = +0.17$ in this example; the interval happens to include 0, a fact we wouldn't ordinarily know.

Section 14–3

1. $3/4$ 3. $E(Y_i) = 3/6$, Var $Y_i = 1/28$ 5. About 0.88
7. Since $P_1 + P_2 + \cdots + P_{n+1} = 1$, Var $[P_1 + P_2 + \cdots + P_{n+1}] = 0$. Then

$$\sum \text{Var } P_i + 2 \sum_{i<j} \text{Cov } (P_i, P_j) = 0.$$

Var $P_i = n/(n + 1)(n + 2)$, and there are $n(n + 1)/2$ covariances, all equal. Therefore Cov $(P_i, P_j) = -1/(n + 1)^2(n + 2)$, $i \neq j$.

9. $E(Y_{10}) = 1/10$, $\sigma_{Y_{10}} \approx 0.03$, and the normal approximation gives

$$P\left(\left|\frac{Y_{10} - 0.1}{0.03}\right| > 1\right) \approx 0.32.$$

11. Var $Y_{16} = 0.00143$
13. Using data from Exercise 12 and the normal approximation, we get

$$P(Y_{64} \geq 0.60) \approx P(Z \geq -1.46) = 0.928,$$

where Z is a standard normal random variable.

15. $P(0.2 < Y_3 < 0.8) = 1 - 2P(0 \le Y_3 \le 0.2)$

$$= 1 - 60 \int_0^{0.2} y^2(1 - y)^2 dy = 1 - 60 \left[\frac{y^3}{3} - \frac{y^4}{2} + \frac{y^5}{5} \right]_0^{0.2}$$

$$\approx 0.884$$

Section 15–1

1. (a) 2/3, (b) 1/2, (c) 1, (d) 1.1

Section 15–2

1. $\mu_3 = -0.524, \sigma_3^2 = 0.19$. For normal with mean 3 and standard deviation 5,

$$\mu_3 = +0.38, \sigma_3^2 = 4.8.$$

3. $P \approx 0.98$

5. Compute normal deviate

$$z = \frac{(\tilde{x}_1 - \tilde{x}_2) - 0}{\hat{\sigma}_{\tilde{x}_1 - \tilde{x}_2}}$$

with

$$\hat{\sigma}_{\tilde{x}_1 - \tilde{x}_2}^2 = \frac{\pi\sigma^2}{2n_1} + \frac{\pi\sigma^2}{2n_2}.$$

7. $g(u) = n(1 - e^{-u})^{n-1} e^{-u}$

9. $g(u_{n+1}) = \frac{(2n + 1)!}{n! \, n!} [F(u)]^n [1 - F(u)]^n f(u)$

Section 15–3

1.
n	M_n	H_n	M_n/H_n
9	1.14	1.49	0.77
19	0.52	1.13	0.46
29	0.34	1.03	0.33
39	0.26	0.96	0.27
49	0.20	0.92	0.22

3. $23\frac{3}{8}$ pounds   5. 4   7. Computer problem   9. 2.1

Section 15–4

1. (a) 0.128, (b) 0.158   3. $\sigma_4^2 \approx 0.22$ from formula (2)

5. $n = 2: 0.466; n = 6: 0.165; n = 10: 0.078; p$ tends to 0 as n grows large.

7. Trimmed mean $= -0.28$

9. $\sigma_T^2 = 0.184\sigma^2$; efficiency $= \frac{0.167\sigma^2}{0.184\sigma^2} = 0.91$

11. (a) 2.06, (b) 0.43

13. Test with standard normal deviate

$$z = \frac{\tilde{x}_r - \tilde{x}_p}{\sigma_{\tilde{x}_r - \tilde{x}_p}}, \text{ with } \sigma_{\tilde{x}_r - \tilde{x}_p} = \frac{\sigma}{0.635\sqrt{n}}.$$

Section 15–5

1. (a) 1.33, (b) 1.35   3. $t = 6.80$, $t' = 6.92$
5. (a) 0.58, (b) $t' = 2.24$, $P(t \geq 2.02) = 0.95$, (c) 1.54, 0.32

Index

Index

Numbers in parentheses refer to exercises on the indicated pages.